THEORIES OF LANDFORM DEVELOPMENT

G. K. Gilbert
(*ca* 1898)

THEORIES OF LANDFORM DEVELOPMENT

Wilton N. Melhorn

Ronald C. Flemal

Editors

A Proceedings Volume of the
Sixth Annual Geomorphology Symposia Series
held at Binghamton, New York
September 26-27, 1975

London
GEORGE ALLEN & UNWIN
Boston Sydney

First published in 1975
First published by Allen & Unwin in 1980
Second impression 1981

GEORGE ALLEN & UNWIN LTD
40 Museum Street, London WC1A 1LU

British Library Cataloguing in Publication Data

Geomorphology Symposium, *6th, State University of New York, 1975*
Theories of landform development. – (The Binghampton symposia in geomorphology: international series; no. 6).
1. Landforms – Congresses
I. Title II. Melhorn, Wilton N.
III. Flemal, Ronald C.
IV. Series
551.4'01 GB400.2 80-41400

ISBN 0-04-551039-3

Printed in Great Britain by Short Run Press Ltd., Exeter.

TABLE OF CONTENTS

INTRODUCTION

In December 1939 a meeting was sponsored by the Association of American Geographers to review the status of Walther Penck's contribution to geomorphology and its relationship to the Davisian system. The objective of that meeting, as stated by its organizer, O. D. von Engeln, was "to present the two approaches [of Penck and Davis] in sharp contrast in order that the differing viewpoints and objectives should be clearly revealed" (AAG, 1940, p. 220). The proceedings of the meeting well served this objective. However, they did not serve what might well have been considered a more basic objective, to resolve the question of which of the disparate theories of landform development was to gain eventual acceptance. Thirty-six years later the Davisian and Penckian systems, both in their original and derivative forms, as well as several additional systems, continue to vie for general acceptance.

The purpose of the Sixth Annual Geomorphology Symposium and this, its proceedings volume, are nearly identical to those of the 1939 meeting. We also hoped to present disparate approaches to landform theory, in contrast, such that their viewpoints and objectives should be clearly revealed. And we also hope, although less tacitly, that the material presented at the meeting and contained herein, will serve to resolve the merits of the contending systems.

The year of 1975 represents also a landmark in the history of geomorphology. It was just a century ago, in 1875, that G. K. Gilbert commenced field work on the project that resulted in publication of the simple yet classical report on the Henry Mountains of Utah (1879). The impact of this monograph, Gilbert's subsequent work on landforms and his pioneering quantitative studies of fluvial processes have earned for him a sort of unofficial status as the patron saint of geomorphology. We are, therefore, fortunate indeed to include in this proceedings volume a paper by a modern biographer that casts new light on Gilbert and his contributions to geomorphological thought.

We have purposely refrained from any editorial changes in the various papers that might distort any author's intent or philosophical thrust. However, we have aspired to provide as broad a range of views as time and availability of contributors would allow. To the authors we express our deep appreciation for making this aspiration a reality.

THE EDITORS

REFERENCES CITED

Association of American Geographers, 1940. Symposium on Walther Penck's Contribution to Geomorphology: AAG Annals, Vol. 30, No. 4, p. 219-284.

Gilbert, G. K., 1879. Report on the geology of the Henry Mountains: U. S. Geog. and Geol. Survey of the Rocky Mountains Region, Washington, 170 p.

ACKNOWLEDGEMENTS

These are the proceedings of the Sixth Annual Geomorphology Symposium, which was held as is customary in the Department of Geological Sciences at the State University of New York at Binghamton, New York on September 26-27, 1975. This is the first symposium in the series organized and published by other than the originators of these symposia, Donald R. Coates and Marie Morisawa. We are grateful to Dr. Morisawa for allowing us to develop the theme of the symposium, for help in procuring the speakers, and for logistical arrangements at SUNY-Binghamton.

We also thank those who helped by acting as session moderators: Donald Doehring, Dale Ritter, and Lawrence Onesti. Lastly, appreciation is expressed to Richard P. Mroczynski for his able assistance in formatting of manuscripts, drafting, and general management that has contributed to completion of this published volume.

THE EDITORS

CHAPTER 1

THEORIES OF LANDSCAPE DEVELOPMENT A PERSPECTIVE

Charles G. Higgins

While the geologist may often be in error
the Earth is never wrong.

— L. C. King, 1967, p.v.

ABSTRACT

Theories of landscape development serve to stimulate communication between the interdependent descriptive, genetic-historical, and process-oriented lines of inquiry in geomorphology, and thereby help to unify them into one science. The theory that has dominated the modern era of geomorphology, which may be said to have begun in 1877 with the publication of G. K. Gilbert's "Land Sculpture", was first formulated by W. M. Davis in 1889 and 1899. The prolonged popularity of Davis' concepts of landscape evolution despite severe criticism of most aspects of his system and despite the availability of at least two major alternative theories, proposed by W. Penck in 1924 and by L. C. King about 1953, suggests that there are compelling and perhaps unrecognized reasons for its appeal. One of these reasons seems to stem from a belief deeply rooted in the origins of geologic thought that natural phenomena progress through cycles. When we begin to recognize and to evaluate such assumptions in the criteria by which we assess theories of landscape development, and when we can recognize that many existing landforms are now relic and that only those in harmony with the processes now affecting them show true relationships between geomorphic form, process, and climate, then we may be able to synthesize a new and long overdue acceptable body of theory for the genesis and development of landscape.

THE ROLE OF GEOMORPHOLOGY IN EARTH STUDY

According to tradition, the first Western scholar to ask, "Of what, and in what way, is the world made?", was the Greek philosopher, Thales of Miletus (about 580 B.C.). His answers to this question would not be meaningful to us today, but the question itself was a significant one, for it represented the birth of the spirit of disinterested curiosity that is necessary for true scientific inquiry. This spirit, disciplined by the requirement that all speculations or hypotheses must be tested, and coupled with the belief that nature operates in a cause-and-effect manner so that nothing in nature is capricious or arbitrary, has led to the development of the modern natural sciences.

The branch of science that today attempts to answer Thales' question most directly is *geology* — the study of the Earth — while a sister science commonly called *geomorphology* is concerned with the nature and origins of the Earth's surface features. Geomorphology — literally, a study of Earth form — is generally understood to embrace the study of landforms and landscapes, although some authors have defined the term so broadly as to include the Earth's configuration as a whole (e.g., see von Engeln, 1942, p. 1) and the dynamic processes that shape its surface. Other writers have restricted geomorphology to the study of subaerial landforms of erosional or depositional origin, or only to the genesis and history of such landforms.

Geologists tend to regard such studies of the Earth's surface as a branch of geology, whereas geographers tend to regard them as part of physical geography, and many geomorphologists themselves regard their work as part of a separate interdisciplinary science that overlaps or interfingers with both geology and physical geography as well as with geophysics and other studies (e.g., see Dury, 1966, p. 2).

THREE LINES OF GEOMORPHIC INQUIRY

Confusion about the nature of geomorphology reflects the fact that, historically, geomorphologists have at one time or another attempted to answer three different sorts of questions about the Earth's landforms and landscapes:

1) How can these features best be described?
2) How have they formed and changed through time?
3) What processes are responsible for them and how do these processes work?

The first of these lines of inquiry would seem to be a function of physical geography, one goal of which is commonly understood to be an accurate, comprehensive and comprehensible description of the Earth's surface. The term *physiography* has come to be used for such descriptive studies through the mistaken impression that the word was originally coined as a contraction of "physical geography" (e.g., see Bryan, 1941, p. 3, 6). Instead, Huxley

(1877) originally used this term in its literal sense for the study of natural phenomena in general. Later Powell (1895) restricted it to the surface features of the Earth, with an emphasis on their mode of origin, and as such the term is approximately synonymous with *geomorphology*, which has largely superceded it. W. M. Davis used *geomorphography* for the descriptive study of landscape (Bryan, 1941, p. 6), and this clear, uncorrupted term deserves to be resurrected (see Gary *et al*, 1972, p. 293).

The third line of inquiry, into the nature and the results of the processes that shape the Earth's surface, is sometimes called *process geomorphology* (see Ruhe, 1975, Preface), although the principles and methods of study are borrowed directly from soil science, soil and rock mechanics, hydrology, and geophysics. Kirk Bryan termed such studies "Dynamic Geology", and emphatically excluded them from his "Geomorphology" (1941, p. 5). Ironically, of the seventeen papers that have won for their authors the Kirk Bryan Award, given annually by the Quaternary Geology and Geomorphology Division of the Geological Society of America, from 1958 through 1974, at least five are devoted primarily to process studies, or to what Bryan himself would have termed "Dynamic Geology".

To Bryan, "Geomorphology" was concerned primarily with the second line of inquiry — the study of landform origin and change — which Davis (see Bryan, 1941, p. 6), Lawson (1894), and other early authors sometimes called *geomorphogeny* (see Gary *et al*, 1972, p. 293). Bryan argued that the "essential and critical" distinction between Geomorphology and Dynamic Geology is:

> "the recognition of land forms or the remnants of land forms produced by processes no longer in action. Thus, in its essence and in its methodology [Geomorphology] is historical . . . This is the true function of the study of land forms within the generous and inclusive arms of the mother science of Geology" (Bryan, 1941, p. 5).

Despite Bryan's restricted definition, geomorphology is nowadays generally understood to cover all three lines of inquiry or aspects of landform study — description, genesis and history, and process. In part this reflects the interdependence of all three sorts of investigations. For example, a sound reconstruction of the history of a particular landform requires both a clear picture of what that landform is today and a clear understanding of the operation and results of the various processes that may have shaped it. On the other hand, a sound description of a modern landform should take into account not only its present form and structure but its antecedents as well. Finally, geomorphic studies of Earth-forming processes necessarily include the effects such processes have on Earth materials and landforms. Such studies, in turn, provide information needed to describe and interpret the histories of existing landscapes that may have been affected by these processes. Thus, each line of geomorphic inquiry serves the others and in turn depends on the others for fresh input of observations and ideas.

THE ROLE OF GEOMORPHIC THEORY

The descriptive, genetic-historical, and process-oriented aspects of geomorphology are further bound together by geomorphic theory — the name given to speculations about the general development of landforms and landscapes under varying conditions and circumstances. Although such speculations are closely related to the genetic-historical aspect of geomorphology, they are intricately interwoven into the entire fabric of the science. Not only must sound theory be based on information from all lines of geomorphic inquiry, but theory itself serves to provide fresh concepts and the stimulus for further research in all aspects of geomorphology.

W. M. Davis, nominally a geographer, formulated a general theory of landscape evolution in order to create an improved system for describing existing landforms. Walther Penck, nominally a geologist, proposed a very different theory of land sculpture in order to interpret the history of dynamic crustal movements. Conflict between these general theories in turn stimulated valuable new investigations into slopes, slope development, and slope-forming processes, while Davis' concepts of the history of streams have helped focus the attention of several generations of investigators on problems of stream processes and channel development, (e.g., see Leopold and Maddock, 1953, p. 51). Geomorphic theory thus functions as a vital and effective means of maintaining communication among the several aspects or branches of geomorphology and thereby serves to integrate these several lines of inquiry into one science.

Geomorphic theories come in all sizes and shapes. They may consist of empirical generalizations about forms and/or processes or of explanations for observed phenomena. Such generalizations and explanations may involve either short- or long-term effects and changes, they may be based on studies of real landforms and processes or on various sorts of models, and they may apply to landforms of tiny or of continental dimensions in circumstances that may be either very specific or very general.

The most challenging theories — and those that seem most likely to provoke the most communication and growth in the science — are those that are the most general; that is, that essay to hypothesize the development of whole landscapes from their inception to some final stage under fairly general conditions. Such theories of landform development are the subject of this Symposium.

THE BEGINNINGS OF MODERN GEOMORPHOLOGY

As long as Man has walked the Earth, tilled the ground, and sailed the seas he has formed ideas about the origins of the Earth and its landscapes. Before the middle of the nineteenth century such ideas about landscape development were incomplete or were based on faulty assumptions or observations or both. Some modern theories share these same shortcomings. However, by the mid 1870's much of the foundation for a modern geomorphic outlook had been laid. James Hutton and others had removed the constraints of catastrophism by postulating that the Earth's features are formed by ordinary processes like those at work today, acting slowly through a series of cycles of uplift, erosion, and deposition in a greatly expanded time scale of Earth history. Louis Agassiz and others had recognized the erosional effectiveness of glaciers, and had established their former existence over much of northern Europe. Playfair, Jukes, Ramsay, Geikie, and geologists working in the American West had established what Davies (1969) has termed the "fluvial doctrine", which recognized the pre-eminent effectiveness of running water to shape the land. To be sure, the important role of downslope movement, or mass wasting, and the minimal role of wind in land sculpture was not yet fully appreciated by the 1870's, but these and other aspects of the nature and relationships of geomorphic forms and processes would come to be much better understood in the succeeding century. Thus if one were to assign a specific date for the beginning of the modern era in geomorphology, one would most likely choose 1877, the year of publication of T. H. Huxley's "Physiography" and of G. K. Gilbert's chapter on "Land Sculpture" in his report on the geology of the Henry Mountains. All of the major theories of landscape development with which we are here concerned were formulated since that date, within the past one hundred years.

G. K. GILBERT AND GEOMORPHIC THEORY

Grove Karl Gilbert, who has been called "the first true geomorphologist" (King, 1966, p. 3), characterized "theorists" as those who fail to test their hypotheses adequately, and he seems to have regarded himself instead as an "investigator" — one who "seeks diligently for the facts which may overthrow his tentative theory" (1886, p. 286). Among the few broad generalizations about landform development that he allowed himself are a group of principles that he formulated to explain certain general tendencies observed in landscapes formed by running water. These he called the "law of uniform slope", the "law of structure", the "law of divides", and the "tendency to equality of action, or to the establishment of a dynamic equilibrium" (1877, p. 115-116, 123). Later he renamed his "law of divides" as the "law of increasing acclivity", and designated a new "law of the interdependence of parts" (1884, p. 75), which seems to include both his concept of equality of action and his earlier "law of structure".

In recent years, John Hack (1960, 1965) has sought to revitalize Gilbert's general principles and to generate new interest in them. By themselves, these principles do not constitute a complete theory of landscape development, although they may be used as tests or as conditions of such theories. For himself, Gilbert — the investigator — does not seem to have been interested in formulating a general theory, and there is no evidence that he ever attempted to apply his own principles as a test of W. M. Davis' concept of landscape evolution as Hack has attempted to do. Gilbert did, in fact, use some of Davis' terminology in his later papers (e.g., 1909, p. 344), and on one occasion he publicly supported the general sense of Davis' theory in rebuking a critic of Davis' terminology (Gilbert, 1905, p. 29). Gilbert evidently sent separates of his papers to Davis (Fig. 1), and Davis freely credited Gilbert's contributions even where he could not find in them support for his own conclusions (1899b, 1909 revision, p. 351).

Figure 1. Inscription presumably written by G. K. Gilbert on the cover of a separate of his 1884 paper, "The sufficiency of terrestrial rotation for the deflection of streams" (Nat. Acad. Sci., Memoir, v. 3, p. 7-10).

Throughout his career Gilbert was chiefly interested in relationships between process and form. This interest culminated in his monumental work on stream processes (1914, 1917) that entitles him to recognition as one of the great pioneers in fluvial hydrology and fluvial geomorphology. While we may lament that this great mind did not engage itself with broad theories of landscape development, we may be grateful for the inheritance that Gilbert did leave us. What he did endures, and without his example, process studies might today be entirely a branch of geology, to the impoverishment of geomorphology.

W. M. DAVIS' GEOMORPHIC THEORY

The first truly general theory (actually a group of theories) of landscape development was proposed by William Morris Davis. In 1889 he described a

"Complete Cycle of River Life" along with elements of a concept of progressive regional denudation. In 1899 he expanded the latter into an ideal "Geographical Cycle" of landscape evolution. The so-called Davisian system thus includes two separate and distinct cyclical concepts, one for the progressive development of erosional stream valleys and another for the development of the overall landscape. Harris and Twidale (1968, p. 238) believe his system may be interpreted to include a third distinct cycle governing the development of slopes.

Davis intended his "Geographical Cycle" to apply to humid temperate regions of uniform resistance to erosion that are uplifted relatively rapidly and then progressively worn down by running water and mass wasting. In subsequent works he and his followers extended this general concept of cyclical development of landforms through progressive stages to apply to landscapes in arid regions (Davis, 1903, 1905, 1930), to those formed by glaciers (Davis, 1900, 1906) and by shore processes (Davis, 1912; Johnson, 1919), and to karst landscapes (Beede, 1911, Cvijić, 1918). As recently as 1950 the concept was further applied to landscapes formed by periglacial processes (Peltier, 1950).

From its inception Davis' system was widely acclaimed and adopted although it also received severe criticism from many quarters. Davis wrote several papers specifically in reply to these objections (e.g., 1899b). Through the years, especially after publication of Walther Penck's concepts of landscape development, Davis further developed and revised elements of his system (1922), particularly as it applied to arid regions (1930, 1932).

Further modifications were made by Davis' followers. For "Geographical Cycle" they used "Normal Cycle", "Fluvial Cycle", "Erosion Cycle", "Geomorphic Cycle", or even "Humid Cycle" (White, 1968), and some of them changed "structure, process, and time", the "trio of controls" on which "all the varied forms of the lands are dependent" (Davis, 1899a, p. 481), to "structure, process, and stage". Recently J. C. Brice added a needed fourth control, which he designates "fundamental form" (1964), and which determines the forms of upland surfaces in early stages of a cycle.

Despite repeated severe attacks on virtually all aspects of Davis' "Geographical Cycle" by early critics, such as R. S. Tarr (1898) and W. S. T. Smith (1899; see also Tilly, 1968), and later ones, such as C. H. Crickmay (1933 and this Symposium), and despite the later introduction of two other major concepts of landscape development, by Walther Penck and Lester King, Davis' system came to dominate both teaching and research in the descriptive and genetic-historical aspects of geomorphology. Its continued viability is attested in part by continuing objections to it by recent critics, such as R. C. Flemal (1971) and C. R. Twidale (1975, in press). That such an obviously flawed doctrine could have enjoyed such prolonged popularity among a large segment of the geomorphic community suggests that there must be compelling reasons for its appeal. Before we consider what these reasons might be it will be helpful to examine briefly the major alternative theories that have been proposed.

CHALLENGES TO THE DAVISIAN SYSTEM

Walther Penck's Geomorphic System.

Walther Penck's "Die Morphologische Analyse" was published posthumously in 1924, one year after his untimely death at age 35. Opinions about his work vary widely, but all agree that it is difficult to read and to understand. A letter quoted by his translators (Penck, 1953, p. v) suggests that, had he lived, Penck would have enlarged the scope of his treatment and possibly would have made it clearer than it now stands. Although an English translation was not published until 1953, Davis had reviewed parts of the work in 1932, and summaries in English and translations of certain chapters had circulated privately among American geomorphologists during the 1930's and '40's. One of the earliest, and still one of the best, published summaries of Penck's concepts is that by O. D. von Engeln (1942, Chapter 13).

In the United States, Walther Penck's system was met with considerable resistance, particularly by geologists, who focussed their disdain on his concepts of slope retreat and long-continued crustal movements. At a symposium organized in 1939 to consider "Walther Penck's Contribution to Geomorphology" (Penck Symposium, 1940) most of the participants were severely critical of Penck's ideas or dismissed them outright, while only a few seemed to find any merit in them. These few included Kirk Bryan, John Leighly, Oscar von Engeln, and Howard Meyerhoff.

To illustrate the extent to which some American geomorphologists impugned Walther Penck's ideas, I was once told that his "peculiar notions" owed to his incomplete recovery from a head wound suffered in World War I. I know of no basis for such a story unless it was contrived, with malice, from von Engeln's remark that "He wrote very tortured German. I have been putting out the theory that it is because he was seriously ill and physically distressed that he wrote such very involved sentences" (Penck Symposium, 1940, p. 280). Penck died of cancer of the mouth (Penck, 1953, p. viii).

Despite widespread rejection of Penck's views, especially in the United States, some young geomorphologists set out to test them in the field, and began to find and describe *knickpunkte* and *piedmont treppen* where an earlier generation had blanketed the countryside with remnants of Davisian peneplains. In England such studies helped lead, in turn, to new methods of landscape study, including morphological mapping. Penck's work led other investigators to reconsider Davis' assumptions about diastrophic activity and about the nature of slopes, and this tended to redirect attention from purely descriptive and theoretical aspects of landform study to the relationships of geomorphic processes and forms.

Lester King's Geomorphic System.

The English translation of "Die Morphologische Analyse" was published the same year as Lester C. King's "Canons of landscape evolution" (1953), which embodied some concepts of landform development that he had been formulating during the previous five to ten years. In a voluminous series of subsequent articles and books, King has elaborated and expanded this system, relating it to the concept of continental drift so that landscape study in his hands becomes a tool for deciphering the origin and history of continents. Although "Canons" is the most often quoted of his works, King's ideas are most fully presented in his massive volume, "The Morphology of the Earth" (1962, 1967), and most lucidly summarized in his shorter book, "South African Scenery" (1963).

Despite King's firm rejections of Davis' system, his own resembles it in many respects. It is evolutionary and cyclical — he refers to his system variously, as "The Landscape Cycle", "The Epigene Cycle of Erosion", and "The Pediplanation Cycle". Each cycle is initiated by a relatively rapid burst of diastrophism followed by a long period of diastrophic inactivity. The end product of long continued subaerial erosion during the inactive period is a *pediplain*, an erosional surface analogous to Davis' *peneplain* to the extent that it is old, has low relief, and is of regional extent. King retains Davis' terms "youth", "maturity", and "old age" for recognizable stages of landscape development in each cycle. However, where Davis' peneplain is worn *down* by erosion, King's pediplain forms by the integration of *pediments* that are enlarged by headward recession of scarps. For this picture of pediment development King credits the earlier work of Kirk Bryan (1922) and Walther Penck (1924), and he further borrows from Penck the concept of retreating fluvial nickpoints (King, 1967, p. 159-160). Another major distinction of King's system is that, whereas Davis' peneplains cease to develop and actually become rejuvenated when a new cycle is inaugurated by crustal uplift, King's pediplains continue to grow headward even while their distal ends are being consumed by new receding scarps. With multiple uplifts, then, King's landscapes are characterized by a number of headward-growing pediplains that closely resemble Walther Penck's *piedmont treppen.*

Some years ago I had the audacity to chide King about his concepts, and his reply was so gracious that it deserves a larger audience:

> "I note that you do not agree with my 'Penckian notions of landscape formation'. Nor did I at one time. I had as good a Davisian training at the hands of C. A. Cotton as any geomorphologist could desire and like many another went about happily interpreting landscape in Davisian terms, which failed more and more to explain what I was seeing. It was not until I was content to forget about both Davis and Penck and learn to stop being clever and interpreting landscape, but just to sit

silently on the hillsides with my chin in my hands and *let the landscape teach me* that I really began to understand it. It is very like being in big game country. So long as one goes on walking one sees nothing; when one sits and becomes part of the landscape then the animals come out all about one." (L. C. King, personal communication, 29 January 1958).

King's general theory of landscape development would seem a likely candidate to challenge the Davisian one since it combines some of the best or most popular features of the latter with some of the more accepted aspects of Penck's system, tempered with well-observed studies of pediment forms in Africa and in the American Southwest. However, response to his system in the United States seems to have been relatively cool. To be sure, many geologists and geomorphologists who formerly described "peneplains" began calling them "pediments", but King's work has not had the influence in this country that it seems to have had in England, Australia, and elsewhere. One reason for this failure is that by the time King published "Canons", in 1953, geomorphic theory was no longer the central topic that it had been in geomorphology for over half a century. Interests had shifted.

Dynamic Geomorphology and the "Quantitative Revolution".

In 1952 Arthur Strahler had called for a new outlook in geomorphology in a paper that is widely regarded as the manifesto of dynamic, or process, geomorphology. In it, Strahler visualized "a system of geomorphology grounded in basic principles of mechanics and fluid dynamics, that will enable geomorphic processes to be treated as manifestations of various types of shear stresses . . ." (1952, p. 923). This emphasis on process and quantification was not new. R. A. Bagnold, W. W. Rubey, and others had been pursuing such studies during the 1930's and '40's, but the end of World War II brought a new surge of interest. Cailleux and Tricart began the *Revue de Géomorphologie Dynamique* in 1950, and by 1952 scores of investigators were attempting to understand the mechanics of streams, glaciers, wind, waves, and other Earth-shaping processes. Strahler's paper served to give focus and encouragement to such efforts, and to affirm that process studies were a legitimate — nay, essential and central — aspect of geomorphology.

In other work, Strahler (1950, 1958) and his students revivified the quantitative description of landforms that Bryan had earlier deplored (1941, p. 8) by using statistical and dimensional analyses intended not only to provide quantitative descriptions of individual landforms, but also to detect and quantify subtle and complex interrelationships between the variables that Davis had designated as "structure, process, and time". In hindsight it seems that although Strahler was neither the first nor the only investigator to call for quantitative methods in studies of landforms and geomorphic processes in the 1950's, his work and writings served to spearhead the so-called "quantitative revolution" that followed.

As a result of this change in intellectual climate, interest in general theories of landscape development waned. Perhaps post-war investigators were dissatisfied with the inconclusive rhetoric of the older generation of general theorists and simply wanted results that at least had the *appearance* of fact. Perhaps, too, they agreed with Strahler that to be a science geomorphology must adopt the methods of other sciences, preferably of physics. In any case, during the next two decades a great deal of effort was devoted to empirical studies of forms and processes, and much has been learned about both. The specific results of such studies do not concern us here, although such information is essential to the formulation of a satisfactory body of geomorphic theory. As C. A. M. King has written, "until the operation of the processes that modify the landscape is understood it will not be possible to understand the genesis of landforms" (King, 1966, p. 17).

These post-war trends in geomorphology help to explain why Lester King's concepts have not generated the kind of widespread discussion that was provoked by earlier theories. However, recently there seems to be a revival of interest in geomorphic theory, as evidenced by the organization of this Symposium, which may be the first such meeting devoted to general theories of landscape development since the Penck Symposium, organized for the Chicago meeting of the Association of American Geographers in 1939.

There are several reasons for such a revival. After prolonged concentration on process studies the science needs to be reintegrated through the communication provided by general theory. Also, enough has been learned about stream networks and slope and channel forms and processes that it may now be possible to synthesize a body of theory more firmly based on a sound understanding of geomorphic processes. Further, after two decades of log-log distribution curves, linear regressions, and morphometric analyses, the urge to make broad generalizations — always a strong force behind any scientific inquiry — may simply have become irresistible. More importantly, the advent of plate tectonics theory in geology has stimulated renewed interest in ancient erosion surfaces as possible aids in interpreting and dating continental drift and interactions between plates. Here, geomorphic study is being asked to serve as Walther Penck thought it should, to supply information about crustal movements and the Earth's diastrophic history. Finally, there is an increasing demand for evaluations and predictions of the effects of man-induced changes on the environment and landscape, and we are discovering that in many cases we are unable to make sound judgments for lack of a reliable body of general theory that could provide a basis for prediction.

To gain some understanding of what such a general theory may eventually consist of one must know what needs it must satisfy, what conditions and constraints will be imposed upon it, and what assumptions may underlie it. I believe we can answer some of these questions by analysing the answers to two other questions: What made Davis' system so appealing, and why do we still lack a satisfactory theory of landscape development?

WHAT MADE DAVIS' SYSTEM SO APPEALING?

First, let's consider why Davis' system was so immediately and widely accepted and then why it has remained popular.

1. It is simple. Davis based his "Geographical Cycle" on a very general case: a landmass is uplifted from beneath the sea into a climate humid enough to sustain perennial streams. The surface is uniform but inclined. Once the relatively rapid uplift is completed there is a long period of crustal stability with no change in either the climate or the regime of streams. During this period streams and downslope movement reduce the landmass through an orderly progression of stages, which are simply, and memorably, named "youth", "maturity", and "old age". Moreover, it would appear that only simple adjustments are needed in this basic model to explain the development of stream-eroded landscapes in more complex conditions and circumstances — where rock structure is inhomogeneous, for example, or where there is repeated uplift or climatic change. These concepts are easily grasped by students and scholars alike, who might otherwise be bewildered by the diversity of landforms. Davis' system makes sense of this diversity.

2. It is applicable. Regardless of the genetic implications of Davis' stage names, students can easily apply them to many existing erosional landscapes with suitable allowances made for special conditions of structure, process, and Brice's fourth genetic factor, fundamental form. The Davisian scheme thus provides a graphic and useful nomenclature for describing the degree of dissection of erosional topography.

3. It was presented in a lucid, compelling, and disarming style. Davis' rhetorical style is justly admired, and several generations of readers became, in Bryan's words, "slightly bemused by long, though mild intoxication on the limpid prose of Davis' remarkable essays" (Penck Symposium, 1940, p. 254). Moreover, Davis illustrated his texts with drawings and diagrams so apt and artfully executed that many are still reproduced or copied in textbooks.

4. It appears to be based on careful field observations. Davis illustrated his conclusions with examples of actual landscapes based on his own and others' field studies in eastern United States and elsewhere. He has been criticized for his "armchair" deductive methods of reasoning, but in fact he travelled widely and his work abounds with well-observed field studies. From what we now know of these areas, errors in Davis' work arose not from faulty observations, but from his misinterpretations of what he saw.

These four reasons for the early successes of Davis' system account in part for its ready acceptance by scholars in general, including physical geographers and geologists. The following reasons relate primarily to the attractions of his work to the geological community, which all along has shown the greatest interest in it and has defended it the most vigorously.

5. It filled a void. Before James Hutton and his followers successfully established the doctrine that came to be known as uniformitarianism, there had been several general theories of Earth history that also served to explain the genesis of landscape. After Hutton, however, there had been no general explanation for the development of landscape in uniformitarian terms, and Davis' system appeared to fill this need.

6. It synthesized current geological thought. Davis masterfully combined the ruling ideas of his time into one great system. This system is clearly uniformitarian and embraces the fluvial doctrine. It incorporated and further developed a number of concepts that were currently being discussed in the literature, such as Powell's "base-level" and genetic classification of stream valleys, Gilbert's "graded stream", and the French engineers' "profile of equilibrium". No one had previously related all of these concepts to each other, and in this way too, Davis' system made sense of diversity.

7. It provides a basis for both prediction and historical interpretation. By following Davis' scheme forward through time one can predict the sequential development of virtually any stream-eroded landscape. By following it backward one can trace the history of the landscape. Turn-of-the-century geologists may not have valued the predictive capability of Davis' system, but they clearly appreciated the ability it appeared to give them to use landform study as a tool for deciphering the later stages of Earth history. To a large extent Bryan's views of the "true function" of geomorphology as a part of historical geology reflect those of the whole geological community, and this attitude has directly influenced the acceptance or rejection of geomorphic theories in the United States. Of all the reasons why geologists accepted Davis' system this one appears to be the most outstanding.

For the foregoing reasons Davis' system won ready acceptance in the scientific community because it satisfied a number of evident needs in both geology and geography. However, one can hypothesize some additional reasons for the success of his system that may be even more important, but that are more difficult to evaluate since they relate to appeals that are more emotional than intellectual and to needs that are generally not recognized or acknowledged. To the extent that these have influenced the popularity of Davis' concepts one must consider them as possible influences on the acceptability of other geomorphic theories and thus take them into account in formulating any future theory.

8. It is rational. Davis' system was consistent with late nineteenth and early twentieth century expectations that "Natural Laws" are above all simple and rational. Geologists may have welcomed the system as a confirmation of their hope that Earth processes could be explained by simple rational models.

9. It is "vivified by evolution". The phrase is I. C. Russell's, quoted by Sheldon Judson, who added:

> "The entire concept is developmental, evolutionary. It fitted well with the new concepts of organic development which swept the scientific world of the time. The very use of the terms Youth,

> Maturity, and Old Age dove-tailed with organic evolutionary thought". (Judson, 1960, p. 197).

By applying the concept of evolution to the physical world at a time when evolution keynoted popular interest in science, Davis' work would have appealed to biological evolutionists as a confirmation of their general concept and to physical geographers and geologists who welcomed the illumination of their fields by the limelight of the current vogue.

10. It seemed to confirm current stratigraphic thought. By its reliance on a tectonic model of rapid diastrophism followed by a long period of crustal stability and rest, Davis' system seemed to confirm orogeny and unconformity as a basis for subdividing geologic time. In the late nineteenth century stratigraphers sorely wanted such confirmation.

11. It set a standard for climate. Davis' system may even have appealed to an unconscious regional chauvinism of geologists living in central-eastern United States and western Europe by designating as "normal" the moderately humid, temperate climates of their homelands.

12. It is cyclic. I have long been puzzled by a peculiar feature of the treatment of Walther Penck's work in the American literature. Although his system is essentially non-cyclic and is dependent not on time but on rates of uplift, this is rarely acknowledged by his critics. Instead, his system seems to have been attacked and rejected chiefly on the grounds that diastrophism does not behave as he thought it does and that hillslopes of the sort he thought should develop are absent in "normal" regions and therefore must result from special conditions in the now-arid regions where they do occur. It is as though the idea of cyclic development of landscape, with its deeper implications about the nature of diastrophism and Earth history, was (and in many quarters still remains) so necessary, central, and ingrained in American geologic and geomorphic thought that the very idea of non-cyclic development was not only incomprehensible but *taboo*. In many cultures taboo subjects are treated as though they do not exist. Similar treatment seems to have been given to the non-cyclic aspect of Penck's concepts, and this may help explain why Gilbert's non-cyclic treatment of landscape also received so little attention.

Lester King's geomorphic system, on the other hand, is avowedly cyclic although it also includes concepts of slope retreat very like those for which Penck's system was attacked. The relative degree of acceptance of King's theory further suggests that the cyclic nature of Davis' system had a powerful appeal to the geological community. Since a desire for a cyclic interpretation of landscape has rarely been acknowledged directly; it may stem from needs that are more emotional than intellectual in nature, including several of those mentioned earlier, such as a desire for confirmation of current stratigraphic thought and reassurance that Earth processes and history are simple and rational. Beyond this, however, a desire for cyclic interpretation may be a legacy from the early development of geologic and geomorphic thought that may still retain force as an unacknowledged fundamental postulate of which we are unaware.

A doctrine of cyclic Earth history predates Davis and uniformitarianism, and can be traced back to the Bible. G. L. Davies (1969) has documented the influence of this doctrine on Western geologic thought, from its Renaissance beginnings through Hutton's "Theory of the Earth".

To sixteenth and seventeenth century Western theorists Nature and Scripture were equally revered sources. Their studies were therefore constrained not only by the limited time span allowed for Earth history by the Bible and by a need to account for effects of the Flood, but especially by a concept derived from the Book of Genesis of a great cycle of Earth history. Davies writes that in the sixteenth and seventeenth centuries:

> "The history of the world was widely regarded as divisible into three distinct phases: firstly, there had been a period of generation extending from the Creation up to the Fall of Man; secondly, there was the prolonged and present period of degeneration initiated by the Fall; and thirdly, there was the eagerly awaited period of regeneration that would be ushered in by Christ's Second Coming." (Davies, 1969, p. 6).

Rugged mountains and deep canyons were seen as evidence of this degeneration and decay. Early theorists, including Steno and Werner, thus accounted for the formation of the Earth's rocks and surface features within the framework of a great, but brief, single cycle of development. Part of James Hutton's great contribution to geology was to discard the too-brief Biblical chronology and to replace the single cycle of creation and destruction with an endless series of cycles, each one beginning with uplift followed by weathering, erosion, and deposition and lithification of the resulting sediments. The cycle ended when the continental mass was worn down and the new sedimentary rocks were elevated to form a new continental mass.

Clearly, Davis' "Geographical Cycle" is a direct descendant of Hutton's "Theory", with the emphasis on landform development in the light of the fluvial doctrine. This further illustrates Davis' masterful synthesis of the ruling ideas of his time. Even though the cyclic aspects of earlier doctrines were not much discussed in late nineteenth century geologic literature, the general concept may have remained as a tacitly accepted or even unconsciously adopted premise in geological thought. This may help explain why Davis formulated a system that was cyclic and why this system was so widely and immediately accepted.

Of the foregoing reasons for the early and ready acceptance of Davis' system, which are responsible for its continued popularity? Certainly its simplicity (#1) and its easy applicability (#2) still make it attractive for inclusion in elementary textbooks, laboratory manuals, and classes in physical geography and geology. But what of the other reasons?

By now we should no longer be disarmed by Davis' "limpid prose" (#3), even though I have observed that students tend to be less critical of Davis' concepts after they have read his own presentations of them. Also, we now have a heritage of carefully observed field studies (#4) much more varied and detailed than those that were available to Davis. Indeed, at least two other general theories of landscape development are based on such studies, so that Davis' system no longer fills the void in geomorphic or geologic thought (#5) that it once did. An evolutionary view of landscape development (#9) is no longer needed to support the concept of organic evolution, in part because biologists have since developed less deterministic, more dynamic concepts of evolution. Nor is it needed to cast a favorable light on the Earth sciences; on the contrary, with growing public controversy over evolution, Earth scientists would tend to avoid such publicity.

Unquestionably Davis' system *is* still valued for providing a basis for predicting and interpreting landform changes and the later stages of Earth history (#7). King's system also provides such a basis, and is accordingly accepted in some quarters. Systems, such as Penck's, that are not time-dependent do not so readily provide such a basis, and are accordingly rejected or ignored. It is hard to know how many scholars may still be swayed by Davis' chauvinistic definition of "normal" climate (#11), but one may safely regard any such appeal as trivial.

It is more difficult to assess the current effectiveness of the remaining four reasons for Davis' original success.

It would seem that Davis' tectonic model of short-lived diastrophism followed by a long period of crustal stability and rest is no longer needed to confirm stratigraphic thought (#10), because most stratigraphers long ago outgrew their need for short-term world-wide orogenies. However, the plate tectonics models now popular in geology seem to entail, at least in part, just the sort of diastrophic activity that Davis and King postulated. Hence their theories have gained added popularity among geological theorists eager to find in landform studies confirmation of deductions based on plate tectonics. The concept of rapid diastrophism followed by long stability also resembles that postulated by Hutton, and thus is one of the several ruling concepts that Davis incorporated in his system.

To the extent that the ruling ideas in geologic thought have changed since 1899, Davis' system is no longer desirable as a synthesis (#6). But this is difficult to evaluate, because for many conservative scholars there has been little or no change in these concepts. For them, geologic and geomorphic thought is still guided by the principle of uniformitarianism, the fluvial doctrine, and by the concepts of grade and profile of equilibrium in much the same form as that in which they were first presented. For less conservative scholars some or all of these guiding principles have undergone slight to great modifications since Davis' time or have been augmented or replaced altogether by other principles, such as Gilbert's "law of interdependence of parts" and concepts of randomness and allometric growth. Conservatives, then, may find little reason to alter the basic framework of Davis' system. For them the system may still hold considerable appeal as a masterful synthesis of basic principles. Until they become convinced that these principles must be modified or replaced, they will remain understandably scornful of any general theory that does not incorporate them.

The appeal of the rationality of Davis' system (#8) is closely related to its appeal as a grand synthesis, since the guiding principle of uniformitarianism is essentially a special geological case of the philosophy of a rational universe, a fundamental belief that underlies much of modern science, and that states in effect that "every detailed occurrence can be correlated with its antecedents in a perfectly definite manner, exemplifying general principles". The phrase is A. N. Whitehead's, quoted by J. W. N. Sullivan (1949, p. 11). It is echoed in G. K. Gilbert's statement that "Phenomena are arranged in chains of necessary sequence" (1886, p. 285). Sullivan went on to comment:

> "It may be, as Eddington has hinted, that the universe will turn out to be finally irrational. This would mean, presumably, that science would come to an end." (1949, p. 12.)

From this it would appear clear that scientists themselves have a large personal interest in the continuing appearance of rationality of the universe, and that such rationality is evidenced by linear and deterministic cause-and-effect relationships.

In certain branches of physics and astronomy this fundamental doctrine has already yielded to concepts of a relativistic universe conceivable only in terms of probability. There are not yet signs of such a revolution in geology, but studies of geomorphic processes by L. B. Leopold, S. A. Shumm, and many others have suggested that the resulting actions and forms can best be described and understood in probabilistic terms. John Mann has reviewed such studies under the title, "Randomness in nature," and he comments that new research methods "have freed scientists from the restraint of deterministic methods and related concepts of strict causality in the analysis of natural phenomena" (1970, p. 95). To many geologists, however, such concepts seem to imply the end of their science, at least as they know it, and they have understandably tended to reject or to ignore the conceptual revolution that seems to be occurring within geomorphology. The appeal of the rationality of Davis' system thus remains very powerful in the geological community.

The cyclic nature of Davis' system (#12) also seems to retain a powerful appeal. This can be deduced from the cool reception given in many quarters to John Hack's attempts to arouse interest in Gilbert's non-cyclic concept of "dynamic equilibrium" and to other so-called "open-system," non-cyclic models of landform development. If the desire for a cyclic, time-dependent model stems from an unacknowledged fundamental postulate that the history of the Earth itself is cyclic, then no non-cyclic theory of landscape development can win general acceptance until this postulate is unearthed, examined, and possibly rejected.

To summarize, Davis' system continued to attract favorable attention for several reasons. Its simplicity and easy applicability make it attractive as a teaching tool in both geology and physical geography. To many geologists it has additional virtues: it provides a rational basis for prediction or historical interpretation of landform change; it synthesizes a number of concepts that retain the conviction of principles in many quarters; and it is cyclic. Since Davis' system still seems to possess so many favorable attributes, it is appropriate to ask at this point:

WHAT'S WRONG WITH THE DAVISIAN SYSTEM — OR — WHY DO WE STILL LACK A SATISFACTORY GENERAL THEORY OF LANDSCAPE DEVELOPMENT?

To take the second question first, the answer depends on what one means by "successful" and on the person giving the answer. Staunch supporters of Davis and King might deny that we lack a successful theory, but "successful" should mean that the theory is widely acceptable, and this is clearly not the case, as critics of the major systems have shown. The objections raised through the years by Davis' many critics, from Tarr (1898) to Twidale (1975, in press), serve to illuminate areas of disagreement. R. C. Flemal (1971) has ably reviewed these objections, finding that they include Davis' concepts of uplift, grade, baselevel, peneplain, the development of slopes, stream networks and channels, the normalcy of climate, and time. This list seems to comprise all the major elements of Davis' "structure, process, and time". With modifications many of the same objections have been raised to King's system. It would seem that one reason we lack an acceptable theory of landscape development is that there is as much diversity of opinion about structure, process, and form as there is diversity among structures, processes, and landforms themselves.

Generally in long-standing controversies of this kind it eventually appears that there is some kind of truth on all sides. Much of this controversy has stemmed from a desire for simplicity — for a single concept of landscape development. But there may be simply too much diversity, not just among opinions but in nature itself, to be accommodated within a single simple conception. True simplicity here may best be served by breadth and flexibility rather than by narrowness of outlook, and resolution may ultimately be achieved only by recognizing that, depending on conditions and circumstances, there may be *several* modes of diastrophic activity, slope and channel development, and penultimate erosional surfaces.

Such flexibility of outlook may eventually enable us to accommodate a wide variety of viewpoints and conditions in one general system. Studies of process and form that have been carried on so intensively during the past two or three decades tend to confirm that multiple rather than single solutions to the problems are required. For example, S. A. Shumm (1966) and British investigators, such as R. A. G. Savigear (1952) and A. Young (1963), have transcended the older Davisian vs. Penckian controversies over slopes by finding that slopes may either wear down and decline or retreat with a constant angle depending on conditions. The task now is to learn just what the conditions are for each mode of development.

In recent years, following Strahler's lead (1952, p. 935), a number of writers (e.g., Chorley, 1962) have attempted to characterize geomorphic theories of landform development in terms of concepts derived from general theory of thermodynamic systems. In these terms Davis' "Geographical Cycle" is identified as a *closed system*, which receives all of its energy at the outset and only reaches equilibrium and a steady state at the end, when the machine has run down, so to speak. Gilbert's concept of dynamic equilibrium, on the other hand, is cited as an example of an *open system*, which requires a continuing flow of energy from the outside in order to maintain a steady state. Such analyses can be very useful, but they have tended to obscure rather than clarify the objectives of the search for an acceptable general theory of landscape development, and they have also been widely misinterpreted.

Some writers, for example, seem to have decided that open systems are "good" and closed systems are "bad", and have accordingly used these terms to express judgments about various geomorphic theories. Others seem to have equated "open system" with "steady state". However, Lee Wilson has pointed out that in open systems there are two forms of dynamic equilibrium — *steady state* and *growth*. Negative growth can result in the kind of evolutionary development of landscape conceived by Davis (Wilson *in* Harris and Twidale, 1968, p. 240). Moreover, as commonly applied, the terms "closed" for Davis' or King's systems and "open" for Gilbert's or Penck's refer only to the tectonic assumptions underlying each system and no more.

Applications of general systems theory to concepts of landform development may be useful in helping to evaluate the dynamics involved and as tests to show whether the concepts are inclusive and consistent — that is, whether they include all the dynamic factors that need to be considered, and whether the resultant change or stability of form is correctly accounted for. Such applications can help us to understand *what happens* at any particular point — in time or in place — in relationships between processes and forms, and they may help us to learn which sequences of landforms are probable and which are improbable, but they cannot be substituted for careful field observations in determining what sequences of forms have actually occurred in various tectonic and climatic settings on the Earth.

To my mind — and this reflects a very biased judgment — the main reason that we still lack satisfactory theories of landform development is that in most geomorphic field studies investigators have misinterpreted the relationships between form and process because they have failed to recognize that in many parts of the world the gross forms of the landscape are relics formed by processes no longer operating there.

Since the work of Peltier (1949, 1950), Clark (1968), and others, it is now generally recognized that landforms of the northeastern United States that Davis believed were formed in a humid temperate climate like that of the present were in fact formed under periglacial conditions. It is less widely recognized that the stream-eroded and karst landforms of many desert regions, which in places are now being buried under wind-blown sand, must be products of a climate different from the present one. If these stream valleys reflect a climate that was formerly wetter, then what of the pediments that occur in the same regions? Almost all writers, including Lester King, seem to have assumed that since pediments now occur in arid and semiarid regions they must form under arid and semiarid conditions. Where pediments have been identified in humid temperate regions it is generally assumed either that these regions must once have been drier or that pediments may form in humid temperate regions as well as in arid and semiarid ones. The latter is King's view.

However, the abundant evidence that arid and semiarid regions were formerly wetter, and that major landforms there, including pediments, are now undergoing dissection or other alterations that are disharmonious or discordant with the forms themselves, can clearly lead to a very different interpretation — namely, that pediments are products of erosion in humid temperate climates and that they are relic landforms in the arid and semiarid regions in which we now find them. L. C. Peltier's field studies in the northern Appalachians and in Missouri are among the few that have attempted to distinguish between landforms that are relic and those that result from present processes. Significantly, in his idealized diagram (1950, Fig. 9) reproduced here as Figure 2, the forms of pre-glacial and present slopes and channels appear to resemble those generally credited to desert regions.

A chief problem with relic landforms is that in most places we know only that they were not formed by the combination of processes now operating there in the present climate. We do now know what combinations of processes *did* form them, nor in what kind of climatic regime, nor when. Such indications of changes of effective processes do not necessarily mean that the relic forms are Pleistocene; Claudio Vita-Finzi (1969) has reported widespread relic alluvial forms now undergoing stream incision and dissection in the Mediterranean Basin, and he relates this change in activity to events, possibly climatic changes, that have occurred within the past 100 to 500 years. Similar evidence of fairly recent changes in processes and associated forms appears to be worldwide.

I expect to develop and document these notions about the significance of relic landforms at some other time and place. For now it is enough to say that failure to recognize them has led to many misconceptions about relationships between landforms and the climates and processes responsible for them, and that these misconceptions are in turn responsible for much of the diversity of opinion about the development of landforms. A discussion of the problems posed by relic landforms in H. F. Garner's new textbook (1974, p.

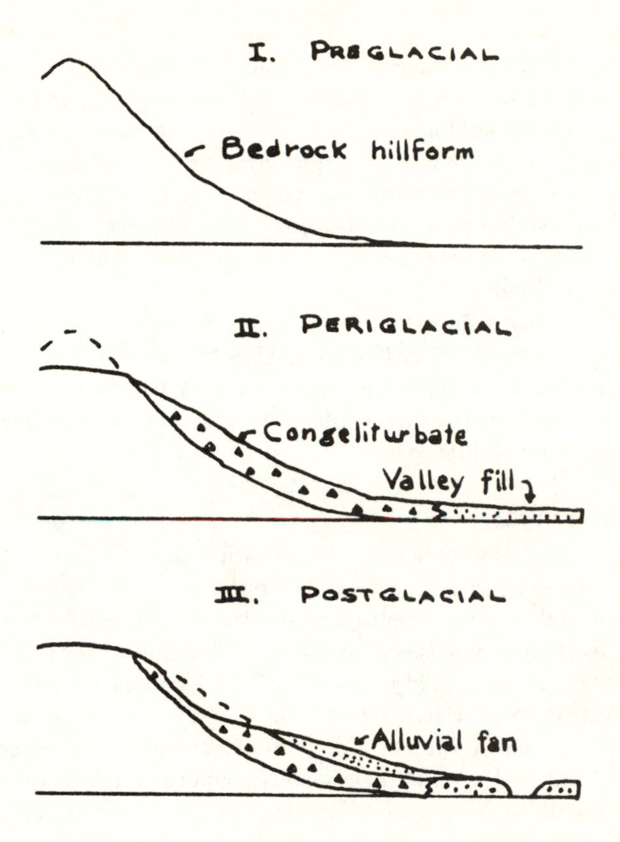

Figure 2. Idealized diagram of the development of composite or polygenetic landscape in Missouri. Slightly modified from L. C. Peltier, 1950, Fig. 9. Relic concavo-convex Pleistocene periglacial slopes still dominate the landscape, leading W. M. Davis to believe that they were formed in the present environment. However both the preglacial slopes and active post-glacial modifications, which *are* presumably formed under humid temperate conditions, resemble landforms generally associated with desert regions.

18-24) gives hope that there may be a growing awareness of relic forms and their significance. If this stimulates new field studies of actual relationships between processes and the forms they are actively producing in today's climatic environments, we may soon have more reliable information on which to base a satisfactory general theory of landscape development.

FUTURE GEOMORPHIC THEORY

A future for geomorphic theory seems assured by the needs of geologists for a sound basis for historical interpretation of landscape, of environmentalists and planners for a sound basis for predicting Man's effects on the landscape, and of the science itself for a means of maintaining communication between its descriptive, genetic-historical, and process-oriented lines of inquiry. Moreover, since geomorphic theories seem to come in their own cycles of about twenty-five years — Davis, 1899, Penck 1924, King (and Penck translated) 1953 — the time seems right for renewed interest in the synthesis of geomorphic thought.

We can even now gain some idea of what any successful future theory of landscape development will be like, at least in general outline. The form that such a theory will take will be determined largely by the needs that it must serve and by the fundamental postulates or assumptions and laboratory and field studies on which it is based.

Needs to Be Served.

There has been widespread and long-continued use of Davis' geomorphic system by teachers — in textbooks and classrooms — who find its terminology simple, memorable, and generally applicable to erosional landscapes. No generally acceptable new theory is likely to fill these needs for simplicity, memorability, and applicability as well as Davis' system appears to do unless it incorporates Davis' classic terminology. This can be done simply by adopting the common current practice of stripping the terms "youth", "maturity", and "old age" of their genetic implications and retaining them solely for describing the degree of dissection of erosional topography.

Other sources of needs are geologists, planners, and geomorphologists themselves. Formulation of any new geomorphic theory must take the needs of geology into account. Most academic and professional geomorphologists in this country are housed in geology departments of various sorts. So, to mix metaphors, since geology pays the rent it expects to call the tunes. That is to say that geologists have tended to support geomorphology because they have certain needs that they think geomorphic studies can help to satisfy. A chief geological need is for a geomorphic system that provides a basis for historical interpretation of landforms and landscapes. Geologists will also want any future geomorphic theory to conform to currently accepted stratigraphic and tectonic models and to other aspects of current geological thought. They may also expect it to explain Earth processes by simple rational models expressed in terms of causalities, and they may object to any new theory that is not evolutionary and cyclic.

Planners need a basis for evaluating and predicting future changes in landforms and landscapes, and will have no use for any theory of landscape development that does not provide such a basis. Geomorphologists themselves will expect such a theory to be consistent with current geomorphic thought and the results of studies of form and process. Finally, any widely acceptable new theory must be presented in clear enough language to be easily and unmistakably comprehensible to a wide audience.

The ideal new theory, then, must include simple terms for describing topography, must conform to current geologic and geomorphic thought, and must provide a basis for historical interpretations and future predictions of landscape changes, preferably in terms of evolutionary and cyclic rational models. Whether or not these conditions can be met depends largely on the assumptions on which the theory is based.

Basic Assumptions to Be Made.

The postulates or assumptions on which any theory is based largely determine its substance, so it is important that they reflect real conditions. It is especially important to identify and examine any assumptions that are normally not acknowledged or that tend to be taken for granted. Such assumptions not only concern the nature of crustal disturbances and the stability of climatic environments, but also involve deeper questions as to whether Earth history and landscape development is cyclic, or indeed whether the general form of some landscapes changes at all. There seems to be at least three major combinations of such assumptions:

1. That the forms of landscapes do change through time and that these changes are progressive and cyclic. These assumptions lead to a closed-system model of landform development or to an open-system model that exhibits growth of some kind. This case seems to require further assumptions of rapid uplift followed by crustal stability, or of progressively changing diastrophic activity and/or climate punctuated by widely spaced episodes of drastic change.

2. That the forms of some landscapes do not change through time but remain constant as continuing uplift balances erosion. These assumptions lead to open-system steady-state models of landform development. Such cases seem to require a constant input of diastrophic activity in an unchanging climatic regime or possibly a delicate interaction between progressively changing uplift and climate.

3. That the forms of landscapes may or may not change through time, depending entirely on the conditions of diastrophic activity and climate. If the latter develop in progressive and cyclic ways, so will the landscapes. These assumptions lead to open-system models that may exhibit either a steady state or growth of various kinds depending on conditions.

What we know of stratigraphic history and plate tectonics and of changing Tertiary and Quaternary climates suggests that the third set of assumptions listed above is the more likely to be adopted in future geomorphic theories. These assumptions lead to a concept of landscape development that accounts in a relatively simple manner for a great diversity of existing landforms and landscapes and of inferred modes of tectonic behavior. Unfortunately this concept is not necessarily evolutionary or cyclic, and because it would yield an equation for landform development with at least three variables — *diastrophic activity, climate,* and *form* — it cannot be used to interpret or predict changes in landform unless the behavior of the other variables is already known or assumed.

More basic postulates concern the uniformity and rationality of Nature. It is most likely that future geomorphic theories will reflect a broadly uniformitarian view of Earth history, at least to the extent of denying any major roles in it to cosmic collisions and other adventitious calamities. However, the relative importance of ordinary catastrophies in land sculpture is still disputed. Although Wolman and Miller have argued that most geological work is done by "events of moderate magnitude which recur relatively frequently" (Wolman and Miller, 1960, p. 72), distinctive long-lasting landforms have clearly resulted from rare catastrophic floods and rockfalls. The part played by such catastrophic events in long-term development of landscape needs further study.

Even more fundamental than uniformity is the question of the rationality of Nature. Past geomorphic and geologic thought has been founded on the assumption that natural processes operate in a rational cause-and-effect manner, that is, in "chains of necessary sequence", to use Gilbert's phrase. Some geomorphologists, however, believe that geomorphic processes are best explained in terms of randomness and probability. Doubtless a general theory of landform development could be based on this concept of Nature and cast in probabilistic terms, but it is doubtful that it would be acceptable to the geological community unless it is also summarized in deterministic terms.

Beside these basic assumptions about the rationality, uniformity, and progression of natural phenomena, some other basic concepts must also be carefully evaluated. These include long-debated ideas about "grade" of streams and slopes, of a "baselevel" of erosion, of the manner in which hillslopes form and change, and of erosional surfaces of low relief. In view of the long-standing controversies over these topics it is likely that no one view is entirely correct or entirely false, and that a satisfactory theory will have to be flexible enough to accommodate a variety of data and opinions.

Input from Laboratory and Field Studies.

The needs to be served by any future geomorphic theory will largely determine its success or failure, and the assumptions on which it is based will determine its general form. Particular features of the theory, or its specificity, will depend on the laboratory and field studies of the interrelationships between forms and processes through time on which it is based.

Because there is a diversity of process interactions resulting from a diversity of climatic regimes there must also be a diversity of modes of landform growth and change, and any successful future general theory must provide a number of alternative models of landscape development in a variety of climatic as well as tectonic settings. To ensure that the theory correctly depicts the development of landscape in any particular climatic setting, field studies on which it is based must carefully distinguish between landforms resulting from combinations of processes now acting on them in the present climate and landforms, now relic, that were produced by combinations of processes that operated there under different climatic conditions.

In conclusion, it seems clear that in the formulation of any widely acceptable general theory or body of theories of landscape development simplicity will best be served by acknowledgement of the diversity of nature, not only of landform types but of climates, climatic fluctuations, crustal movements, and other influential factors. To incorporate such diversity and the wealth of field and laboratory studies that have been made of it into one comprehensive and comprehensible system is a formidable undertaking, but one that appears feasible so long as we recognize that diversity itself is the key to understanding, and that it demands an explanatory system flexible enough to accommodate alternative modes of landform development. Such a system is long overdue.

ACKNOWLEDGEMENTS

I wish to express my appreciation to the Organizers of this Symposium, W. N. Melhorn and R. C. Flemal, and to its Director, Marie Morisawa, for this opportunity to explore further and to present a number of ideas that I have been developing over the past six or eight years but have not until now managed to set down on paper, and to Gordon L. Davies, *in absentia*, for his masterful history of British Geomorphology (1969) which has helped me to understand at least part of the reason why non-cyclic theories of landscape development have so consistently been disdained by the English-speaking geological community.

In the foregoing account I have deliberately and regretfully omitted mention of a number of recently active lines of geomorphic investigation and many able investigators and theorists whose work has contributed greatly to the development of modern geomorphic thought. Their omission here in no way reflects their worth, but simply results from the selective treatment I have chosen, with its special emphasis on the development of geomorphic thought in North America.

REFERENCES CITED

Beede, J. W. (1911) The cycle of subterranean drainage as illustrated in the Bloomington, Indiana, quadrangle. Indiana Acad. Sci., Proc., v. 20, p. 81-111.

Brice, J. C. (1964) Fourth genetic factor for sculptural relief. Geol. Soc. America, Special Paper 82, p. 19.

Bryan, Kirk (1922) Erosion and sedimentation in the Papago Country, Arizona. U. S. Geol. Surv., Bull. 730, p. 19-90.

——————— (1941) Physiography. Geol. Soc. America, 50th Ann. Vol., p. 1-15.

Chorley, R. J. (1962) Geomorphology and general systems theory. U. S. Geol. Surv., Prof. Pap. 500-B. 10p.

Clark, G. M. (1968) Sorted patterned ground: new Appalachian localities south of the glacial border. Science, v. 161, p. 355-356.

Crickmay, C. H. (1933) The later stages of the cycle of erosion. Geol. Mag., v. 70, p. 337-347.

Cvijić, Jovani (1918) Hydrographie souterraine et évolution morphologique du karst, Rec. Travaux Inst. Geog. Alpine, v. 6, n. 4. 56p.

Davies, G. L. (1969) The Earth in Decay: a History of British Geomorphology 1578-1878. New York: American Elsevier. 390p.

Davis, W. M. (1889) The rivers and valleys of Pennsylvania. Nat. Geog. Mag., v. 1, p. 183-253. See also: Geographical Essays (1909), p. 413-484.

——————— (1899a) The geographical cycle. Geog. Jour., v. 14, p. 481-504. See also: Geographical Essays (1909), p. 249-278.

——————— (1899b) The peneplain. American Geologist, v. 23, p. 207-239. Reprinted "with numerous minor changes" *in* Geographical Essays (1909), p. 350-380.

——————— (1900) Glacial erosion in France, Switzerland, and Norway. Boston Soc. Nat. Hist., Proc., v. 29, p. 273-322. See also: Geographical Essays (1909), p. 635-689.

——————— (1903) The mountain ranges of the Great Basin. Harvard Univ. Mus. Comp. Zool., Bull., v. 42, p. 129-177. See also: Geographical Essays (1909), p. 725-772.

——————— (1905) The geographical cycle in an arid climate. Jour. Geol., v. 13, p. 381-407. See also: Geographical Essays (1909), p. 296-322.

——————— (1906) The sculpture of mountains by glaciers. Scottish Geog. Mag., v. 22, p. 76-89. See also: Geographical Essays (1909), p. 617-634.

——————— (1909) Geographical Essays. D. W. Johnson, ed. Boston: Ginn. 777p. Facsimile republication 1954 by Dover Press.

——————— (1912) Die erklärende Beschreibung der Landformen. Leipzig: Teubner. 565 p. 2nd edition 1922.

——————— (1922) Peneplains and the geographical cycle. Geol. Soc. America, Bull., v. 23, p. 587-598.

——————— (1930) Rock floors in arid and humid climates. Jour. Geol., v. 38, p. 1-27, 136-158.

——————— (1932) Piedmont benchlands and Primärrümpfe. Geol. Soc. America, Bull., v. 43, p. 399-440.

Dury, G. H. (1966) The Face of the Earth, rev. ed. Baltimore: Penguin Books, 238p.

Flemal, R. C. (1971) The attack on the Davisian system of geomorphology: a synopsis. Jour. Geol. Educ, v. 19, p. 3-13.

Garner, H. F. (1974) The Origin of Landscapes. New York: Oxford. 734p.

Gary, Margaret, Robert McAfee, Jr., and C. L. Wolf, eds. (1972) Glossary of Geology. Washington: Am. Geol. Inst. 857p.

Gilbert, G. K. (1877) Report on the geology of the Henry Mountains. U. S. Geog. and Geol. Surv. of the Rocky Mtn. Region. 160p.

——— (1884) The topographic features of lake shores. U. S. Geol. Surv., Fifth An. Rpt., p. 69-123.

——— (1886) The inculcation of scientific method by example. Am. Jour. Sci., v. 31, p. 284-299.

——— (1905) Style in scientific composition. Science, n.s., v. 21, p. 28-29.

——— (1909) The convexity of hilltops. Jour. Geol., v. 17, p. 344-350.

——— (1914) The transportation of débris by running water. U. S. Geol. Surv., Prof. Pap. 86. 263p.

——— (1917) Hydraulic-mining débris in the Sierra Nevada. U. S. Geol. Surv., Prof. Pap. 105. 154p.

Hack, J. T. (1960) Interpretation of erosional topography in humid temperate regions. Am. Jour. Sci., v. 258-A, p. 80-97.

——— (1965) Geomorphology of the Shenandoah Valley, Virginia and West Virginia, and origin of the residual ore deposits. U. S. Geol. Surv., Prof. Pap. 484. 84p.

Harris, S. A., and C. R. Twidale (1968) Geomorphic cycles, p. 237-240 *in* Fairbridge, R. W., ed. Encyclopedia of Geomorphology. New York: Reinhold. 1295p.

Huxley, T. H. (1877) Physiography: an Introduction to the Study of Nature. London: Macmillan. 384p.

Johnson, D. W. (1919) Shore Processes and Shoreline Development. New York: Wiley. 584p.

Judson, Sheldon (1960) William Morris Davis — an appraisal. Zeitschr. Geomorph., v. 4, p. 193-201.

King, C. A. M. (1966) Techniques in Geomorphology. London: Arnold. 342p.

King, L. C. (1953) Canons of landscape evolution. Geol. Soc. America, Bull., v. 64, p. 721-752.

——— (1962, 1967) The Morphology of the Earth. Edinburgh: Oliver and Boyd. 699p. 2nd edition 1967, 726p.

——— (1963) South African Scenery. Edinburgh: Oliver and Boyd. 308p.

Lawson, A. C. (1894) The geomorphology of the coast of Northern California. Univ. of Calif., Bull. Dept. Geol., v. 1, no. 8, p. 241-271.

Leopold, L. B., and Thomas Maddock, Jr. (1953) The hydraulic geometry of stream channels and some physiographic implications. U. S. Geol. Surv., Prof. Pap. 252. 57p.

Mann, C. J. (1970) Randomness in nature. Geol. Soc. America, Bull., v. 81, p. 95-104.

Peltier, L. C. (1949) Pleistocene terraces of the Susquehanna River, Pennsylvania. Penn. Topog. and Geol. Surv., Bull. G-23. 158p.

——— (1950) The geographic cycle in periglacial regions as it is related to climatic geomorphology. Assoc. Am. Geog., Annals, v. 40, p. 214-236.

Penck, Walther (1924) Die Morphologische Analyse. Geog. Abh., ser. 2, v. 2. Stuttgart: J. Engelhorn. 283p.

——— (1953) Morphological Analysis of Land Forms. Translated and edited by Hella Czech and K. C. Boswell. London: Macmillan. 429p. Republished 1972 by Hafner.

Penck Symposium (1940) Assoc. Am. Geog., Annals, v. 30, n. 4, p. 219-284.

Powell, J. W. (1895) Physiographic features. Nat. Geog. Soc., Monographs, v. 1, n. 2, p. 33-64.

Ruhe, R. V. (1975) Geomorphology. Boston: Houghton Mifflin. 246p.

Savigear, R. A. G. (1952) Some observations on slope development in South Wales. Inst. Brit. Geog., Trans., Publ. 18, p. 31-52.

Schumm, S. A. (1966) The development and evolution of hillslopes. Jour. Geol. Educ., v. 14, p. 98-104.

Smith, W. S. T. (1899) Some aspects of erosion in relation to the theory of peneplains. Univ. Calif. Publ., Bull. Dept. Geol., v. 2, p. 155-178.

Strahler, A. N. (1950) Equilibrium theory of erosional slopes approached by frequency distribution analysis. Am. Jour. Sci., v. 248, p. 673-696, 800-814.

——————— (1952) Dynamic basis of geomorphology. Geol. Soc. America, Bull., v. 63, p. 923-938.

——————— (1958) Dimensional analysis applied to fluvially eroded landforms. Geol. Soc. America, Bull., v. 69, p. 279-300.

Sullivan, J. W. N. (1949) The Limitations of Science. New York: Mentor Books. 192p. Republication of 1933 edition.

Tarr, R. S. (1898) The peneplain. American Geologist, v. 21, p. 351-370.

Tilly, P. (1968) Early challenges to Davis' concept of the cycle of erosion. Prof. Geogr., v. 20, p. 265-269.

Twidale, C. R. (1975, in press) On the survival of palaeoforms. Am. Jour. Sci.

Vita-Finzi, Claudio (1969) The Mediterranean Valleys. Cambridge Univ. Press. 140p.

von Engeln, O. D. (1942) Geomorphology. New York: Macmillan. 655p.

White, S. E. (1968) Humid cycle, p. 538-541 *in* Fairbridge, R. W., ed. Encyclopedia of Geomorphology. New York: Reinhold. 1295p.

Wolman, M. G., and J. P. Miller (1960) Magnitude and frequency of forces in geomorphic processes. Jour. Geol., v. 68, p. 54-74.

Young, A. (1963) Deductive models of slope evolution. Nachr. Akad. Wiss. Göttingen, II. Math.-Physik, K1. 5, p. 45-66.

CHAPTER 2

EVOLUTION OF APPALACHIAN TOPOGRAPHY

Sheldon Judson

ABSTRACT

The orogeny which closed the proto-Atlantic ocean in mid-Ordovician time continued spasmodically through the Paleozoic and produced northwesterly drainage onto the North American craton. Drainage continued to the northwest through the Jurassic. The northwestward regional tilt was reversed to the southeast as the continental margin subsided as a result of the divergence of the North American plate from the spreading center now seen in the Mid-Atlantic ridge. Beginning with that reversal the modern streamways and associated topography have developed as rivers adjusted themselves to a changing base level in ways outlined by W. M. Davis and others.

Present-day seismic activity, evidence of Cenozoic and Cretaceous faulting, regional gravity anomalies, and very large present-day vertical movements all suggest that the Appalachian orogen is still to come to equilibrium.

GENERAL STATEMENT

William Morris Davis published in 1889, "The Rivers and Valleys of Pennsylvania". This was not the first statement of the Geographical Cycle, more generally known as the Davisian Geographical Cycle, but is the strong, enduring and most quoted basis for the Cycle. Davis' purpose in the introduction to the Pennsylvania study was "to see the causes of the streams in their present courses; to go back, if possible to the early date when" the area "was first raised above the sea, and trace the development of the several river systems then implanted upon it from their ancient beginning to the present time." (Davis 1889, p. 183).

The purpose of this paper is the same. But some caveats are in order. First, Davis' 1889 paper ran to some 70 pages, and even if this review were of similar length it could not do justice to the original. Second, in the 86 years since Davis wrote, our data on Earth history, structure, tectonics, and processes have grown enormously. In one way this eases the task. Actually, however, there are so many data now available that marshalling them becomes formidable indeed. Third, the more often and more closely that one reads Davis the more impressed one is. His grasp of time, space, and change; his command of detail; and his ability to order his information and frame his argument remind us again that we are in the presence of a giant. Certainly no strand stretched by one Lilliputian will fetter him.

FROM EARLY PALEOZOIC TO EARLY JURASSIC TIME

Cambrian to Early Ordovician Time.

The clastic sedimentary rocks of early and mid-Cambrian time are followed by carbonate deposits which continued to accumulate into lower Ordovician time in the New Jersey-Pennsylvania area. The early sources of detrital sediments were to the west and north reflecting drainage to the east and southeast. The resulting rocks are interpreted along with other evidence as marking the shelf area of a continental margin which faced on the Proto-Atlantic. (See for instance Williams and Stevens, 1975.)

Mid-Ordovician to Permian Time.

By mid-Ordovician time sediments changed in nature and source. In the New Jersey-Pennsylvania area the change is first seen in the sediments of the Martinsburg shale, which become increasingly coarse upward. Similar sedimentary activity, as well as volcanism, also occurs to the northeast and southwest. As the old Atlantic closed, compressional stresses developed between the converging plates and produced the Taconic orogeny of the Ordovician. As we progress through the Paleozoic section the basic pattern remains the same. Sediments carried from the east become progressively finer westward giving way to carbonate deposits the farther we move onto the North American craton. Rock units thin westward and units which in the geosyncline to the east reached hundreds of meters in thickness thin westward to a few meters in the relatively stable continental interior.

During this time the area was close to the border between converging plates. This proximity accounts for the westward flowing streams carrying sediments from the east. It accounted, also, for the tectonic activity that characterized the Paleozoic from mid-Ordovician through Permian times. These displacements not only produced the folding and thrusting of the Appalachians but also the metamorphism and igneous activity now recorded in the crystalline belts of the Piedmont and Blue Ridge. This tectonic activity appears to have had three major peaks, namely the Taconic Orogeny (Ordovician), the Acadian Orogeny (Devonian), and the Appalachian Orogeny (Permian).

Davis had noted this westward drainage during the greater part of Paleozoic time. The cause for reversal of this drainage to the present southeastern flowing streams was one of his major concerns, as it is ours.

Triassic to Early Jurassic Time.

Davis postulated the reversal of Appalachian drainage from northwest to southeast as a result of what he called the "Newark Depression". The evidence he cited lay in subsiding Triassic fault-block valleys and the sediments which accumulated in them. He was somewhat troubled by what he called the "Jurassic tilting".

In New Jersey and Pennsylvania Davis interpreted the northwestward dip of the rocks of the Newark System as due in part to uptilting by compressional pressures directed from the southeast. He saw this "tilting" as operating against the southeastward reversal of the old northwestward flowing streams. He was forced to postulate that the disturbance was not effective in halting the eastward drainage which he believed began in the Triassic.

We can perhaps throw some light on this problem. In the first place we can now consider that there was uplift during the Triassic rather than subsidence. The Triassic basins indeed reflect downward motion but only locally along rift valleys. Our modern analogue is with the East African rift valleys and the Red Sea rift. The heights flanking the rift valleys serve as divides to turn water east and west. Thus the Congo River heads on the western side of the African rift highlands and empties into the Atlantic over 1,700 km to the west. From the western escarpment of the Red Sea surface water is diverted westward until intercepted by the northward flowing Nile 200 to 600 km distant. From the eastern escarpment of the Red Sea the slope is eastward to the Arabian Gulf 1,000 km away. The Triassic rift belt of eastern North America must have operated in a similar fashion. By analogy the rivers of New Jersey and Pennsylvania can be pictured as flowing westward toward the continental interior. How far to the west we cannot say.

We can look upon the Triassic of eastern North America as marked by a thermally induced arching approaching 1,000 km in width and 2 km or more in elevation above the surroundings. The effect of the arching must have diminished east and west from the axis to the limit of the thermal effect.

Meyerhoff (1972) has suggested that drainage to the west ceased after the deposition of the Permian beds and that it began to flow eastward as a result of late Paleozoic tectonism. He inferred that the divide of the time approximated that of the present. I do not find Meyerhoff's argument on this aspect of Appalachian drainage persuasive. Certainly there has been more erosion in eastern Pennsylvania than in central, and more in central than in western, a point referred to later. It is reasonable that this was in part due to a higher country in the east. We find, for instance, that the Triassic basins of New Jersey and Pennsylvania were filled largely from the east and that the coarse conglomerates of Paleozoic clasts accumulated along the western margin as the result of short, steep drainages along the fault scarp (Van Houten, 1969).

I think it is very reasonable to picture the divide as related to the Triassic thermal rise superimposed on whatever Paleozoic elevations may have been present. From this divide the drainage flowed east and west. Recent studies indicate that Newark deposition extended into the early Jurassic time (Cornet, Traverse and MacDonald, 1973; Cornet and Traverse, 1975).

Later Jurassic to Cretaceous Time.

The reversal of stream flow to a southeasterly direction into the present Atlantic seems to have begun in the Jurassic as the Atlantic began to widen. The basic cause for the reversal lies in the subsidence of the continental margin. The subsidence occurred as a result of several factors, namely thinning of the continental crust, decay of the thermal rise as the continental margin diverged westward from the spreading axis, and loading of the subsiding continental margin with sediments and water. Kinsman (1975) has developed a model for the subsidence of a trailing edge continental margin from the pre-rifting stage to a stage comparable to that of the present margin of eastern North America (Fig. 1).

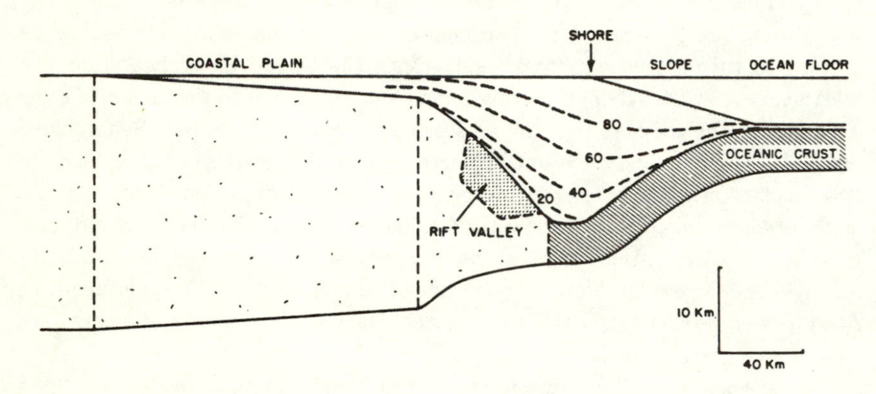

Figure 1. A model of a fully loaded, mature, divergently rifted, continental margin and adjacent oceanic crust 100 million years after rifting. The numbers refer to the evolving sediment pile in millions of years after continental rupture. (Kinsman, 1975). It is thought that, in a general way, the eastern margin of North America reflects this pattern during the first 100 million years since rupture. Subsequent history would little change the model.

Figure 2 is a generalized geologic cross-section through the New Jersey-Pennsylvania area and out to the continental rise. It represents the mature trailing edge of a continent on a diverging plate and in its development very probably followed the sequence of stages outlined by Kinsman (1975). Thus, the outer, rifted margin of the continent is shown as attenuated over a 60 to 80 km width. Thinning was most probably structural but sub-crustal thinning has also been suggested. In addition erosion prior to subsidence assisted in the thinning process. Inland from the attenuated margin thinning has also taken place but primarily as the result of supra-crustal erosion. This thinning of the continental crust was eventually expressed in subsidence of the crust as the continental margin diverged from the spreading axis. Cooling of oceanic and continental crust allowed for a large component of subsidence. And, as Kinsman points out, if the continental crust is 30 km or less in thickness it will submerge eventually beneath the sea and water loading can then add another component. Finally sediment load along the new slope and shelf caused still further subsidence in that area.

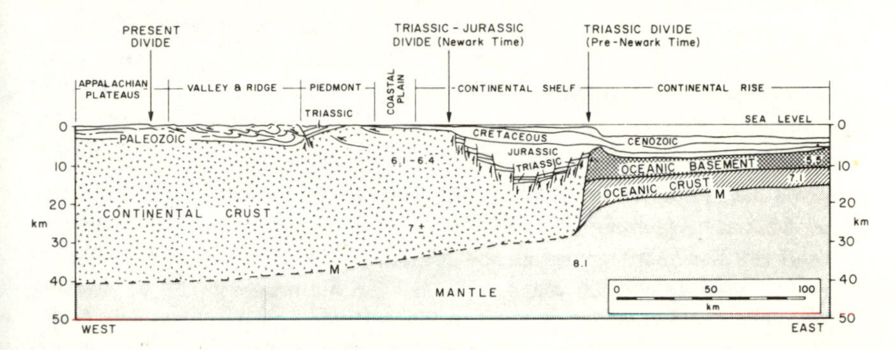

Figure 2. Geologic section of the eastern margin of the North American continent and adjacent oceanic crust in the New Jersey-Pennsylvania area. The presumed westward migration of the main east-west divide is shown. Geology compiled from various sources.

Before the subsidence, the divide between the east- and west-draining streams must have lain somewhere along the crest of the thermal ridge. With initial rifting the divide moved toward the western edge of the rift valley. With subsidence it moved farther and farther westward and an increasingly larger area was added to the drainage flowing toward the widening Atlantic.

With the widening of the Atlantic a sequence of deposits, beginning with Jurassic carbonates and terrigenous sediments, was laid down on the developing slope and shelf. These were followed by pelagic and hemi-pelagic sediments of the Cretaceous and by a thin wedge of Cretaceous shelf sediments lapping onto the thinned edge of the continent. Erosion on the subsiding continental margin ceased along the westward migrating shoreline which today is marked approximately by the position of the most westerly Cretaceous outcrop. Erosion continued to the west of this shoreline as now eastward flowing streams brought the waste of weathering to the Atlantic. The position of the present divide is discussed in a later section.

SOME OBSERVATIONS ON EROSION

We have found that streams were westward flowing from early Paleozoic time into the Jurassic, at which time the divide shifted westward and eastward drainage began. Through most of the Paleozoic the belt of erosion lay east of the present prism of Paleozoic sediments which date from mid-Ordovician to Permian. In post Paleozoic time these sediments were subject to erosion even as they are today. Davis recognized that erosion had been extremely extensive, although he had few data on which to base quantitative estimates. Some information is now becoming available on which to begin assessment of the amount of erosion and, to some extent, their rates.

Wagner and Crawford (1975), in a study of the Precambrian Baltimore gneiss in the Piedmont of southeastern Pennsylvania show that the Taconic metamorphism has been impressed on rocks that had already been modified in an older (1,000 my) metamorphic event. The later phase is dated at 440 my (Taconic) and temperatures of 650°-700°C at 7 to 8 kb are deduced. Allowing 3.2 km burial per kilobar then burial at time of metamorphism was about 24 km. Assuming that erosion and not structural movement has exposed the Baltimore gneiss at the surface, then the rate of erosion has averaged about 5 cm/1,000 years since the Taconic orogeny. R. V. Amenta (1974) maps a metamorphic succession of staurolite-kyanite-sillimanite facies in the nearby Wissahickon rocks of the Philadelphia area. This, too, argues for a similar amount of denudation if the metamorphism is Taconic, although Amenta suggests that it may be Acadian in age and if so then the average erosion rate was more rapid.

In the southern Blue Ridge, Dallmeyer (1975), on the basis of field relations, metamorphic grades, and radiometric ages, has outlined an uplift and erosional history dating from the Ordovician. He points out that metamorphism approximately 480 my ago testifies to a burial of 25 km. If erosion has been constant since that date then we again have a rate of denudation of 5 cm/1,000 yrs. Dallmeyer introduces evidence, however, to suggest that the rate of denudation was twice this (10 cm/1,000 yrs.) up to the Triassic and presumes that erosion was slower since that time and particularly in Cretaceous and Cenozoic time.

We know that in Pennsylvania and New Jersey the crystalline rocks were exposed to the east of the Triassic basins by the beginning of Newark time. On the western side of the basin a Paleozoic cover dominated throughout Newark deposition and few crystalline rocks were exposed until post-Newark time. Stratigraphic relations of the Newark series with older rocks indicate considerable erosion since Newark time. Thus on the basis of data derived from the Pennsylvania Geologic Map 1:250,000 (1960) we find that approximately 2.5 to 3.0 km of erosion has taken place in post-Newark time in the area southeast of the Susquehanna River.

It is clear that with more precise information on the grade and date of metamorphism coupled with the structural relationships of the metamorphic rocks, much information on depths and rates of erosion will be available along the Appalachian orogen. The examples given, however, suggest the great depths of denudation that have taken place.

In the Plateau district post-Paleozoic erosion has been generally less than 1 km and close to 0.25 km. In the Valley and Ridge the amount of stratigraphic section removed varies with the structure. Over the Nittany arch, erosion has removed a stratigraphic section of some 6 km. In the Broadtop syncline immediately to the east and south, the depth is 1 km or so. But in general the wedge of material eroded increases eastward, and along the border of the Valley and Ridge and Piedmont a minimum of 10 km of sediments has been removed.

Gilluly (1964) estimated that sediments of Triassic time and later, on the east coast shelf, represented a denudational rate of a little over 2 cm/1,000 yrs., a figure that he felt was conservative. Matthews (1975) computed the volume of clastic sediments off eastern North America above reflection zone "A". His figures indicate sediments equivalent to 2 km of erosion since the Eocene. This is equivalent to a denudation rate of a little over 3 cm/1,000 yrs. between the Coastal Plain and the main divide. Judson and Ritter (1964) on the basis of sediment load records of streams felt that the present rate of denudation along the basins draining to the Atlantic was about 5 cm/1,000 yr. Judson (1969) suggested that the rate of denudation before human interference in the landscape would have been about 3 cm/1,000 yrs. in this area.

Present Seismic Activity, Gravity, and Crustal Movement.

The seismicity in eastern United States and Canada, as compiled by Oliver, Isacks, and Barazangi (1974) and York and Oliver (in press), is shown in Figure 3. The area, traditionally thought of as aseismic, in reality shows zones of pronounced seismic activity. For our purposes we note that the zone in Figure 3 labeled "Appalachian Foldbelt" is very active. It is paralleled, perhaps, by a zone running from the head of the Mississippi embayment northeastward to the St. Lawrence lowland. The coastal plain and shelf are quiet except for the activity centered on Charleston, South Carolina, and two quakes recorded on the Grand Banks. The foci of quakes are shallow and probably lie within the upper crust. Data on focal mechanisms are few but Sbar and Sykes (1973) feel that the few available mechanisms in eastern North America are consistent with east-west compressional stress and may be related to the movement of the North American plate.

Regional gravity anomalies show a pattern generally parallel to the coast, as is to be expected. A widespread negative anomaly coincides with the Appalachian system and is defined in Figure 4 by a -50 mgal contour. The strongest and widest portion of the anomaly lies in the southern section where at its minimum it is less than -100 mgal in the Blue Ridge and measures over 200 km across. One can interpret the anomaly in several ways. But the most reasonable conclusion is that it represents a thicker zone of continental crust than the zones surrounding it. Seismic refraction studies yield some information on this point. Warren (1968) shows that there is little difference in the depth to the Moho at Burgaw, North Carolina, near the

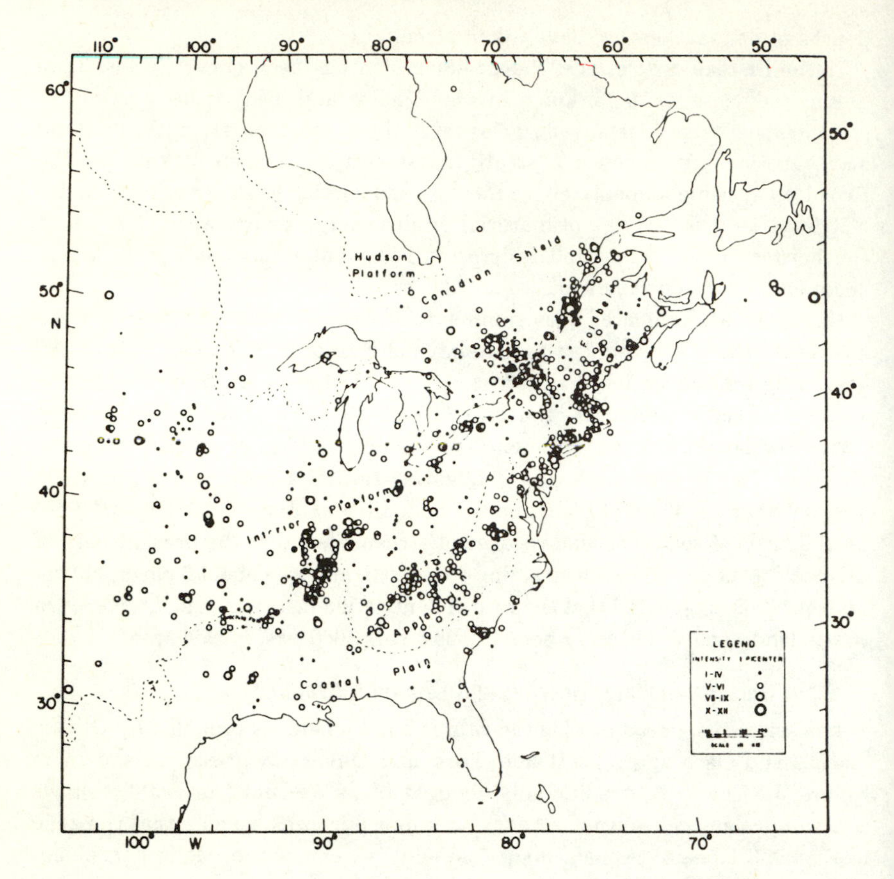

Figure 3. Distribution of reported earthquakes in eastern North America from historic and instrumental data, 1534-1971. (York and Oliver, in press)

coast on the Cape Fear arch, and at Erwin, Tennessee, near the North Carolina line and on the eastern boundary of the Valley and Ridge. But he does report a significant change in the thickness of the upper, lower-velocity (6.1 km/sec) crust. At Erwin it is 35 km thick and at the coast it is 25 km.

Leveling data obtained by the National Geodetic Survey provide some startling data on relative crustal movement in the eastern United States. An extensive analysis of these data by Brown and Oliver (in press) shows that the Appalachian belt is rising relative to the Central Lowlands and the Gulf and Atlantic Coastal Plains. Relative to the Atlantic Coastal Plain the rate of rise is up to 6 mm/yr and the rate is the greatest in the southern section of the Appalachians where the topography is the highest. The Atlantic and Gulf Coastal Plains are tilting away from the interior of the continent at $1\text{-}5 \times 10^{-8}$ rad/yr and the Central Lowlands are tilting down to the east at the rate of $1\text{-}3 \times 10^{-8}$ rad/yr.

These rates are extremely high. They represent rates nearly two orders of magnitude greater than present erosion rates. If real, they are (as Brown and Oliver point out) either episodic or oscillatory over periods much less than one million years.

THE PRESENT TOPOGRAPHY

The Present Divide.

We have argued that the divide between eastward and westward draining streams lay someplace near the rift valleys in Triassic time and, as the continental margin subsided, this divide moved westward. Subsidence was at an exponential rate and the greater part of the migration was probably accomplished by the beginning of the Cretaceous.

Figure 4 shows the position of the present divide from New York southward. Note first that it swings from the Appalachian Plateaus in New York and Pennsylvania, follows in general the western boundary of the Valley and Ridge to the vicinity of Blacksburg, Virginia, more or less in the area where tectonic style changes from one dominated by folds to one dominated by thrust faults. In this area it crosses the Valley and Ridge and the Blue Ridge and then follows the eastern margin of the Blue Ridge and swings westward with that boundary in northern Georgia. It crosses the Valley and Ridge again south of Chattanooga, Tennessee, and, as the boundary between the Gulf drainage and the Tennessee-Mississippi drainage, continues across the Plateaus to the Mississippi embayment. Through its length it roughly parallels the present Atlantic shoreline as well as the line of Cretaceous outcrop on the inner edge of the Coastal Plain.

Hack (1973) suggests that the location of the divide is in part due to the fact that streams are in disequilibrium along the southeastern border of the Blue Ridge Highland south of Roanoke. Another somewhat analogous suggestion arises out of consideration of the regional gravity pattern.

Figure 4 also shows the relation of the present divide to the negative gravity anomaly. In New York and Pennsylvania the divide has pushed to the western edge of the anomaly (-50 mgal) and beyond. Southward the divide begins to cross into the anomaly and at Blacksburg, Virginia, where the anomaly reaches -80 mgal it takes an abrupt turn toward the southeast side of the strengthening and broadening gravity low. Once across the Blue Ridge it follows along the strongest portion of the negative anomaly (less than -100 mgal). It follows around the southern end of the Blue Ridge and across the Valley and Ridge roughly coincident with the -50 mgal contour. Brown and Oliver (in press) add another bit of information that needs explanation. They point out that some peaks in relative vertical uplift correspond to the divide, and that one possible correlation of uplift patterns could call for a trend which cuts across the grain of the structure in a manner similar to the divide.

Why does the present divide lie near or west of the main anomaly in New York and Pennsylvania and cross into the southeastern side of the anomaly in the Blue Ridge section? An appealing suggestion is that the continental crust was thicker in the south than in the north. As a result the streams, reversed eastward in the Mesozoic, were able to push their divides in New York and

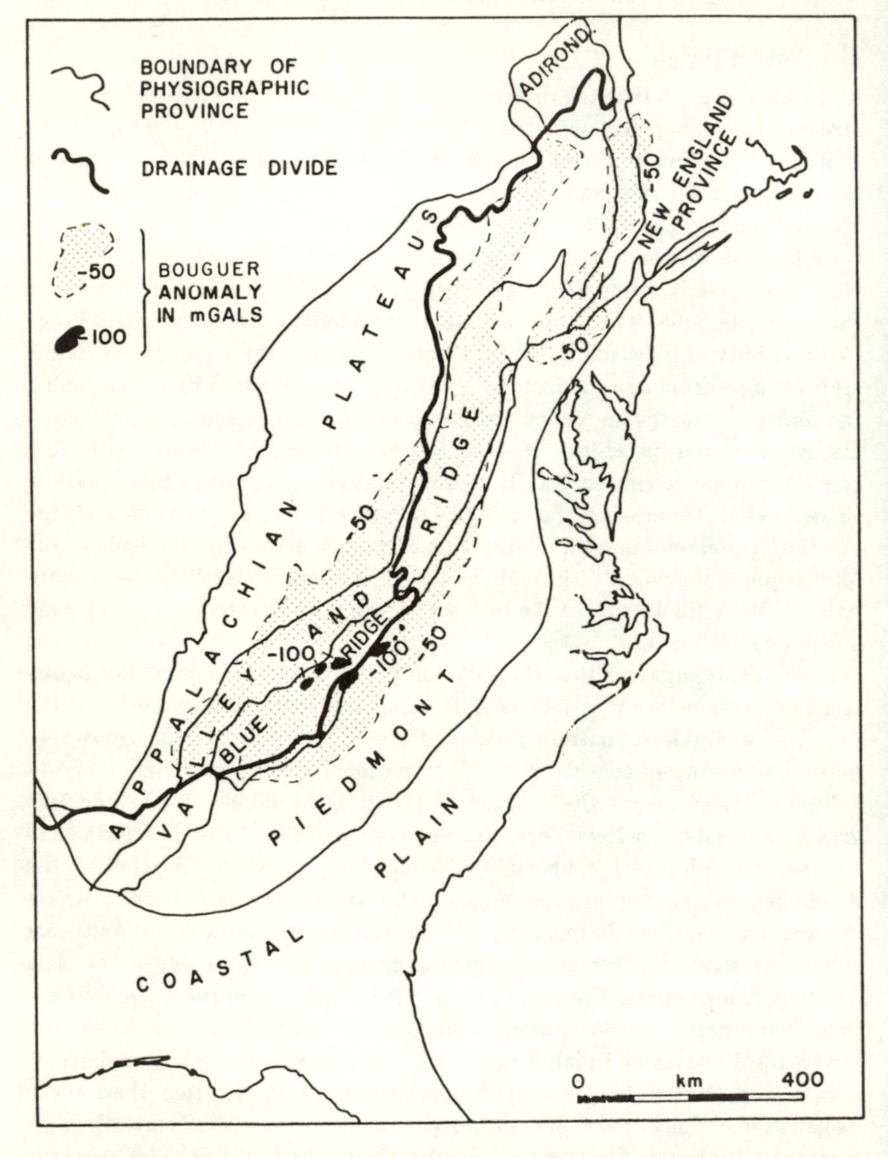

Figure 4. The main Appalachian drainage divide does not coincide with regional geologic structure physiographic provinces, or gravity data. Gravity data from Woollard and Joesting (1964).

Pennsylvania into the Plateaus. In the southern section, on the other hand, the increased crustal thickness has held up the migration of the divide along what is now the Blue Ridge escarpment.

Whatever the relation of the divide to the physiographic provinces and regional gravity low as shown in Figure 4, one can imagine that it arrived in its present position by a combination of processes. These include new streams which became consequent upon the regional tilt of the continental margin in post-rifting time, reversal in their original courses of stream or of stream segments from west to east, successive capture of westward flowing streams by streams on new, easterly directed gradients, superposition from a sedimentary cover and superposition structurally controlled. [On this last mentioned mechanism see Meyerhoff 1972.] Capture is evident along the Shenandoah which has worked its way southwestward along the strike of non-resistant rocks west of the Blue Ridge. Whether it will eventually capture the headwaters of the James River still farther southeast is not clear. But certainly it would not take much for the Roanoke, next to the south, to capture the headwaters of the westward draining New River near Blacksburg.

The Courses of Present Streams.

The direction of drainage across structure is not a problem for us. Nor was it with Davis. Only a regional tilt of surface is necessary. Surface runoff will find its way asking only the line of the lowest way to the sea. This Davis recognized in the case of the streams draining westward from Appalachia. His discussion of the origin and adjustment of streams across the newly-created folds of the Appalachians is ingenious, well-argued, and reasonable. Oberlander (1965) offers us a modern analogue of the process from the Iranian fold belt across which streams pick their way westward toward the Arabian Gulf.

Davis saw also that the present eastward drainage demanded an eastward regional tilt toward the Atlantic. He thought the tilt formed in the Triassic and we now can say, with some confidence, that it formed after the rifting of Gondwana, becoming effective sometime in the Jurassic. Davis outlines in detail the processes by which more and more area was brought into the Atlantic drainage. We may wish to change some of the details, or question some of the specifics, but the principles of the adjustment of streams, as they established essentially their modern courses, remain unchanged.

The adjustment which streams in our area went through, both during their westward-flowing history and their subsequent history as they developed their eastward courses, lies in the late youth and early maturity in Davis' Geographical Cycle. Davis postulated that thereafter the streams and the region they drain pass through later maturity and on into old age until the end product, the peneplain, is achieved. And this brings us to our real problem with Davis' Appalachian erosional history.

Davis, we will remember, felt that after initial and rapid uplift occurred stillstand persisted for a very long period of time to allow the old age stage to develop. In the Appalachians, Davis saw the evidence of the old age stage — the peneplain — on the ridges and uplands held up by the most resistant rocks. This was the Schooley-Kittatinny surface. It's age need not concern us although he saw it as having developed as the result of erosion in Jurassic and Cretaceous time. It was from this surface, regardless of its age, that he developed the details of the present landscape in our area. He called upon Tertiary uplift (rejuvenation) to start renewed erosion etching out the less resistant rocks and to form a lower level, the Harrisburg surface. Erosion in this second and partial cycle did not have time to go to completion and erase the more resistant ridges carrying the evidence of the Schooley peneplain. Still a third surface, the Somerville, came into being as the result of another uplift which allowed even less time for erosion than did the uplift which brought the Harrisburg into being.

Inherent in this process is another assumption, namely that after uplift and differential erosion left stranded the remnants of a surface it was essentially immune to further lowering. I will not discuss this here and merely assert that I find the assumption difficult to accept.

We may now ask whether we have any independent information that would lead us to accept or reject the long period of stillstand needed for peneplain development. We know now that the Appalachians throughout their extent are the site of, not intense, but appreciable, seismic activity. We know, too, that geodetic leveling has shown that there is vertical crustal movement from the coast across the Appalachians with very remarkable figures for relative vertical uplift along the Appalachians being suggested. We must admit that the uplift figures of the magnitude given (millimeters per year) do not represent a steady state that has persisted over a long period of time. If they did we would be studying the equivalent of the Himalayas. For instance, 6mm of uplift per year will produce 6km of uplift if continued for a million years. Assuming reality of the motion, regardless of its magnitude, we still do not know its cause.

But I think we can assign some possible causes to the seismic activity and the vertical crustal movement in the Appalachians. First, the site is within a plate and also within its continental segment. We are now convinced that the great bulk of tectonic activity occurs along plate boundaries. But here we have activity in an intraplate location. It seems reasonable to assume that this must in some way be due to plate movement. And if this is borne out we have no reason to believe that the activity has not been with us since the North American plate began to move. In other words we can advance the assumption that crustal unrest has been continuous from rifting time to the present.

Second, gravity data suggest that the Appalachians are underlain by an excess of continental crust. We would then expect isostatic adjustment upward. Added to this should be a similar adjustment due to continued erosion.

To me it seems more reasonable to assume that we have had long-continued crustal movement in the Appalachians than that the area was stable over long, uninterrupted periods. The geologic record lends some support to this thought. Thus York and Oliver (in press) discuss the Cretaceous and Cenozoic faulting in eastern United States. It is clear that, although the motion has been modest, there is enough of a sample to suggest that it has been widespread and continuous. Faulting in the eastern United States has been long continued and, by extrapolation, so has the seismic activity.

CONCLUSIONS

I think that we can safely say that the origin of transverse drainage in the Appalachians is not a real problem. We have the mechanism (orogeny attendant upon converging plates) that gave us the regional northwestward tilt and resulting in drainage to the northwest across our Appalachian orogen until the Jurassic. We have a reasonable mechanism to invoke to reverse that tilt to the southeast in the Jurassic (subsidence of the continental margin as a result of divergence of the North American plate from the spreading center now seen in the Mid-Atlantic Ridge).

Davis and others have outlined the ways in which streams adjust themselves to changing baselevel. This involves streams taking a consequent course as a result of regional tilting, reversal of stream segments for the same reason, capture of streams by adjustment to differing lithologies and structural geometries presented to the streams as they deepen their valleys, and superposition from sedimentary or structural covers.

We have a reasonable basis for challenging the Davisian assumption of long periods of stillstand and the resulting development of the peneplain in the Appalachian region. Present day seismic activity, although modest by the standards of activity found along plate boundaries, is real. Furthermore, the geologic record suggests that intraplate activity has been with us from the Cretaceous on. Vertical movements, up in the Appalachians, and down in the flanking Coastal Plain and Interior Lowlands, take place at the present time. Isostatic adjustment may explain some of this motion but by no means all (or even a small part) of its apparent magnitude. All in all, however, we have real reason to hypothesize motion in the Appalachians from post-rifting time to the present.

If you allow me the assumption that there has been more or less continuous crustal movement in the Appalachians rather than long periods of stillstand then it seems to me that seeking a better understanding of Appalachian landscape lies in two directions. First, we need to know more about the nature of the crustal movement. We need to know its distribution in time and space, its rates, and its causes. Second, it seems we are searching for surficial mechanisms and processes which will account for the details of the

Appalachian landscape. These details include Davis' Schooley peneplain, and the newer surfaces described by him. Succeeding authors discuss the mechanics of landscape development and from quite different points of view. What they have to say will be applicable to the origin of the present Appalachian topography, a question that nearly a century after it was posed by Davis is still to be answered to our satisfaction.

It would seem fitting to close this report with a quotation from William Morris Davis taken from his paper on the "Rivers and Valleys of Pennsylvania":

> "If the postulates that I (have used) seem unsound and the arguments seem overdrawn, error may at least be avoided by not holding too fast to the conclusions that are presented, for they are presented only tentatively. I do not feel by any means absolutely persuaded of the correctness of the results, but at the same time deem them worth giving out for discussion."

ACKNOWLEDGEMENTS

I have benefited from comments by William E. Bonini, Kenneth S. Deffeyes, Alfred G. Fischer, John T. Hack and Howard Meyerhoff on an early draft of this report.

REFERENCES CITED

Amenta, R. V., 1975, Multiple deformation and metamorphism from structural analysis in eastern Pennsylvania: Geol. Soc. America Bull., vol. 86, pp. 1647-1660.

Brown, Larry D., and Oliver, Jack E., (in press), Vertical crustal movements from leveling data and their relation to geologic structure in the eastern United States: Review of Geophysics and Space Science.

Cornet, Bruce, Traverse, Alfred, and McDonald, N. G., 1973, Fossil spores, pollen, and fishes from Connecticut indicate early Jurassic age for part of the Newark Group: Science, vol. 82, no. 4118, pp. 1243-1246.

Cornet, Bruce, and Traverse, Alfred, 1975, Palynological contributions to the chronology and stratigraphy of the Hartford basin in Connecticut and Massachusetts: Geoscience and Man, vol. 11, (April) pp. 1-33.

Dallmeyer, R. D., 1975, Incremental $^{40}Ar/^{39}Ar$ ages of biotite and hornblend from retrograde basement gneisses of the southern Blue Ridge: their bearing on the age of Paleozoic metamorphism: Amer. Jour. Sci., vol. 275, pp. 444-460.

Davis, W. M., 1889, The rivers and valleys of Pennsylvania: Nat. Geogr. Mag., vol. 1, pp. 183-253.

Gilluly, James, 1964, Atlantic sediments, erosion rates, and the evolution of the Continental Shelf: some speculations: Geol. Soc. America Bull., vol. 75, pp. 483-492.

Hack, John T., 1973, Drainage Adjustment in the Appalachians: pp. 51-69, *in* Morisawa, Marie, (ed.) Fluvial Geomorphology: Publications in Geomorphology, State University of New York, Binghamton, N. Y., 314 p.

Judson, Sheldon, 1968, Erosion of the land, or what's happening to our continents?: American Scientist, vol. 56, pp. 356-374.

Judson, Sheldon, and Ritter, Dale F., Rates of regional denudation in the United States: Journal of Geophys. Research, vol. 69, pp. 3395-3401.

Kinsman, David J. J., 1975, Rift valley basins and sedimentary history of trailing continental margins: pp. 83-126, *in* Fischer, A. G., and Judson, Sheldon (eds.) Petroleum and Global Tectonics: Princeton University Press, Princeton, N. J., 322 p.

Matthews, W. H., 1975, Cenozoic erosion and erosion surfaces of eastern North America: Amer. Jour. Science, vol. 275, pp. 818-824.

Meyerhoff, Howard A., 1972, Postorogenic development of the Appalachians: Geol. Soc. America Bull., vol. 83, pp. 1709-1728.

Oberlander, Theodore, 1965, The Zagros Streams: a new interpretation of transverse streams in an orogenic zone: Syracuse Geographical Studies No. 1, Department of Geography, Syracuse, N. Y., 168 p.

Oliver, Jack E., Isacks, B. L., and M. Barazangi, 1974, Seismicity and continental margins: pp. 85-92 *in* Burk, C. A., and Drake, C. L., The Geology of Continental Margins: Springer-Verlag, New York, 1009 p.

Sbar, M. L., and Sykes, L. R., 1973, Contemporary compressive stress and seismicity in eastern North America: an example of intraplate tectonics: Geol. Soc. America Bull., vol. 84, pp. 1861-1862.

Van Houten, F. B., 1969, Late Triassic Newark group, north-central New Jersey and adjacent Pennsylvania and New York: pp. 314-347 *in* Subitsky, (ed.), Geology of selected areas in New Jersey and eastern Pennsylvania: Rutgers University Press, New Brunswick, N. J., 382 p.

Warren, David, 1968, Transcontinental geophysical survey (35°-39°), Seismic refraction profiles of the crust and upper mantle from 74° to 87° W. Longitude: U. S. Geol. Surv. Investigations, Map I-535, 1:1,000,000.

Wagner, Mary Emma, and Crawford, Maria Luisa, 1975, Polymetamorphism of the precambrian Baltimore gneiss in southeastern Pennsylvania: Amer. Jour. Sci., vol. 275, pp. 653-682.

Williams, Harold, and Stevens, R. K., 1974, The ancient continental margin of eastern North America: pp. 781-796 *in* Burk, C. A., and Drake, C. L., (eds.) The Geology of Continental Margins: Springer-Verlag, New York, 1009 p.

Woollard, George E., and Joesting, H. R., 1964, Bouguer Anomaly map of the United States; U. S. Geol. Surv., 1:2,500,000.

York, James E., and Oliver, Jack E., (in press), Cretaceous and Cenozoic faulting in eastern North America: Geol. Soc. America.

CHAPTER 3

THE PENCKIAN MODEL – WITH MODIFICATIONS

Howard A. Meyerhoff

ABSTRACT

Field and map analysis of western New England's upland surface demonstrated the presence of a succession of erosional terraces or straths, not a single surface or peneplane as postulated by W. M. Davis. The surfaces were formed by successive vertical changes in sea level, with progressive headward entrenchment of streams and partial cyclical development of valley floors whose width was a function of stream volume, magnitude of uplift, rock resistance and structure, and duration of cycle. Each change of level is recorded by a headward retreating knickpoint in mainstream and tributaries and by lateral "migrating" or retreating scarps and strath surfaces. Excellent correlation with terraces of marine origin, identified by Barrell in southern New England, was established.

Subsequently, field studies in the central and southern Appalachians confirmed the work of five other authors who identified cyclical levels in these sections of the Appalachian system. Observations have been extended to the mobile belts, notably in the Caribbean. There the concept also applies wherever late Cenozoic tectonics and/or volcanism have not materially altered the fluvial regimes.

It is concluded that the concept, which is basically the Penck model, has worldwide application for fluvially dissected landforms. Although it relegates peneplanes (peneplains) to theoretical status, it recognizes the validity of incomplete erosion cycles, and it attributes scarp retreat and strath formation to the processes involved in pedimentation.

INTRODUCTION

The invitation to present a paper on Penck's Treppen concept at this symposium came as a welcome surprise. I had just submitted a brief discussion on the so-called treppen concept to the *Journal of Geological Education* (1975), taking issue with an author (Rahn, 1971) who, on the basis of dubious "evidence", cavalierly dismissed Joseph Barrell's analytical study (1920) of the southern New England upland. In preparing the paper I realized that the Barrell-Penck concept of landform analysis and interpretation had receded into the dim pages of history. Notwithstanding the spread of quantitative methods in geomorphology, landform analysis, if undertaken at all, is still qualitative; in its neglect, geomorphologists are missing a timely opportunity to make a major contribution to other fields of earth science, notably to tectonics and potentially to stratigraphy.

My invitation to participate in this session can be traced to a small part I played in a Penck symposium sponsored by the Association of American Geographers and organized by O. D. Von Engeln. The fact is, I got into that 1939 symposium on the mistaken impression that I was a follower of Walther Penck, of whom I had scarcely heard and whose tortuous German I had not read. There was no English translation of his *Analyse*, and I had to find out what Penck and I had in common before I could present my discussion on the migration of surfaces (Meyerhoff, 1940). Apparently, I had discovered by serendipity the principles which Penck enunciated in his controversial work (1924). The discovery was made in western Massachusetts and in Vermont in a futile attempt to trace William Morris Davis' New England peneplane (1895) into the Green Mountains. There wasn't any peneplane. Instead there was a series of straths which we (Meyerhoff & Hubbell, 1929) called cyclical terraces, but which Penck would have termed "treppen". Our analytical work was inspired not by Penck but by Barrell, with whom we differed solely in our interpretation of the erosional terraces in western New England as fluvial, in contrast to his equally valid interpretation of those he studied in Connecticut as marine.

In this way I sidled into the Penck school of thought, into the 1939 symposium and, now, into this conference. In my acceptance to participate, I did warn Dr. Flemal that Penck and Meyerhoff might become inextricably mingled in my presentation, but this seemed not to daunt him. I can merely hope that there are no Penck purists present to take me to task for deviating from the precepts of the master. In what follows I may appear dogmatic, and on certain points I am. On others I recognize qualifying factors but may slight them in the interest of brevity, but even a summary presentation of a complex concept cannot be too brief without becoming obscure.

This draft of the Penck or mobile landform model was started in Guatemala. The advance summary, submitted in January, was prepared in Tulsa, Oklahoma. No greater contrast in tectonic backgrounds can be imagined than between the youthfully faulted, volcanically active, ash-covered upland and the lowlands of Central America and the stable cuesta-lowland topography of the gently dipping late Paleozoic strata of the Osage Plains.

What the two regions have in common, however, is precipitation and running water — the basic agency, or process, to be considered in this symposium. In the beginning, to be sure, is the tectonism that created the types of landforms on which the water operates, as well as the rock types exposed to the water treatment. Atmospheric and biologic activities are secondary agencies of importance, but whether rocks are resistant or non-resistant, weathered or unweathered, the victims of peneplanation, pedimentation, or the "treppen degradation" of Penck, water is the dominant force to be dealt with. To a degree, we are myopic, for Davis, Penck, and King — and I — have been more concerned with denudation than with nature's balance of degradation and aggradation. A great deal of the aggradational phenomena we wisely leave to the sedimentologists, but alluvial and lacustrine phenomena are landforming processes that are well within the province of the geomorphologist. Glaciation and deflation have their own milieus, and they infringe on the province dominated by running water. Where they do, they cannot be ignored. If, in what follows, they are barely mentioned, do not think they are forgotten.

THE MOBILE EARTH

For any student raised on the Pacific rim or in Middle American-Mediterranean countries, the concept of a mobile earth is readily grasped. For those of us born, raised and trained in the staid and steady East or Mid-West, the concept comes harder. Yet, even the true believers in the New England-Schooley and Harrisburg "peneplanes" accept the fact of their uplift and presumed warping in Tertiary time. And who can contemplate the remarkably straight scarp bounding the Ozark Plateau and Mississippi floodplain, together with the comparatively youthful dissection of the plateau, without realizing that uplift was geologically recent. Movement along the Balcones escarpment in Texas has been so recent that, locally, there is little adjustment between the initial drainage and karst of the Edwards Plateau and the mature topography of the Gulf Coastal Plain. These geologically young tectonic features are part of the staid and steady East and Mid-West.

Non-resistant sediments may undergo considerable planation within limited areas, but even so limited and provincial a landform as the Appalachian Piedmont is *not* a peneplane. Bascom (1920) and Knopf (1924) have shown its composite form. Barrell (1913, 1920) and Hatch (1917) did as much for the New England upland in Connecticut. The so-called Fall Line is more than a simple contact between the non-resistant strata of the Atlantic Coastal Plain and the resistant crystalline rocks of the Piedmont. Few of the falls which gave it its name are located at or even near the contact. The Potomac, for example, has worked headward from the coastal plain in two distinct pulses to Little Falls and Great Falls, each recording a separate uplift, although Little Falls has had its problems with the submergence of Chesapeake Bay and at the hand of man.

THE FLUVIAL SYSTEM

The literature on fluvial phenomena is all but limitless. A substantial amount of it now is quantitative, but I shall neither clarify nor befog my thesis with formulas, even though there are some that apply. The most important definitions for our purposes are those of baselevel, gradient and grade (Davis, 1902).

The baselevel for any drainage system is its terminus, whether it be the sea, a lake or playa in an undrained basin, or a dry bolson in a desert valley. The fact that the last two are variables and may be ephemeral is generally accepted, but the fact that the sea itself is subject to changes of level in relatively short periods of geologic time is rarely stressed. Whether the changes be eustatic and worldwide, epeirogenic and regional, or tectonic and comparatively local is immaterial. The sea is no more permanent in level than the saline lake that rises, falls, and even evaporates with cyclical variations in precipitation. Sealevel changes relative to specific drainage systems may take place more slowly and are likely to be more enduring once made. But they affect the inflowing drainage no less than the local and less enduring baselevels of interior drainage systems. I am sure that few systems, beyond quite recently formed first- and second-order consequent streams, have been unaffected by sealevel changes. I am almost as sure that few theoretical plots of stream or stream-system profiles recognize the composite characteristics which the profiles actually possess.

Most river systems — and all old systems — are composed of streams that have not one but $n + 1$ gradients, where n equals the number of vertical uplifts and/or monoclinal deformations the drainage area has undergone since its inception. I realize this usage calls for a definition of gradient which, if properly defined, is the normal angle of declivity achieved by a stream. It differs therein from a stream profile, which may or may not exhibit a norm. What is a norm? A norm is the hydrodynamically determined minimum declivity required to maintain uniform flow for a given volume of water. In nature this is modified by variables which, however, are secondary in their effects. The more important variables include sedimentary load, seasonal or diurnal variations in stream volume and channel cross-section. Sedimentary load and channel cross-section are related to lithology and rock structures which, in themselves, are not participating variables.

Although it is not pertinent to our thesis, it might be noted that the gradient, or norm, for a given stream is not a smooth curve, but a succession of straight lines that slope at progressively lower angles downstream with the input of tributaries. The underflow, or groundwater, does not increase volume between tributaries; it merely sustains volume.

If I seem to belabor these points and to narrow some definitions, I have reason. Although geomorphologists have taken to the hydrodynamics of drainage basins as elaborated quantitatively by Horton (1945), the classroom and the literature still abound in such vagaries and inaccuracies as "gradients steepened by uplift", "hard rocks as the cause of waterfalls and rapids", and other deviations from hydrodynamic theory and fluvial fact. The actual fact, in conformance with hydrodynamic theory, is that, with simple vertical uplift, a stream instantly seeks to re-establish its gradient, or norm, *at the point of uplift* and does so progressively headward (Fig. 1). Upstream from the limit of headward entrenchment, the knickpoint, no critical factor has been altered. Water volume is the same, and the mainstream and tributaries continue in the old regime, oblivious of the change in elevation.

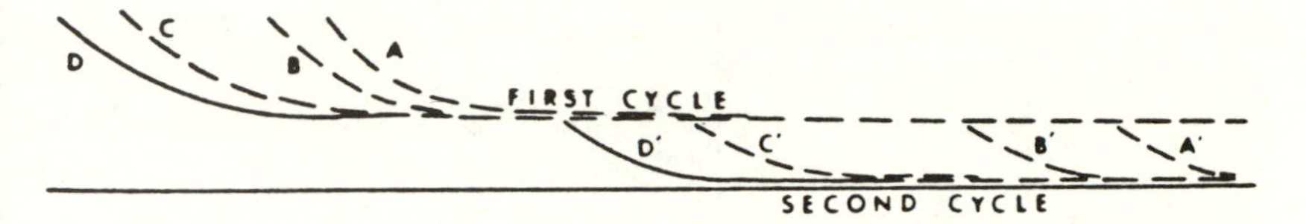

Figure 1. Stream profile development following simple vertical uplift.

No more convincing illustration can be found than the terrace above the Great Falls of the Potomac, where the comparatively placid river, graded in a flat-floored valley within the Piedmont upland, contrasts with the roaring stream entrenched in a young gorge below the knickpoint (Fig. 2). The headward retreat of the knickpoint is evident from the potholes and other fluvial features that can be identified along the upper margin of the gorge until they fade and disappear as the valley walls weather, widen and mature toward Georgetown.

Like any other river draining a land that has been raised vertically with respect to baselevel, the Potomac is developing a new gradient below the break, while the old gradient remains unaffected until tapped by headward entrenchment. The profile is compound — actually composite. There is no change in rock types at, immediately above, or below the falls, although entrenchment obviously is proceeding more slowly in the crystalline rocks of the Piedmont than it did in the less resistant Coastal Plain sediments downstream. Similarly, older knickpoints upstream are working headward. Between each knickpoint in the Potomac, as in other composite rivers, the river is developing its normal gradient without reference to what goes on at higher or lower levels. Each reach migrates headward, never overtaking its predecessor or being telescoped or merged with its successor until the ungraded headwaters are reached. Each tributary also undergoes progressive entrenchment with correlative knickpoints that register, in elevation, their smaller water volumes. The spacing between knickpoints reflects (a) the magnitude of uplift; (b) differential speed of entrenchment in rocks of different types; (c) diminishing water volume upstream in mainstream and tributaries.

Although the incidence of simple vertical change without deformation is high, many uplifts entail arching or monoclinal tilt. Post-orogenic movements in the Appalachian region-Atlantic slope have been dominantly of this type. Whereas there is no change in the fluvial regime at any level with vertical uplift, arching steepens stream profiles, thereby accelerating velocity without

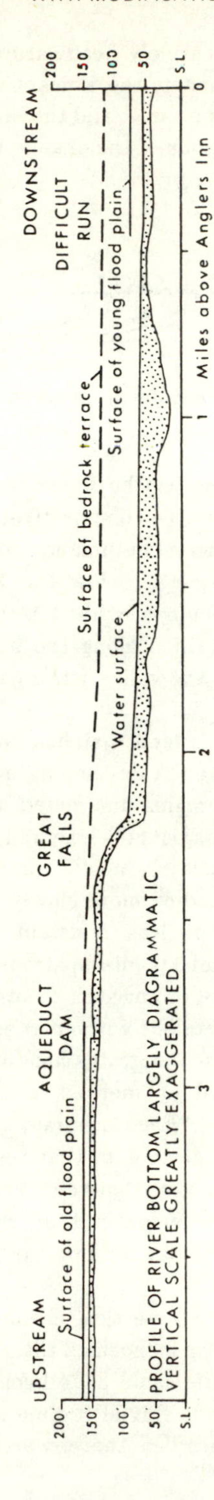

Figure 2. Profile of Potomac River showing knickpoint and relation of old flood plain above Great Falls to bedrock surface and young flood plain below Falls. Vertical scale exaggerated. (From U. S. Dept. Interior, 1970, p 20-21)

affecting volume. The inexorable hydrodynamic laws of erosive-transportative energy immediately start to work through the entire system. In the least probable case, in which there is no elevation at the point of drainage discharge into the sea, gradation will entail entrenchment through the entire system and restoration of the gradient norm progressively headward from the point of discharge, as shown in the profile of the San Joaquin River in the Sierra Nevada of southern California (Fig. 3; Wahrhaftig, 1965).

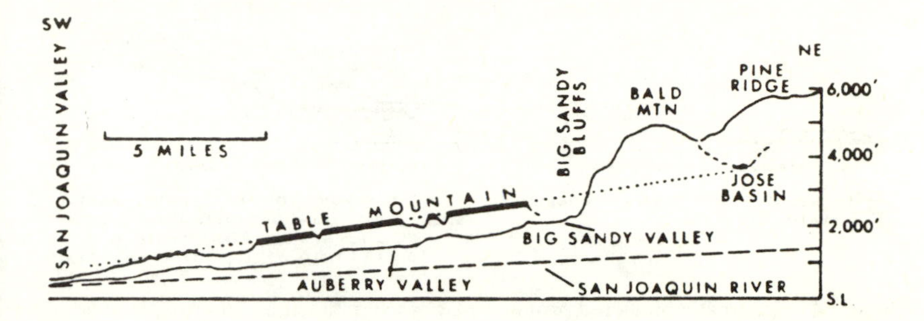

Figure 3. Projected longitudinal profile along lower San Joaquin River, southern Sierra Nevada, Cal. (After Wahrhaftig, 1965, Geol. Soc. Amer. Bull., v 76, p 1170, Fig. 5; reproduced with permission of the Society and author.)

More commonly, there is effective vertical uplift at the prior point of discharge. There may be an exposed apron of unconsolidated delta and shoreline deposits which give the semblance of smooth arching, but these sediments can be trenched in such a brief span of time that the practical effect of arching is vertical uplift at the prior drainage terminus. The result is regional entrenchment to tilt and the initiation of gradient adjustment to two levels — the pre-arching level and the new lower level of discharge. The simultaneous development of two contemporaneous gradients initiated by a single tectonic event such as arching, doming, or tilting probably would escape detection in regions of complex structure and prolonged erosional history. Hubbell and I failed to identify any clear case of dual development in New England. In Puerto Rico, however, where the middle Tertiary coastal plain acquired a 4°-5° tilt to the north in middle or late Miocene time, compound gradients were identified in several north-flowing rivers, notably Ríos Bayamón and Arecibo (Fig. 4). Although Wharhaftig (1965) has shown rather convincingly that the cyclical treppen concept does not apply to post-tilt treppen surfaces in the Sierra Nevada, he has demonstrated that high-level steps formed by differential weathering and erosion of plutons have surface gradients "corrected" for tilt (Fig. 3). In their lower reaches the main rivers have gone far toward regradation to the Central Valley, but progress is slowed upstream by the rising elevation (Fig. 3). I have not found any definitive study of river profiles above mainstream knickpoints to relate gradients either to tilt or to the pre-tilt surfaces described by Lawson (1904), Matthes (1930, 1937) and other later workers. Post-tilt glaciation has made such studies difficult but probably not insoluble.

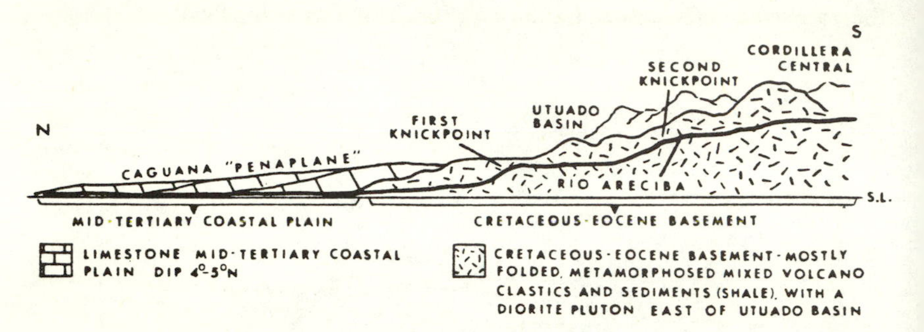

Figure 4. Schematic projected profile east from Rio Arecibo, Puerto Rico, vertically exaggerated.

SCARP AND SURFACE RETREAT

Development of the typical V-cross section of a valley may start during the early stages of entrenchment, depending upon the interaction of lithology and climate. Valley widening, however, does not become a systematic process much before the downstream reach of a stream attains its normal gradient. Stream undercutting may initiate or accelerate the process locally; but scarps, or valley walls, recede without reference to any auxiliary lateral cutting the stream may do. The mechanism takes a miscellany of wasting forms, from the alternation of vertical sandstone or limestone cliffs and shale slopes in a horizontal sedimentary sequence, to the talus of a lava palisade or granite batholith. Although frost action, water saturation, and biologic factors contribute to the result, gravity is the primary force involved, and for the retreating scarp I have proposed the term "gravity slope (1940, p. 251). The surface at the base of a gravity slope, which is seeking a pluvial gradient with respect to the graded stream, I have called a "wash slope" (1940, p. 251). Both terms are defined in the *Glossary of Geology* (1972, p. 311, 785).

There is little argument about the retreat of scarps in arid, semi-arid and sub-humid regions, for mesas, buttes, and inselbergs permit no other explanation. Nor can there be any argument about the complementary wash slopes whether they be fans or pediments. The notion still persists among some geomorphologists, however, that gravity slopes and wash slopes lose their identities when the climate is humid (Garner, 1974). It should be obvious that geomorphic forms developed by two distinct forces cannot merge. Garner (Ch. 5) contends that the vegetation of humid regions reduces slope corrasion toward zero as a limit; but, in so doing, he ignores two theoretical and observable facts: wash slopes seek grade with respect to stream gradient and water table controls; gravity slopes have no deterrent other than the magnitude of the vertical drop between summit level and base on the wash slope.

A gravity slope, ideally, approximates the 30°-35° angle of repose for unconsolidated material, but actually it may vary from a vertical wall to a chaotic slump mass at the base of a clay or shale outcrop. Regardless of vegetation, it is subject to creep if soil covered, and in heavy rains it can undergo actual flow. In Puerto Rico, with 80 inches of rainfall, and in western Massachusetts with 41 inches, I have seen soil slurries, formed by excessive precipitation on saturated gravity slopes, flow from beneath and around bush and tree roots. On one Puerto Rican slope scarcely a tree remained upright after the slide. Vegetation may retard erosion by gravity, but the idea that it inhibits wasting and retreat loses sight of the forces that may operate during relatively short spans of time — almost "overnight" in the two examples mentioned. Cumulatively, if spasmodically, these brief episodes effectuate scarp or valley-wall retreat.

The wash slope, in contrast, is a "controlled" surface that rarely exceeds 15°, and is usually less than 10°. The variation depends largely upon the capacity of sheet wash to move gravity slope detritus. On a macro-scale the process may be seen in the fans and pediments in undrained grabens or depressions in the Basin and Range Province, where debris from adjacent horsts or ranges exceeds the carrying capacity of deficient rainfall to cope with stream- and gravity-contributed material. Fan and interstream or inter-distributory pediment slopes steepen (I have measured 15° angles), yet are not steep enough to develop a balance between sediment supply and removal.

On a mini-scale, graded wash slopes evolve quite systematically with respect to graded master streams in humid valleys. Gradation takes place with respect to graded tributaries as well as to the master stream, but the latter, maintained not only by runoff but also by its relation to the water table, is the base line of control for gradation. Although the wash-slope angle tends to be low, it depends upon the height and lithology of the retreating scarp. At Trebisacce, in southern Italy facing the Gulf of Taranto, a young fault scarp, nearly 1000 meters high, is contributing limestone blocks and calcareous mudstone beyond the limit of sheet wash capacity to carry them into the Coscile River even at an 8-to-10°-slope, but the wash slope does approach grade in relation to the stream.

I assume that the orthodox view of valley widening starting at graded stream mouth and progressing upstream needs neither further exposition nor justification (Fig. 5). The extent to which widening occurs depends upon the height of the land undergoing dissection, its lithology, and the duration of the stillstand of baselevel. Ample illustration can be provided from innumerable fluvial systems. The treppen concept assumes that every stillstand is limited in duration and that uplift will initiate a new partial cycle of development. Analysis of the effects of uplift are in order.

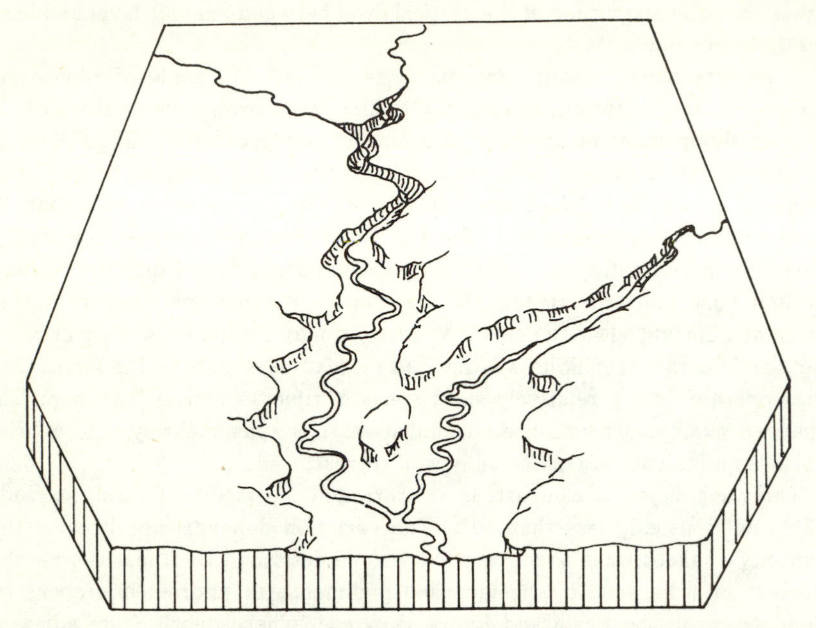

Figure 5. Progressive headward stream entrenchment, valley-wall retreat and valley-floor widening, first partial cycle.

THE TREPPEN CONCEPT

To limit exposition, the simplest, yet commonest, case of vertical uplift will be assumed, with little or no arching or other deformation inland from the line of uplift, where stream entrenchment begins. Headward incision and development of a normal gradient to the newly established base line is a relatively rapid process. To be sure, the rate of retreat varies drastically with rock types, giving rise to the myth that waterfalls merely indicate resistant rock rather than cyclical knickpoints. I have ridden horseback for 12 miles on a graded stream bed through a vertical-walled gorge cut 60 feet in basalt in western Chihuahua. Valley widening had not started, yet upstream in less resistant country rock wash slopes were developing. This type of seeming inversion is, of course, common. Again the Potomac River illustrates the point. The graded, flat-floored valley above Great Falls widens from a scant half mile to nearly two miles upstream where it flows across the Triassic sandstone and shale of the Leesburg Basin (Fig. 6).

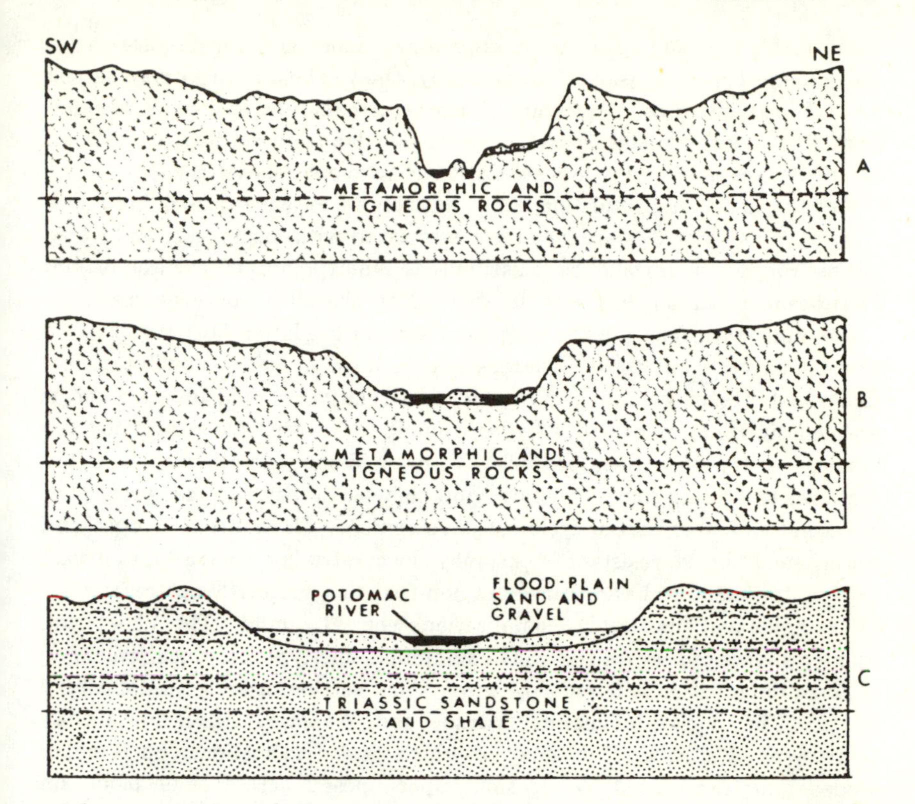

Figure 6. Cross sections of Potomac River valley. A, Four miles below Great Falls near Carderock. B, Five miles above Great Falls near Violets Rock. C, Leesburg basin 9 miles above Great Falls (From U. S. Dept. Interior, 1970, p 7)

Excavated soft rock lowlands behind resistant barriers across which master streams were superposed or to which they were antecedent are numerous — in the Southern and Central Rocky Mountains, Connecticut Valley, Asheville Basin, and innumerable other localities. Structure is commonly a causal factor, as in alternating hard and soft rocks in cuesta-lowland, ridge-and-valley, anticlinal-valley, domal-basin topography.

Rock resistance is strictly relative. Solution, for example, commonly proceeds more slowly in the humid tropics than combined chemical weathering and mechanical erosion of so-called less (or in-) soluble rocks. Limestones are common ridge-formers and cap rocks; granitoid-igneous rocks, which undergo rapid "granulation" by mineral-contact decay, may form basins as the crystal grains are washed out and deposited as arkose. In the Lago Enriquillo trench of the Dominican Republic, a 90-foot bed of halite forms a prominent hogback, not because it is in the rain-shadow of the Cordillera Central, but because the sub- and superjacent sandstone, shale, and gypsum are easier preys to mechanical erosion. Paradoxically, the weakest of rocks may persist in, and actually form, highlands if masterstream entrenchment has not worked far enough headward to tap them. Lowlands do not form on rocks of low resistance until and unless stream entrenchment reaches them.

To be sure, stream gradation is not entirely insensitive to rock hardness. Waterfalls and cascades are generally associated with hard rocks, and by no means can all waterfalls be classified as knickpoints of cyclical origin. Pleistocene glaciation has seen to that. Nor are all — or even many — knickpoints waterfalls. In the process of stream gradation falls are likely to develop on resistant rocks, whereas rapids form in easily erodible rocks. Headward stream gradation, however, is a relatively rapid process, and it proceeds in conformance with hydrodynamic law with little regard for the lithology of the underlying bedrock. Valley widening, in contrast, is highly sensitive to lithology and structure, and this sensitivity becomes prominent in second and later partial cycles of development. Downstream gorges and water gaps cut in resistant rocks may form striking contrasts to broad valleys, lowlands, or basins carved in non-resistant materials upstream.

Examples and, inevitably, complications come to mind. The Valley and Ridge province of the Appalachians exemplifies one type, in which prolonged regional degradation is burrowing into a thick section of folded sedimentary strata. The Rocky Mountains "parks" and Wyoming Basin present an assortment of situations, in many of which post-deformational fill is being eroded and transported by streams superimposed across older mountain axes. The cuesta-lowland (or inner lowland) type may appear to be the simplest of landforms in this category, but its simplicity almost invariably is deceptive, for the temptation is strong to correlate the lowlands between cuestas. Likewise, in the Valley and Ridge province all the valleys have been uncritically and erroneously lumped into the "Harrisburg peneplane". Landform analysis, even with the limitations of the old USGS 20-foot contour maps, demonstrates that the Harrisburg surface is a composite of several erosional levels, separated by knickpoints and scarps, and forming low steps at successively higher elevations in upstream areas. These successive elevations exceed the normal gradient requirements of stream control; and, though related to arched or monoclinal uplift, they actually record a chronology of intermittent uplifts and stillstands.

CRITERIA FOR CYCLICAL PLANATION

Hubbell and I discussed analytical methods in our study of Vermont (1929), borrowing from, and expanding on, Barrell (1920). Both publications are old and, apparently, forgotten; but for anyone not familiar with the rigid techniques of constructing linear (Fig. 7), zonal, and projected profiles, and with the uses and limitations of the three types, consultation of those references is

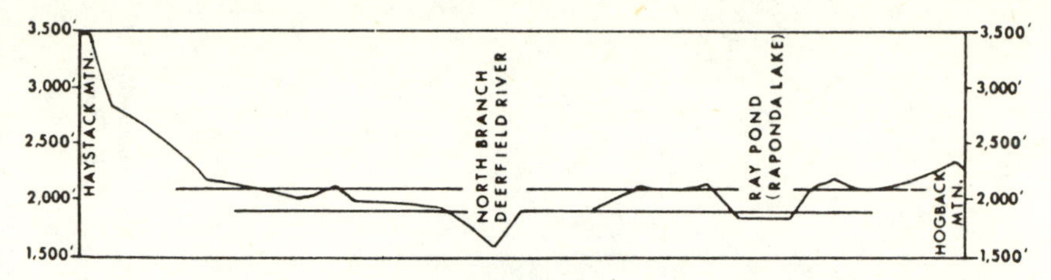

Figure 7. Linear profile showing 2100- and 1900-foot surfaces between Haystack and Hogback Mountains, Vermont. (From Meyerhoff and Hubbell, 1929)

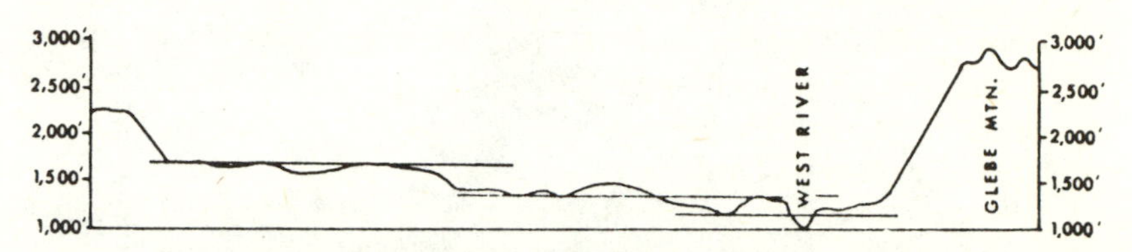

Figure 8. Zonal profile across Londonderry basin, Vermont, showing levels at 1680, 1360, and 1200 feet. (From Meyerhoff and Hubbell, 1929.)

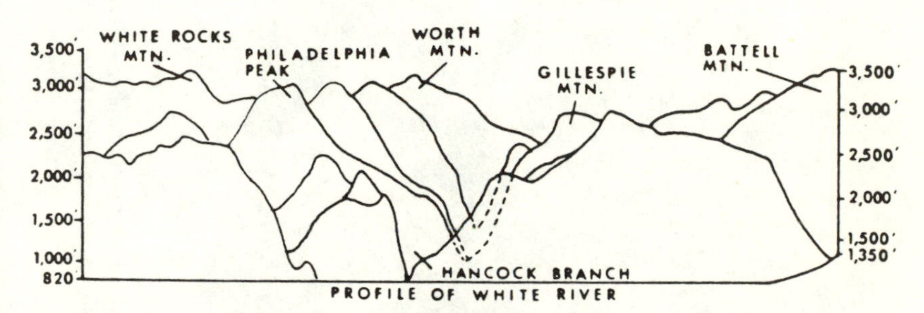

Figure 9. Projected profile west from White River between Rochester and Granville, Vermont, showing accordant peaks and limited surface preservation at 3200 feet. (Meyerhoff and Hubbell, 1929)

recommended. Here I shall restrict my remarks to theory and to field criteria, but I might add that Barrell and I — and our contemporary landform analysts — were born too soon. We labored hours over topographic maps to produce results such as those shown in the zonal profile of Figure 8 and in the projected profile of Figure 9. Now, with judicious human aid, the computer can turn out such spectacular results as the backprop of Vancouver, British Columbia, shown in Figure 10.

After a stream has established its normal gradient in any reach of its course, valley-wall or scarp retreat begins (Bryan, 1940). Ideally, the valley

Figure 10. Perspective view of Greater Vancouver, British Columbia. View from S.W. (Reproduced by permission from the Association of American Geographers' Committee on College Geography Resources Paper No. 17, "Computer Cartography," T. K. Peucker, 1970.)

walls are gravity slopes approximating 30°-40°; actually, they may be vertical sandstone or limestone cliffs or chaotic slump masses. At their bases wash slopes are formed. Whereas the normal gradients of a masterstream and its tributaries are controlled by water volume, with suspended and traction load as a modifying variable, sediment input is a much larger factor in determining the gradient or declivity of the wash slope. Nonetheless, the profile of the water table is the ultimate control and, as a general rule, the water table is accordant with the permanent streams. Theoretically, therefore, for any stillstand of baselevel, masterstreams and tributaries will seek normal gradients to which wash slopes will acquire accordance. Development of the wash slope may be described as a cut-and-fill process, involving sheet wash, rilling, erosion, transportation, deposition — in short, all the processes associated with pedimentation.

Uplift without deformation initiates stream entrenchment and headward development of a normal gradient, without any alteration of stream or wash slope functions on the upraised surface until headward entrenchment of the masterstream knicks the stream profile. Even when entrenchment occurs, wash slope encroachment on the retreating gravity slope or scarp continues, and a rock terrace or strath is produced. One significant change does take place: entrenchment of tributaries cuts the strath into isolated segments, and gravity slope retreat proceeds from the tributaries as well as from the mainstream. Locally, complete destruction of one or more segments may occur, especially where the drainage texture is fine; but I know of few instances where the destruction of an erosional rock terrace has been regionally complete. In New England, for example, zonal and projected profiles invariably revealed remnants of all the levels which had reached any given part of the Connecticut River and its major tributaries (Fig. 11).

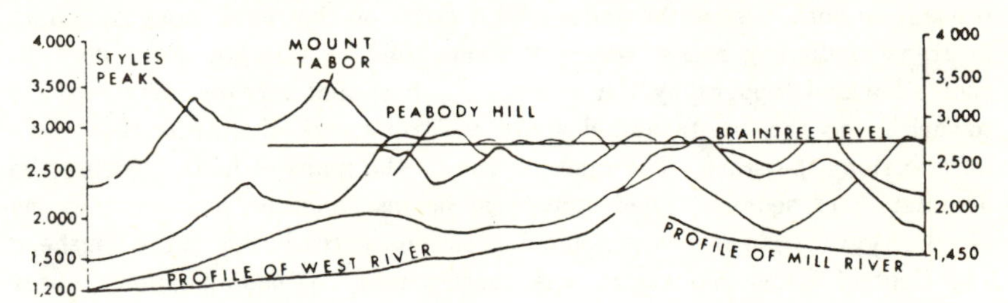

Figure 11. Projected profile west and southwest from West and Mill Rivers between Weston and Tarbelville, Vermont, showing remnant of cyclical surface at 2750 feet (Meyerhoff and Hubbell, 1929).

The identification of cyclical terraces or straths calls for the application of rigid criteria, among which the following are most critical:

1) Generally, every strath should be gradationally correlative with a graded or pediplaned, unentrenched valley floor upstream above a knickpoint.

2) Correlative straths should be present and identifiable in mainstream and major tributary valleys and should maintain approximately the same down-valley declivities as the gradients of the streams with which they are associated.

3) Notwithstanding the sensitivity of wash-slope development or pedimentation to lithology, cyclical terraces should show a marked degree of regional independence of rock types and rock structures. Benching by stripping of resistant sandstone, conglomerate, or limestone strata in areas of flat-lying sedimentary rocks may give the illusion of strath formation, but the normal declivity of rock terraces of cyclical origin should suffice to etch and preserve them on differing lithologic units in a horizontal stratigraphic section. Dipping lithologic units likewise tend to weather and erode subuniformly and to produce ridge crests and terraces which look deceptively like cyclical forms (Fig. 12).

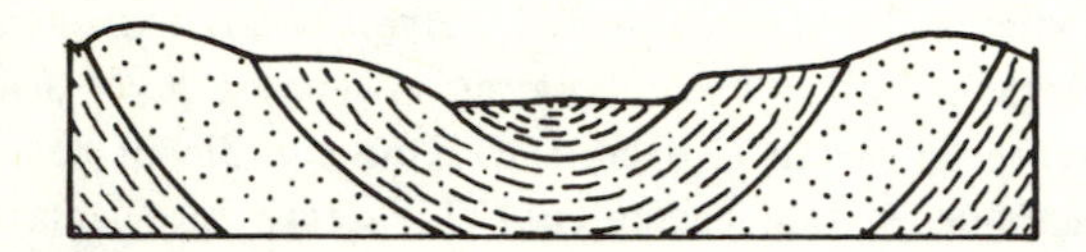

Figure 12. Pseudo-strath development produced by differential erosion.

4) In homogeneous rocks, valley widening takes place symmetrically with respect to normally graded streams, hence matching straths should be present on both sides of the valley. This criterion, however, does not apply in areas of dipping rocks, where streams tend to migrate down dip. In cuesta-lowland topography, for example, subsequent streams may migrate toward or to the cuesta and destroy any remnants of straths that were formed along the scarp. The same is true on the limbs of folds. During the denudation of sediment-filled structural basins, another problem may be faced. Aggradation in such tectonic depressions as the basins of the Southern and Central Rocky Mountains was accompanied by pediplanation. Later removal of the fill probably has exhumed planed surfaces on older rock that have no cyclical significance. Their random distribution in space and elevation should serve to distinguish them, even when they have considerable areal extent. For example, the western-sloping surface of the Mississippian limestone in the Black Hills appears to be a pediment, formed during the early Tertiary aggradation of the Powder River basin, with the Precambrian rocks of the Harney Peak district forming the eastern divide. Generally, pediplanes of this type have steeper profiles than benches of cyclical origin, but this is not always a diagnostic criterion.

WIND GAPS

One interesting corollary evolved from a study of erosional levels in the Appalachian fold belt. The wind gaps were found to fall into sets correlative with cyclical levels (Meyerhoff and Olmsted, 1934). Apparently capture took place as subsequent tributaries of some mainstreams entrenched themselves in soft-rock strike-valleys and lowlands more rapidly than the victimized streams could regrade their courses across resistant ridges. Statistical analysis revealed that, of 84 gaps for which reasonably precise elevations were available, only 10 (12%) had a clearly random distribution, unrelated to identified cyclical levels.

PENEPLANES

The literature is still replete with references to peneplanes, and it will be difficult to expunge them from the written geologic record. In textbooks of my day (circa 1915-1925), much was made of a Precambrian peneplane formed during the so-called Lipalian interval, which is now known not to have existed. The term Lipalian thus has fallen into disuse, but not so the concept of a Precambrian peneplane (see, e.g., Chenowith, 1968). A study I directed at the University of Pennsylvania demonstrated that, in the Mid-Continent, Lower and Middle Paleozoic sediments buried relief as high as 3000 feet (Meyerhoff, 1969, 1970). Attempts to explain the Laurentian Upland or Canadian Shield surface by the stripping of a Paleozoic cover over beveled Precambrian rocks fall flatter than the surface to be explained. The Precambrian Baraboo Range of Wisconsin, for example, rises high above the Upper Cambrian St. Croix sandstone at its base, and the Baraboo is but one of the ranges buried and partly exhumed. Yet no one will deny that the Laurentian upland is a flat erosional surface of huge dimensions. I am willing to call it a peneplane, but how to explain it?

First, it should be noted that the high relief on the early-middle Cambrian basement is almost entirely on middle and late Proterozoic rocks, whereas the flat surface is chiefly on crystalline metamorphic rocks of Archean and early Proterozoic rocks. The former are infolded and infaulted into the Archean shield, and the evidence indicates that deformation and intrusion of some magnitude occurred as late as early-middle Cambrian time, at least, in parts of the Mid-Continent — e.g., the Wichita Mountains. The relief so produced is still preserved where Upper Cambrian and Lower Ordovivian sediments buried the rugged areas. In fact, some of the eminences must have survived as islands in Paleozoic seas until early Devonian time (Meyerhoff, 1969, 1970). Significantly, the relief is absent only where the Archean basement was not buried or where removal of the northward-thinning cover of early Paleozoic strata has left the homogeneous metamorphic rocks exposed since Mesozoic time, and probably at prolonged intervals during parts of the late Paleozoic and Mesozoic eras.

Geophysicists have demonstrated that the Archean shield is a region of remarkable stability, although it has been affected tectonically along the Labrador-Ungava margin in late Tertiary time, and regionally its isostatic equilibrium was upset by the Pleistocene ice sheets. In general, however, it has been a positive area undergoing mild increments of uplift from the Cretaceous period until late Miocene time. During this time span, the climate was warm and humid; weathering of the homogeneous metamorphic rocks presumably was deep. Uplifts, however frequent, were so minor that they were quickly impressed by entrenchment on the residual soils, but without penetrating to unweathered rock. As a guess, cyclical terraces could only have been identified, if at all, by precise leveling, for they were separated only by a few vertical feet.

The early phases of Pleistocene glaciation stripped the deep residual soils, and transported and deposited them on the Till Plains of western Ohio, Indiana, and Illinois, and in the Dissected Till Plains of Iowa. The shallow cyclical terraces were rubbed out or "bulldozed" into a visually flat surface, the Laurentian "peneplane", and the final phase of glaciation had chiefly a flat, bare-rock surface to pluck. Glaciation was the final effective agent in producing a peneplane.

The only first-hand evidence I can advance in support of this postulate admittedly is circumstantial. In the western Adirondack Mountains, where Precambrian rocks have been moderately elevated, glaciation has produced a bare-rock "peneplaned" upland surface on the dominantly resistant, intricately folded and metamorphosed Grenville sedimentary sequence. Pre-Pleistocene entrenchment, however, was well advanced and has been preserved in the sinuous valley of Indian River, which follows the outcrop of a low-resistant carbonate formation. Thus, there was dissection to bedrock and observable relief where doming raised Precambrian crystalline rocks above the general low level of the shield.

Happily, it is as impossible to disprove this interpretation of the Laurentian Upland as it is for me to prove it conclusively; but, if it is accepted as a plausible and rational postulate, it can be claimed that the Penck treppen concept has universal application except in those desert, high-latitude and high-altitude regions where, for climatic reasons, normal fluvial processes have long been inoperative.

MARINE INTERRUPTIONS OF THE CYCLE

Up to this point, it has been assumed that continental landforms are generally on the "up-and-up". Although continents are positive, with uplift most consistent and pronounced along their mountain borders, marginal and interior submergence has occurred, and in the mobile belts both positive and negative movements of considerable magnitude have taken place. Many have involved deformation, but some have been essentially vertical. Without getting into the complications of mobile belt tectonics, let us examine briefly the effects of marine transgression in a comparatively simple case. None has been as well documented as southern New England, which was analyzed by Barrell (1920).

Davis postulated a marine overlap of southern New England, primarily to explain the departure of the Connecticut River from the Triassic Lowland at Middletown. That the spot marked the limit of the Cretaceous coastal plain deposits seemed logical at the time; but, in spite of a lack of direct sedimentary evidence, a much later date — early or middle Miocene — seems incontrovertible from geomorphic considerations. It was left to Joseph Barrell (1913, 1920) and Laura Hatch (1917) to demonstrate that the upland from Middletown south to Long Island Sound is not a simple slope but a composite set of steps, each marking a regression and short stillstand. The pauses in marine retreat were long enough for the cutting of a cliffed shoreline and the grading of an offshore slope. Subsequently Olmsted and Little (1946) traced the Middletown marine limit eastward to Rhode Island and westward to Sparkill Gap in the Palisades west of Hudson River. Map analysis, as well as visual inspection, shows that the old shoreline serves as the head for a series of consequent streams (Meyerhoff, 1946), although the marine advance was not prolonged enough to obliterate some structural control exerted on pre-submergent drainage. Especially striking is the abrupt change from what might be described as rugged micro-relief in the interfluves north of the marine scarp to singularly flat skylines in the interstream areas south of the Middletown marine boundary line.

Table 1. Comparison of erosional levels in southern New England and Vermont.

Southern New England Barrell		Taconic Section Pond	Eastern and Central Vermont Meyerhoff and Hubbell
Name	Elevation	Elevation	Elevation
Dorset	****	3,200	(?) 3,000(?)-3,200
Braintree	****	2,700	2,700-2,800
Becket	2,450	2,500	2,300(?)-2,500
Canaan	2,000	2,000	2,000-2,150±
Hawley	****	****	(?) 1,820-1,920
Cornwall	1,720	1,700	1,660(?)-1,720
Goshen	1,380	****	1,360-1,420
Litchfield	1,140	****	1,180-1,240
Prospect	940	900±	(?) 980-1,080(?)
Towantic	740	700+	780- 900(?)
Appomattox	540	****	540- 620(+)
New Canaan	450	****	420- 480
Sunderland	240	****	(?) 260- 300
Wicomico	120	****	(?) 160- 180

Unfortunately, no precise study has been made of the area between Barrell's marine terraces south of Middletown and the fluvial terraces identified by Meyerhoff and Hubbell in northwestern Massachusetts and southern and central Vermont and by Pond (1929) in the Berkshires of western Massachusetts and Vermont. Hence the effects of Miocene (?) marine transgression north of the Middletown boundary have not been determined. They appear to have been minimal and temporary, for extrapolation (Table 1; Meyerhoff and Hubbell, 1928, p. 325) indicates good correlation with Barrell's levels. Further study is needed, and it should embrace the effects of Pleistocene changes in sealevel, as well as post-glacial isostatic rebound. This is but one item of unfinished business in the "treppenization" of the New England "peneplane". Analysis of New Hampshire and Maine are others.

DATING OF SURFACES

In my presentation I have mingled theory and observation. It must be evident that in theory the operation of the treppen concept is perfect: uplift is followed by stream headward entrenchment and the development of a normal gradient to the new baselevel. A discordant knickpoint separates the new gradient from the old, which continues its headward gradation on the upraised surface oblivious of the change in elevation. As the normal gradient is achieved, valley walls recede and pedimentation at their base carries out

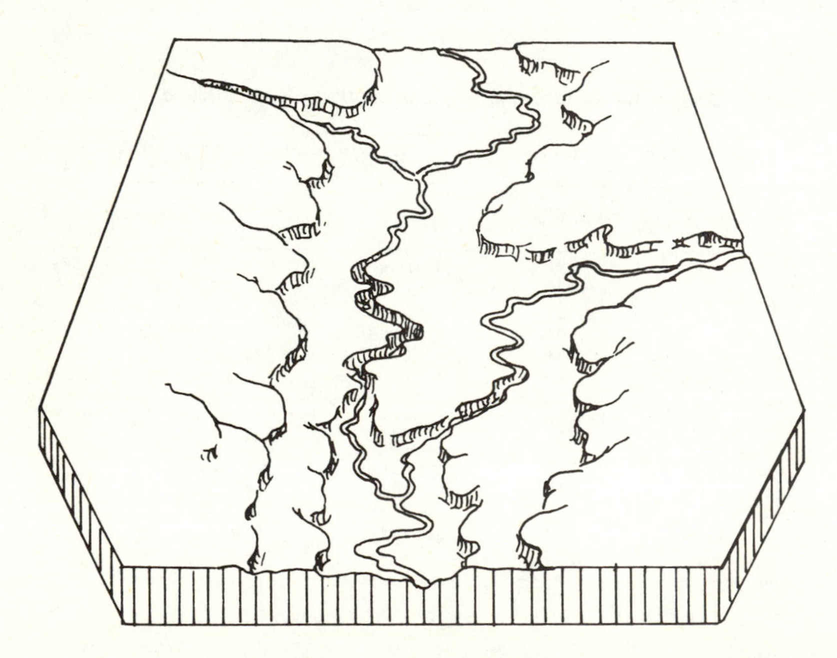

Figure 13. Second cycle stream incision and wash- and gravity-slope retreat.

the process of valley widening. Lithology and structure permitting, the scarps or gravity slopes continue their retreat, as do the wash slopes at their base, even after stream entrenchment and valley floor encroachment occur at the lower level (Figs. 13, 14). Tributary streams may dismember the straths or erosional terraces, thereby attacking them with retreating scarps from two or even three directions, but complete destruction is slow, and it is rare that a cyclical terrace leaves no surviving remnants.

The implications of this progressive development must be obvious. There is no such thing as a Cretaceous peneplane, or even a Cretaceous strath (cf. King, 1933). The Braintree surface we see at elevation 2750 feet west of the Connecticut River (see Fig. 11) or in upstream segments above the Deerfield River and other tributaries of the Connecticut may have been initiated by uplift in Cretaceous time, but it has taken a substantial part of the 70 million years which have elapsed since the close of the Cretaceous Period to migrate headward and form in its present position. The surfaces or surface remnants we actually see are younger than the dates of their inception. To a degree, all are only slightly less than contemporary.

There are many corollaries to the thesis presented herein. For example, in a region with oceanic drainage, old divides, like old soldiers, never die. In fact, they haven't even faded away in areas of positive movement (see Meyerhoff and Olmsted, 1938; Meyerhoff, 1972, p. 1711). Differential erosion and stream capture may sever an old divide and create a new one, but relics

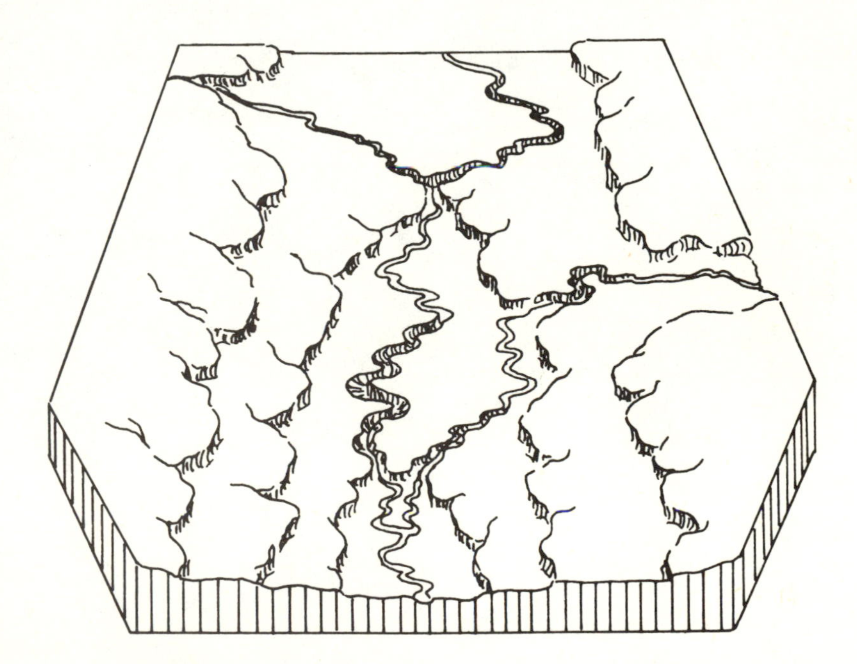

Figure 14. Third cycle entrenchment and strath retreat.

of the old persist as long as deformation does not alter the original topography.

The treppen concept is difficult to apply to mildly positive or neutral areas like the interior plains of the United States, but its application in the mountainous continental borders, whether old or young, offers a field of quantitative study that could and should bring new vitality to regional geomorphology. Landform studies have languished since the USGS published Fenneman's "Physiographic Map of the United States", for it has been assumed that this is the last word in regional geomorphic classification and description, save for details. The fact is, regional geomorphology has much that is new to contribute to moot points in tectonics. I am sure that the organizers of this symposium recognized the latent importance of this neglected field in calling for a review of the concepts by which landforms are interpreted. Have I made it clear that the Davis cycle is not obsolete or erroneous despite its limited application, that pedimentation is operative on scales both large and small, and that the treppen concept cannot do without both of them? Debate on the verity of rival models or concepts is the life of science, but just as surely, the embrace of one to the arbitrary exclusion of others is its death.

REFERENCES CITED

American Geological Institute, 1972, see Gary, Margaret.

Assoc. of Amer. Geographers, 1972, see Peucker, T. E.

Barrell, Joseph, 1913, Piedmont terraces of the northern Appalachians and their mode of origin: Geol. Soc. Amer. Bull., v 24, p 668-690.

Barrell, Joseph, 1920, Piedmont terraces of the Northern Appalachians: Amer. Jour. Sci, v 199, p 225-258, 327-361, 407-428.

Bascom, Florence, 1920, Cycles of erosion in the Piedmont province of Pennsylvania: Jour. Geol., v 29, p 540-559.

Bascom, Florence, and Miller, B. L., 1920, Elkton-Wilmington Folio: U. S. Geol. Sur. Atlas, Folio 211.

Bryan, Kirk, 1940, Retreat of Slopes, *in* Symposium — Walter Penck's contribution to geomorphology: Assoc. Amer. Geographers Annals, v 30, no. 4, p 254-268.

Carter, M. A., and Kirby, M. J., 1972, Hillslope Form and Process: Cambridge University Press, 475 p.

Chenowith, P. A., 1968, Early Paleozoic (Arbuckle) overlap, southern Mid-Continent, United States: Amer. Assoc. Petroleum Geologists, Bull. 52, p 1670-1688.

Commission on College Geography, 1972, see Peucker, T. E.

Davis, W. M., 1895, The physical geography of southern New England: National Geog. Soc. Monograph I, no. 9, p 269-304.

Davis, W. M. 1902, Base level, grade, and peneplain: Jour. Geol., v 10, p 77-111.

Davis, W. M., 1909, Geographical Essays, Boston, Ginn and Co., 777 p.

Garner, H. F., 1974, The Origin of Landscapes: A Synthesis of Geomorphology: Oxford Univ. Press, New York, London, Toronto, 734 p.

Gary, Margaret; McAfee, Robert, Jr.; Wolf, C. L., editors, 1972, Glossary of Geology: American Geol. Inst., Washington, 857 p.

Hatch, Laura, 1917, Marine terraces in southern Connecticut: Amer. Jour. Sci., 4th ser., v 44, p 319-330.

Horton, R. E., 1945, Erosional development of streams and their drainage basins: Geol. Soc. Amer. Bull., v 56, p 275-370.

King, L. C., 1953, Canons of landscape evolution: Geol. Soc. Amer. Bull., v 64, p 721-752.

Knopf, E. B., 1924, Correlation of residual erosion surfaces in the eastern Appalachian highlands: Geol. Soc. Amer. Bull., v 35, p 633-668.

Lawson, A. C., 1904, The geomorphology of the upper Kern basin: Univ. Calif. Pub., Bull. Dept. Geol., v 3, p 291-376.

Matthes, F. E., 1930, Geologic history of the Yosemite valley: U. S. Geol. Survey Prof. Paper 160, 137 p.

Matthes, F. E., 1960, Reconnaissance of the geomorphology and glacial geology of the San Joaquin Basin: U. S. Geol. Survey Prof. Paper 329, 62 p.

Meyerhoff, H. A., 1940, Migration of erosional surfaces — Walter Penck's contribution to geomorphology: Assoc. Amer. Geographers Annals, v 30, no 4, p 247-254.

Meyerhoff, H. A., 1946, Upland terraces in southern New England, a discussion: Jour. Geol., v 54, p 126-129.

Meyerhoff, H. A., 1969, Early Paleozoic (Arbuckle) overlap, southern Mid-Continent, United States — discussion: Amer. Assoc. Petroleum Geologists Bull., v 53, p 1519-1520.

Meyerhoff, H. A., 1970, Basement relief beneath the Upper Cambrian of the Mid-Continent: Geol. Soc. Amer. Abstracts with Programs, v 2, p 396.

Meyerhoff, H. A., 1972, Postorogenic development of the Appalachians: Geol. Soc. Amer. Bull., v 83, p 1709-1728.

Meyerhoff, H. A., 1975, Landform analysis — a lost art? Jour. Geol. Education, v 23, p 47-49.

Meyerhoff, H. A., and Hubbell, Marion, 1929, The erosional landforms of eastern and central Vermont: Vermont State Geologist, 16th Biennial Rept., p 315-381.

Meyerhoff, H. A., and Olmsted, E. W., 1934, Wind- and water-gap systems in Pennsylvania: Amer. Jour. Sci., 5th ser., v 27, p 410-416.

Meyerhoff, H. A., and Olmsted, E. W., 1938, Evolution of the northern Appalachian drainage divide (abs.): Geol. Soc. Amer. Bull., v 49, p 1938.

Olmsted, E. W., and Little, L. S., 1946, Marine planation in southern New England (abs.): Geol. Soc. Amer. Bull., v 57, p 127.

Penck, Walther, 1924, Die Morphologische Analyse: Stuttgart.

Peucker, T. K., 1972, Computer cartography: Assoc. Amer. Geographers, Commission on College Geography Resource Paper No 17, 75 p.

Pond, A. M., 1929, Preliminary report on the peneplanes of the Taconic Mountains of Vermont: Vermont State Geologist, 16th Biennial Rept., p 292-314.

Rahn, P. H., 1971, The weathering of tombstones and its relationship to the topography of New England: Jour. Geol. Education, v 19, p 112-118.

United States Dept. Interior, 1970, The river and the rocks — the geologic story of Great Falls and the Potomac River gorge: U. S. Dept. Int., Geol. Survey — National Park Service, 46 p.

Wahrhaftig, Clyde, 1965, Stepped topography of the southern Sierra Nevada, California: Geol. Soc. Amer. Bull., v 76, p 1165-1190.

CHAPTER 4

EPISODIC EROSION: A MODIFICATION OF THE GEOMORPHIC CYCLE

S. A. Schumm

ABSTRACT

Most models of geomorphic evolution are oversimplified and, therefore, they are unsatisfactory for short-term interpretation of landform change. For example, the extrapolation of average denudation rates from a 10-year record to longer periods of geologic time is based on the assumption of progressive erosional evolution which even without the influence of climate change and diastrophism is probably not correct. In fact, the inherent workings of a fluvial system may prevent progressive reduction of a valley floor.

It is proposed that stream gradients and valley floor altitudes do not change progressively through geologic time, but rather relatively brief periods of instability and incision are separated by long periods of relative stability (grade). Although the climatic and diastrophic history of the Quaternary prevents the identification of unstable periods due to geomorphic controls alone, some field and experimental evidence indicates that such a model is possible. Therefore, a very complex denudational history of a landscape may be geomorphically "normal".

INTRODUCTION

It is stimulating to consider the grand changes of a landscape during the millions of years of its erosional evolution. Unfortunately, this overview provides little assistance to those concerned with the short-term behavior of landforms.

The reason that most models of geomorphic evolution are unsatisfactory for short-term interpretation is that they are oversimplified, and this is largely because they are based on very limited information. For example, the extrapolation of average denudation rates from a ten-year record to a thousand or million years of erosional evolution of a landscape produces a model of landscape evolution that is based on an assumption of progressive slow change, which may not be correct (Gage, 1970).

The criticism of the Davis geomorphic cycle in the writings of Penck and John Hack reflect their concern with this simplistic model. Some years ago, Lichty and I (1965) attempted to resolve some of the controversy by considering the landscape during very different spans of time. The time required for the denudation of a landscape was subdivided into cyclic, graded and steady time periods (Fig. 1). Under the category of cyclic time are time spans of geologic duration, that is, the period of time required for the denudational evolution of a landscape. For example, during this period, one expects an essentially exponential decrease of stream gradients, which is a landscape component that reflects changes in the fluvial system. However, cyclic time can be subdivided into graded time and steady time periods. During graded time, average gradient will remain relatively constant, but there will be, through time, fluctuations about this mean. Graded time, therefore conforms to the definition of a graded stream, as expressed by Mackin (1948). During the very short period of steady time, there is no change. When considering a landscape or its components, it is helpful to think in terms of the time spans and how a landscape is altered during the time span under consideration. In short, the period of time referred to as being cyclic or geologic in duration can be represented by the Davis curve showing the erosional evolution of the landscape (Fig. 2). Of course, it seems unlikely that denudation will continue for such an appreciable period of time without interruption by climate change or by isostatic adjustment. So the smooth curves presented by Davis (Fig. 2) and by Schumm and Lichty (Fig. 1a) can be expected to be complicated, when

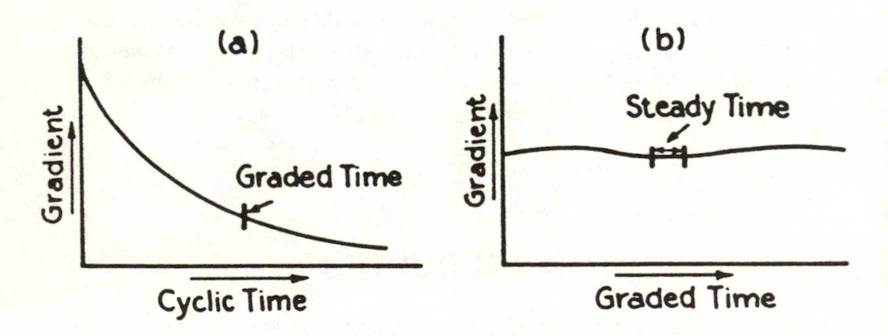

Figure 1. Diagram illustrating change of channel gradient during different time spans (from Schumm and Lichty, 1965).

A. Progressive reduction of channel gradient during cyclic time. During graded time, a small fraction of cyclic time, the gradient remains relatively constant.

B. Fluctuations of gradient above and below a mean during graded time. Gradient is constant during the brief span of steady time.

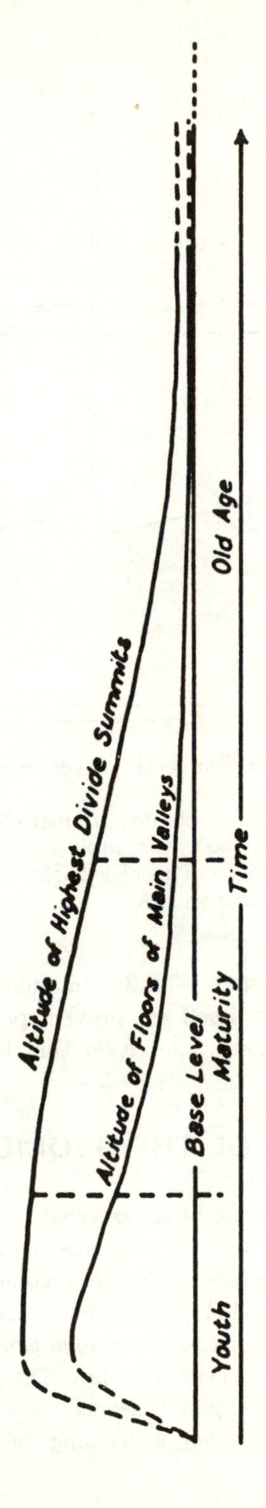

Figure 2. The Davis geomorphic cycle as usually presented.

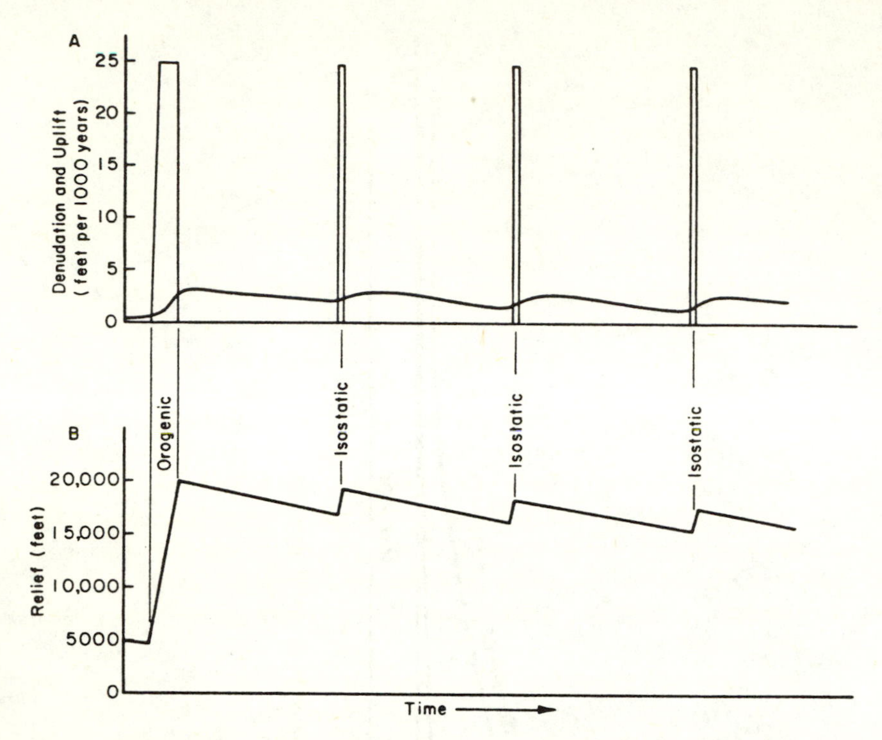

Figure 3. The effect of isostatic adjustment on the geomorphic cycle (from Schumm, 1963a).

A. Hypothetical relation of rates of uplift (25 feet per 1000 years) and denudation (3 feet per 1000 years) to time.

B. Hypothetical relation of drainage basin relief to time as a function of uplift and denudation shown in A.

external forces act on the system (Fig. 3). In addition, it is likely that the workings of the fluvial system itself will prevent progressive reduction of a valley floor or stream gradient, and it is this hypothesis that will be considered here.

COMPLICATIONS OF THE GEOMORPHIC CYCLE

It is understood that a change of an external variable will interrupt the progress of the geomorphic cycle, and changes of stream profile and variations of gradient during graded time are readily understood as reflecting variations of discharge and sediment load (Fig. 1b). Nevertheless, Figure 1 poses a problem. It is difficult to image how the graded time curve (Fig. 1b) can be compatible with the cyclic time curve (Fig. 1a). The progressive reduction of gradient shown on Figures 1a and 2 seems reasonable, but this in turn prevents a graded condition from developing until "old age" (Fig. 2).

Conversely, if graded conditions exist then progressive reduction of the valley floor and stream gradient is impossible.

This line of reasoning apparently requires the elimination of either the concept of progressive erosion or grade. However there is an alternative solution. If valley floors (Fig. 2) and stream gradients (Fig. 1a) do not evolve progressively but rather change rapidly, during brief periods of instability that separate longer periods of grade, then a model incorporating both progressive change and grade can be proposed.

One central aspect of landscape denudation that has attracted little interest is that, as a landscape changes, components of the landscape, hillslopes, tributaries, and main channels, will not necessarily be adjusted to one another or be graded. That is, a channel adjusting to uplift may not be ready to cope with the effects of the rejuvenation in the watershed upstream. Hence, an actively eroding system will be continually searching for a stability that cannot be maintained. Interestingly enough, Davis was to some extent aware of this problem. For example, when the diagram illustrating the geomorphic cycle that was prepared by Davis himself is inspected (Fig. 4), it is apparent that part of his scheme is missing in Figure 2.

A description of Figure 4 in Davis' own words is appropriate here (Davis, 1899, pp. 254-255):

> "the base line represents the passage of time, while verticals above the base line measure altitude above sea level. At the epoch 1 let a region of whatever structure and form be uplifted, B representing the average altitude of its higher parts and A that of its lower parts, AB thus measuring its average initial relief . . . The larger rivers, whose channels initially had an altitude A, quickly deepen their valleys, and at the epoch 2 have reduced their main channels to a moderate altitude represented by C. The higher parts of the interstream uplands, acted on only by the weather without the concentration of water in streams, waste away much more slowly, and at epoch 2 are reduced in height only to D. The relief of the surface has thus been increased from AB to CD. The main rivers then deepen their channels very slowly for the rest of their lives, as shown by the curve CEGJ, and the wasting of the uplands, much dissected by branch streams, comes to be more rapid than the deepening of the main valleys, as shown by comparing the curves DFHK and CEGI. The period 3-4 is the time of the most rapid consumption of the uplands, and thus stands in strong contrast with the period 1-2, when there was the most rapid deepening of the main valleys. In the earlier period the relief was rapidly increasing in value, as steep-sided valleys were cut beneath the initial troughs. Through the period 2-3 the maximum value of relief is reached, and the variety of form is greatly increased by the headward growth of side valleys.

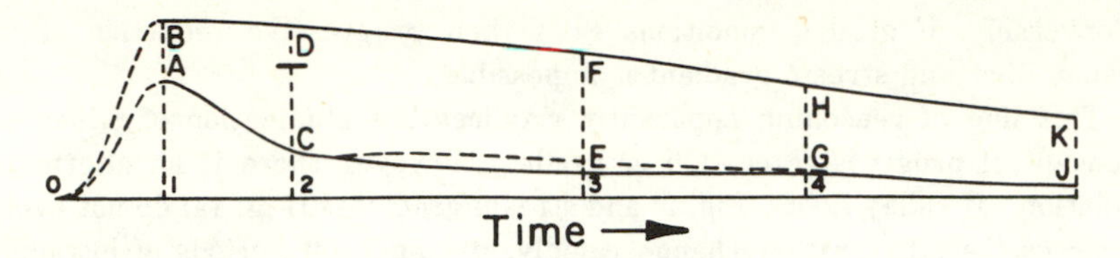

Figure 4. The geomorphic cycle according to Davis (1899).

> During the period 3-4 relief is decreasing faster than at any other time, and the slope of the valley sides is becoming much gentler than before; but these changes advance much more slowly than those of the first period. From epoch 4 onward the remaining relief is gradually reduced to smaller and smaller measures, and the slopes become fainter and fainter so that some time after the latest stage of the diagram the region is only a rolling lowland, whatever may have been its original height."

A few pages further (pp. 260-261) Davis returns to the topic of valley deepening and the effect of sediment load on this process, and he clearly states that downwearing of the valley floor may not be a continuous process. That is, at some stage the main river adjusts by aggradation to the increased quantities of sediment being delivered from upstream, and a valley-fill deposit is formed within the valley. Davis, thus, envisioned a period of aggradation during the progressive erosion of the drainage basin. This concept was stated also in an earlier paper (Davis, 1895, p. 130) as follows:

> ". . . streams proceed to entrench themselves in the slanting plain, and in a geologically brief period, while they are yet young, they will cut their valleys down so close to base level that they cannot for the time being cut them any deeper; . . . When this condition is reached, the streams may be described as having attained a "profile of equilibrium;" or, more briefly, they may be said to be *graded*. It may be noted, in passing, that inasmuch as the work that the stream has to do is constantly varying, it must as constantly seek to assume new adjustments of grade. In the normal course of river events, undisturbed by outside interference, the change in the work is so slow that the desired adjustment of capacity to work is continually maintained. It may be that during the adolescence of river life, the work to be done is on the increase, on account of the increasingly rapid delivery of land waste from the slopes of the growing valley branches; and in this case, part of the increase of waste must be laid down in the valley trough so as to steepen the grade, and thus enable the stream to gain capacity to carry the rest. Such a stream may be said to aggrade its valley . . .; in this way certain flood plains (but by no

> means all flood plains) may have originated. Aggrading of the valley line may often characterize the adolescence of a river's life; but later on, through maturity and old age, the work to be done decreases, and degrading is begun again, this time not to be interrupted."

The deduction of Davis that deposition will naturally follow initiation and presumably rejuvenation of a drainage system was a very astute one, and one that he illustrates by the dashed line CEG in Figure 4. This idea seems to have been ignored; nevertheless, the possibility that deposition takes place naturally within the drainage system is of great importance.

In summary, it appears that an uplifted drainage system will have difficulty in disposing of all the sediment delivered to its major channels from minor tributaries and interfluve areas, and sediment will be stored within the system. There is, of course, a reduction of slopes and stream gradients as well as a widening of the main valley in a downstream direction, which creates a situation whereby sediment delivered from upstream may be stored in the downstream parts of a drainage system. For example, there is a very dramatic downstream increase in the area available to receive deposition within even small drainage basins, as valleys widen downstream (Hadley and Schumm, 1961). That is, within any natural drainage basin there are many places where sediment can be stored permanently or temporarily.

GEOMORPHIC THRESHOLDS AND COMPLEX RESPONSE

Recently I discussed two geomorphic concepts that have potential for aiding in the development of an understanding of the complexity of the landscape; these are geomorphic thresholds and complex response (Schumm, 1975). For example, when a small experimental drainage basin was rejuvenated, the system responded not simply by incising, but by hunting for a new equilibrium by incision, aggradation, and renewed incision. This was referred to as the *complex response* of the system. The concept of *geomorphic thresholds* suggests that there can be changes within the fluvial system that are not due to external influences but rather they are due to geomorphic controls inherent in the eroding system. Field and experimental studies demonstrate that when sediments are stored within a fluvial system they become unstable at a critical threshold slope, and erosion takes place. This seems to be a reasonable explanation for the distribution of some arroyos and gullies in the West (Patton and Schumm, 1975), and it is also a partial explanation for the different morphological characteristics of alluvial fans, that is, the presence or absence of fan-head trenches (Weaver and Schumm, 1974).

When the influence of external variables such as isostatic uplift is combined with the effects of complex response and geomorphic thresholds, it is clear that denudation, at least during the early stage of the geomorphic cycle, cannot be a progressive process. Rather, it should be comprised of episodes of erosion separated by periods of relative stability, a complicated sequence of events. Much of this complexity is the result of a delayed transmission of information through the system. That is, channel changes that take place near the mouth of a drainage basin following incision are responding to the conditions at that time and location. Therefore, the channel is not prepared for the changes that its incision induces within the system upstream; hence downcutting is followed by deposition when the upstream response occurs (Fig. 4).

LANDSCAPE EVOLUTION

When the concepts of geomorphic thresholds and complex response are applied to landscape evolution the model becomes as summarized by Figure 5. Figure 5a is essentially a modification of the left side of the Davis scheme shown in Figures 2 and 4 (youth and early maturity), but the progress of denudation is interrupted by periods of isostatic adjustment. As in Figures 2 and 4, the upper line of Figure 5a represents divide elevations and the lower line valley-floor elevations. The divide elevations probably change as shown by Figure 5a. That is, only major external influences affect the divides, and they are subjected to a relatively uniform downwearing. However, if the valley floor is considered in greater detail over a shorter span of time (Fig. 5b), a stepped pattern of valley floor reduction emerges, as a result of storage and flushing of sediment from the valleys. This model ignores variations due to external influences, and it shows a system that is in dynamic metastable equilibrium (Chorley and Kennedy, 1973).

A steady state equilibrium involves fluctuations about an average, but a metastable equilibrium occurs when an external influence carries the system over some threshold into a new equilibrium regime. The effects of external variables on equilibrium systems are expected, but in the case of landscape denudation the dynamic metastable equilibrium may reflect the response of the system to inherent geomorphic thresholds (Schumm, 1973), in this case, the accumulation of sediment to an unstable condition. When a geomorphic threshold is crossed, the drainage system will be rejuvenated, and the complex response will come into play (Fig. 5c). Figure 5c shows periods of instability separated by longer periods of dynamic equilibrium or grade. Because periods of erosion are followed by periods of deposition, the bedrock floor of the valley is reduced in a steplike manner through time as shown by the dashed lines. Hence, separating the periods of erosion will be periods of deposition and storage of alluvium. In addition, during these periods of relative stability, channel pattern may change from straight to increasingly sinuous, as the nature of the sediment moved through the channel changes (Schumm, 1963b). Hence, sinuosity may also increase to a condition of incipient instability when a large flood can cause an abrupt shortening of the

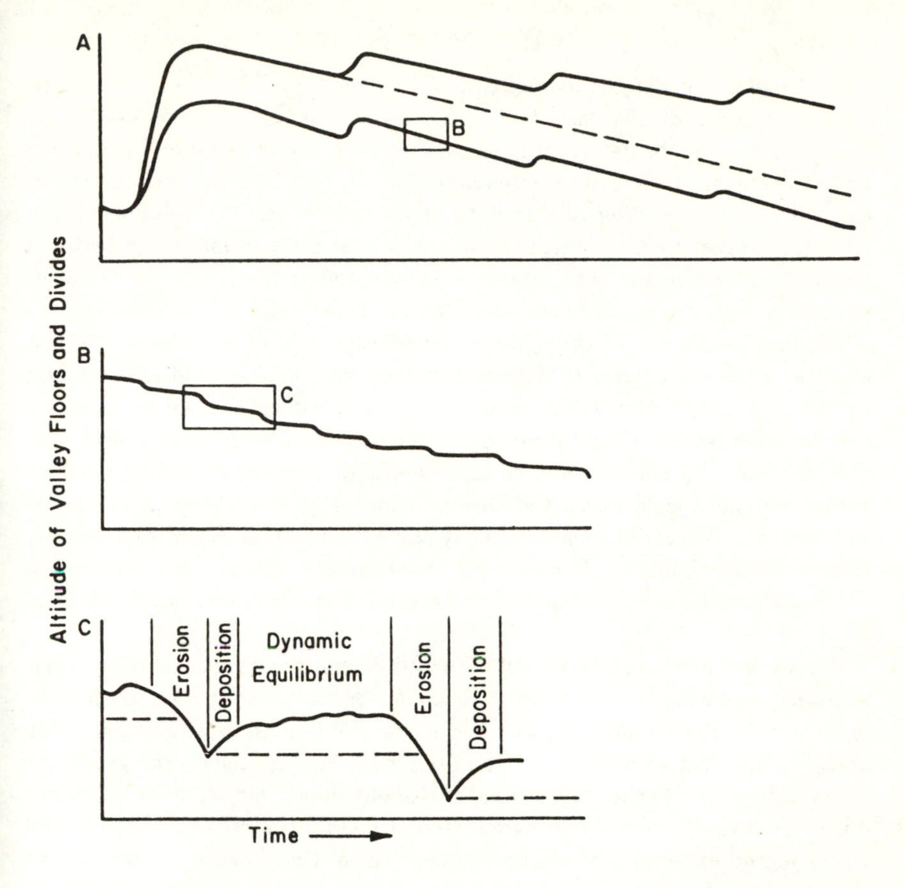

Figure 5. Modified concept of geomorphic cycle.
A. Erosion cycle, as envisioned by Davis (dashed line), following uplift, and as affected by isostatic adjustment to denudation.
B. Portion of valley floor in A above, showing episodic nature of decrease of valley-floor attitude.
C. Portion of valley floor in B above, showing periods of instability separated by longer periods of dynamic equilibrium.

river course of the type documented by Fisk (1944) for the Mississippi River. This type of event could also play a role in the establishment of a period of valley erosion.

If, in fact, this type of episodic erosion takes place many of the details of the landscape, small terraces, and recent alluvial fills, do not need to be explained by the influence of external variables because they develop as an integral part of system evolution. The Davis curve of Figure 2 averages out these variations during cyclic time (Fig. 1a).

EVIDENCE

The reader's obvious response to Figure 4 is "prove it." In view of the climatic history of the last few million years it may be impossible to demonstrate that the details of landscape are not due to climatic fluctuations or to tectonism. In fact, it is extremely difficult to find locations where one can with confidence view the results of an uninterrupted evolution of the landscape rather than the response of the landscape to climatic and tectonic changes. Nevertheless, the evidence is suggestive that, when a drainage system is rejuvenated, incision takes place, and relatively soon thereafter, depending on the size of the system considered, deposition follows, and the complex response occurs, as detected in the experimental studies described by Schumm and Parker (1973). During these experimental studies, following rejuvenation sediment yield from the drainage system decreased rapidly, but this decrease did not continue to an essentially uniform or average value; rather sediment yield fluctuated through time (Fig. 6). These variations of sediment yield occurred during each experiment and under essentially uniform experimental conditions. For example, the quantity of water delivered to the drainage system was constant, and there were no further changes of base level.

The variations of sediment yield occurred during the normal uninterrupted erosional evolution of a very small drainage basin. Additional experiments demonstrate that periods of low sediment yield occur when sediment is stored in the upstream valleys. When sediment storage causes the gradients of the valley floor to exceed a critical threshold slope, this material is flushed out to produce the periods of high sediment yield. The erosion of the valley floors in the experimental studies should follow the scheme as outlined in Figures 5b and 5c.

Further tests of this idea can be obtained in situations where rejuvenation of a basin is so great that progressive downcutting of the stream is expected. It is well established in the geomorphic literature that a major reduction of base level will cause a progressive downcutting and readjustment of the stream gradients until a new graded or equilibrium situation has been developed. However, where such an event has occurred, terraces and evidence of pauses in the erosional downcutting are found, but these are usually attributed to some external influence, such as variations in climate, the rate of base level change, or variations in the rates of uplift of the sediment source area. However, recent studies in Douglas Creek drainage basin of western Colorado support the idea of discontinuous downcutting. The investigation of the recent erosional history of this valley (R. Womack, 1975, unpublished M.S. Thesis, Colorado State University) shows that modern incision of the valley fill began after 1882. Yet, there are four surfaces now present below the two pre-1882 surfaces (Fig. 7). These surfaces are unpaired, discontinuous terraces that elsewhere have been explained by the shifting of a channel laterally across the valley floor, during progressive

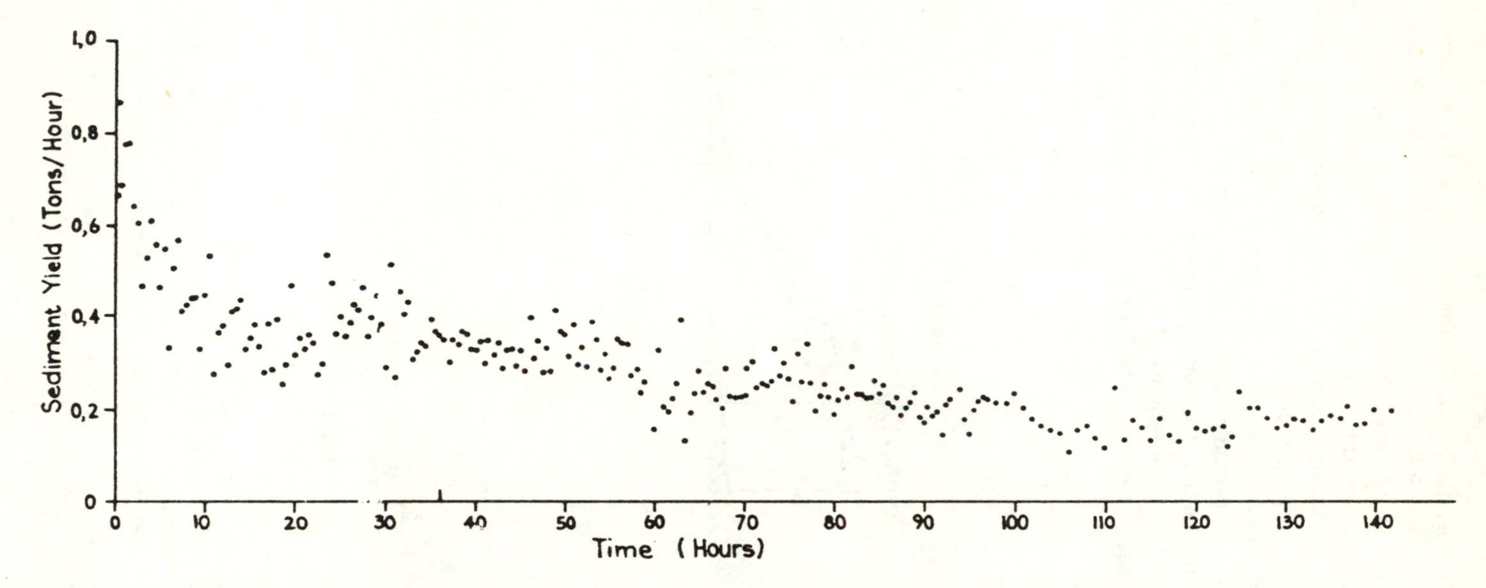

Figure 6. Sediment yield from DERF following rejuvenation of drainage network at time 0. Secondary peaks of sediment production occur at about 30, 50, 75 and 97 hours (Parker, R. S., 1975, unpublished Ph.D. dissertation).

downcutting (Davis, 1902). In the Douglas Creek Valley, however, downcutting was discontinuous. In fact, during pauses in downcutting there was deposition. The denudation scheme as sketched on Figure 5c portrays the sequence of events in this 400 square mile drainage basin in western Colorado. That is, during incision of the main channel, there is rejuvenation of tributaries and a progressive increase in sediment yield from upstream. Sediment loads become so great that downcutting ceases and deposition begins. Deposition continues until it is possible for the channel to incise again and to continue the downcutting process. Multiple surfaces of this sort can be identified on photographs of the Rio Puerco area, and elsewhere. Probably the sequence of events in these areas was similar to that of Douglas Creek.

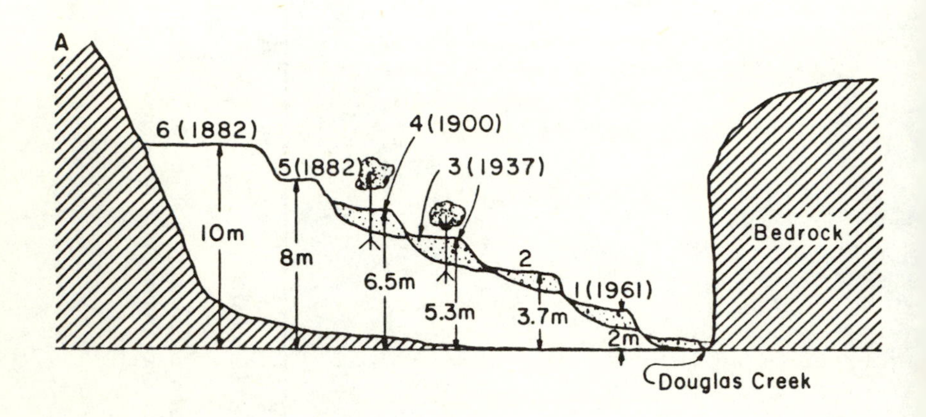

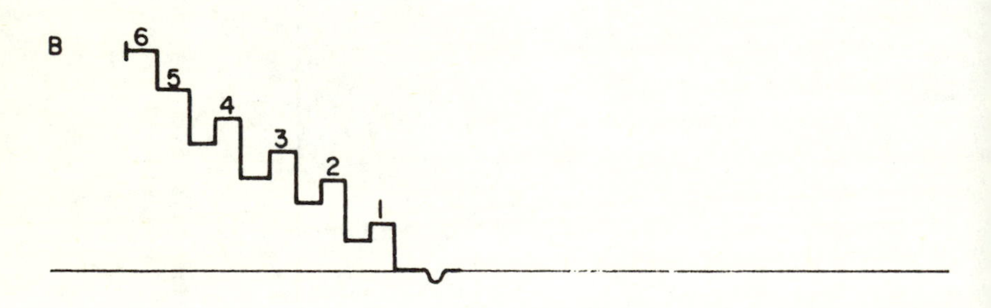

Figure 7. A. Sketch of Douglas Creek valley showing erosion surfaces formed since 1882. Age of surface is based on tree-ring dating and historical data. Note burial of trees by deposition. Surfaces 5 and 6 were present before modern erosion began after 1882 (after Womack, 1975, unpublished M.S. thesis).
B. Summary of behavior of Douglas Creek. Vertical segments indicate incision or deposition, horizontal segments periods of relative stability.

The Douglas Creek situation appears to conform to the observations of Born and Ritter (1970) who have mapped six discontinuous and unpaired terraces at the mouth of Truckee River, where it enters Pyramid Lake in Nevada. A reduction of the water level in Pyramid Lake reduced the base level of the lower Truckee River, but instead of simple downcutting commensurate with the lowering of the base level, the channel in fact paused as many as six times.

Gage (1970) cites an example of rapid deposition, which caused aggradation of from 10 to 80 feet in the Waiho River of New Zealand. This glacier-fed river then proceeded to clear the deposited sediment over a period of a few weeks. The erosion produced a flight of 10-foot terraces. Gage attributed this and similar events to 10-year weather patterns, and he cautioned that if some of these terraces were preserved they could easily be mistaken for surfaces of considerable antiquity.

Other examples of multiple unpaired terraces can be cited that may have formed as a result of rapid but episodic incision similar to that which occurred in the Douglas Creek Valley (Davis, 1902; Small, 1973).

CONCLUSIONS AND DIRECTIONS

The major conclusion is that a very complicated erosional evolution of a landscape can be expected, and the erosional evolution will be complicated even though there are no external influences affecting the system. Richard Hey (1972) has reached similar conclusions on a theoretical approach to channel incision.

This model of landscape evolution may seem to pose problems for the interpretation of some landscape details, but of course the usual appeal to climate fluctuation or tectonics is no solution. In fact, one need not attempt to explain all of the details of a landscape as a result of climate change or distrophism because some of the complications are inherent in landscape evolution. To one concerned with river stability and erosion control this model may be of considerable aid, because it suggests that it is possible to recognize unstable components of the landscape based on the identification of geomorphic threshold slopes, and to take measures to prevent further development of the instability (Patton and Schumm, 1975).

In geologically and geomorphically homogeneous regions the high variability of sediment yield and runoff characteristics of drainage basins may also be explained by this model, as illustrated by the variations of sediment yield from the experimental watershed (Fig. 6). That is, even within the same geomorphic region, similar drainage basins need not be producing comparable quantities of sediment. In effect, the revised model explains why the attempt to relate geomorphic variables to hydrology has been successful but less satisfying than one would like.

Some evidence to support an episodic model of at least the early stages of landscape evolution has been presented based on experimental studies and limited field data. The field data are restricted to the semi-arid and arid regions of western United States and high sediment producing areas elsewhere. The applicability of the model to perennial streams and the humid regions is unknown. In addition, a major difficulty with regard to the field evidence is that it was obtained from relatively small drainage systems (less than 400 square miles), where changes take place at rates sufficiently rapid to be recorded. In large drainage systems the progression of erosion will be sufficiently slow that it will be impossible to detect changes in sediment-yield variations and in valley slope and stream gradient through the life span of any one individual.

To find convincing evidence of the ideas put forth here it seems necessary that the geomorphologist must scrutinize the records of depositional events. The stratigraphic record is complex and fragmentary, and this in turn may be considered to be a natural result of the erosional development of the landscape as described here. Nevertheless, there should be sequences of rocks that, when analyzed, will provide information on the rate and variability of the erosional denudation of the landscape.

Apparently what is required are investigations of the type suggested by Mutti (1974) who concluded from his study of deep-sea fan deposits that an understanding of these features must come from the study of the geologic record rather than the few cores and profiles of modern deep-sea fans. This is true also of an attempt to evaluate the erosion cycle of a fluvial system. The complexities of erosional evolution through time will be preserved for the most part in the depositional basin, whereas the evidence of these events will be destroyed by the denudation of the landscape itself. It seems appropriate therefore to close this discussion with consideration of a topic that has concerned stratigraphers for some time, that is, cyclic sedimentation. A great deal has been written on this subject, not, of course, all related to fluvial processes. The symposium volume edited by Merriam (1964) and the book by Duff, Hallam and Walton (1967) adequately review the problem. The concern, of course, is whether or not such cycles actually exist, and if they do, the reason for them. It appears from the previous discussion on the behavior of fluvial systems that the development of cyclic sedimentation is inevitable in fluvial sedimentary deposits.

Cycles of several dimensions should be found within a fluvial depositional unit (Fig. 8). Referring to Figure 5, which shows the progress of the erosion cycle through time, one can substitute sediment yield for elevation or relief on the abscissa of that figure. The curves then show sediment produced by the fluvial system through time. There should be at least five fining upward cycles of decreasing magnitude in the deposits associated with this erosional evolution. The primary cycle is related to denudation following uplift, with maximum sediment production at the beginning of the erosional event and a progressive decrease in quantity and size of material through time to yield a massive fining upward sequence (Figs. 2 and 4). Interruptions of this,

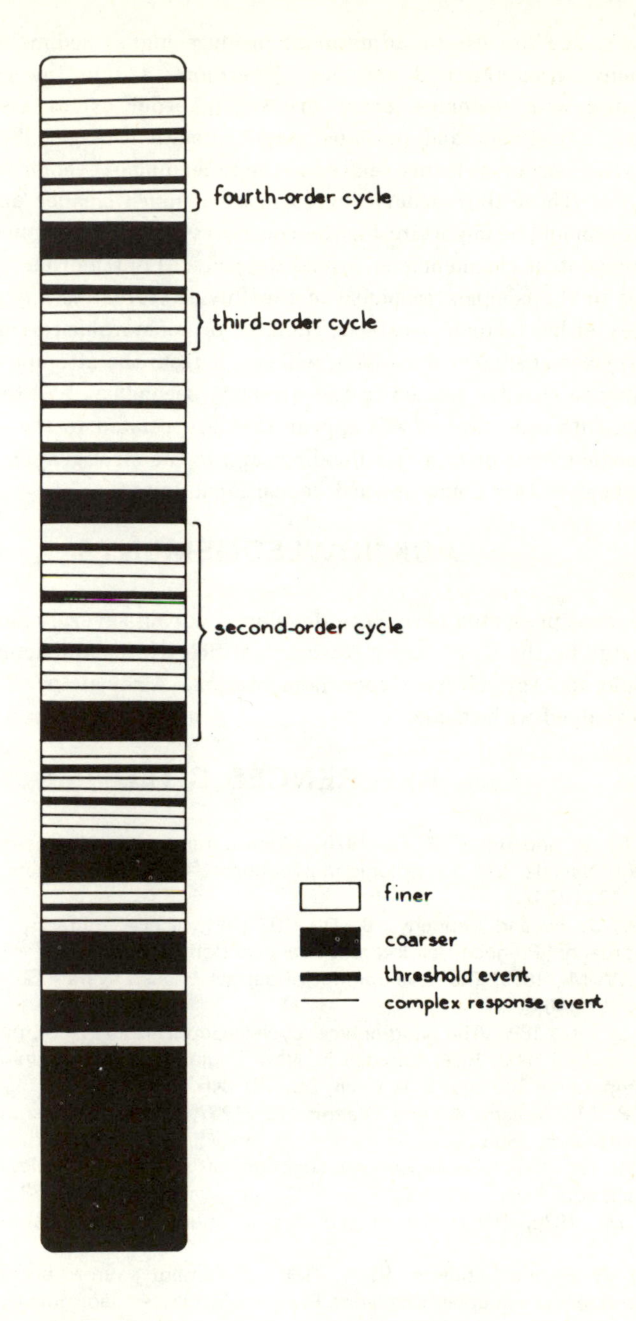

Figure 8. Diagrammatic model of primary sedimentation cycle with higher order components.

however, due to isostatic adjustment produce higher sediment yields from the source area (Figs. 3 and 5a). Therefore, within the primary cycle associated with tectonics, there are second-order cycles associated with isostatic adjustment and perhaps major climatic change. Between these events third-order cycles are related to the exceeding of geomorphic thresholds (Fig. 5b). These third-order cycles will be of much smaller dimension, and yet they should be important for the concentration of heavy minerals and the development of channeling in fluvial deposits. Fourth-order cycles will be related to the complex response of the fluvial system to any of the above changes, either tectonic, isostatic, climatic, or geomorphic threshold (Fig. 5c). These cycles of smaller dimension will result from the attempt of the system to adjust to changes related to the primary, secondary, and tertiary cycles. Finally, fifth-order cycles will appear that are related to the seasonality of hydrologic events or to major flooding, and in the stratigraphic record these will appear as thin fining-upward depositional units.

ACKNOWLEDGEMENTS

The ideas presented here were developed during several research projects supported by the U. S. Army Research Office, National Science Foundation and Colorado Agricultural Experiment Station. A review by M. P. Mosley is acknowledged with thanks.

REFERENCES CITED

Born, S. M. and Ritter, D. F., 1970, Modern terrace development near Pyramid Lake, Nevada, and its geologic implications: Geol. Soc. America, Bull. v. 81, pp. 1233-1242.

Chorley, J. R. and Kennedy, B. D., 1971, Physical Geography — A Systems Approach: Prentice-Hall International, London, 376 p.

Davis, W. M., 1895, The development of certain English rivers: Geogr. Jour. v. 5, pp. 127-146.

———————, 1899, The geographical cycle: Geogr. Jour. v. 14, pp. 481-504.

———————, 1902, River terraces in New England: Harvard Univ., Museum of Comparative Zoology, Bull. v. 38, pp. 281-346.

Duff, P. M., Hallam, A. and Walton, R., 1967, Cyclic Sedimentation: Elsevier, Amsterdam, 280 p.

Fisk, H. N., 1944, Geological investigation of the alluvial valley of the lower Mississippi River: Mississippi River Comm. Vicksburg, Miss. 78 p.

Gage, M., 1970, The tempo of geomorphic change: Jour. Geology v. 78, pp. 619-625.

Hadley, R. F. and Schumm, S. A., 1961, Sediment sources and drainage basin characteristics in upper Cheyenne River basin: U. S. Geol. Survey Water Supply Paper 1531-B, pp. 137-196.

Hey, R. D., 1972, An analysis of the factors influencing the hydraulic geometry of river channels: Unpublished Ph.D. dissertation, Univ. of Cambridge.

Merriam, D. F., 1964, Symposium on cyclic sedimentation: State Geol. Survey of Kansas, Bull. 169, 2 vols., 636 p.

Mutti, E., 1974, Examples of deep-sea fan deposits from circum-Mediterranean geosynclines: Soc. Economic Paleo. Mineralogists, Spec. Pub. 19, pp. 92-105.

Mackin, J. H., 1948, Concept of the graded river: Geol. Soc. America, Bull., v. 59, pp. 463-511.

Patton, P. C. and Schumm, S. A., 1975, Gully erosion, northwestern Colorado: A threshold phenomenon: Geology, v. 3, pp. 88-90.

Schumm, S. A., 1963a, The disparity between present rates of denudation and orogeny: U. S. Geological Survey Prof. Paper 454-H, 13 p.

———————, 1963b, Sinuosity of alluvial rivers on the Great Plains: Geol. Soc. America Bull, v. 74, pp. 1089-1100.

———————, 1973, Geomorphic thresholds and complex response of drainage systems: pp. 299-310 *in* Fluvial Geomorphology (editor M. Morisawa) Pub. in Geomorphology, SUNY Binghamton, N. Y., 314 p.

———————, and Lichty, R. W., 1965, Time, space, and causality in geomorphology: Am. Jour. Sci., v. 263, pp. 110-119.

Small, R. J., 1973, Braiding terraces in the Val D'Herens, Switzerland: Geography, v. 58, pp. 129-135.

Weaver, William and Schumm, S. A., 1974, Fan-head trenching: An example of a geomorphic threshold: Geol. Soc. America Abstracts, v. 6, p. 481.

CHAPTER 5

DYNAMIC EQUILIBRIUM AND LANDSCAPE EVOLUTION

John T. Hack

ABSTRACT

The principle of dynamic equilibrium when used to explain landscape features is not in itself an evolutionary model such as the geographic cycle. However, the tendency toward dynamic equilibrium is a universal principle that can be used to explain specific landscape features and problems, when it is assumed that the landscape has developed during a long period of continuous downwasting. This concept can be tested and compared with the multiple erosion cycle concept by examining a variety of specific features in a landscape, as the writer has done previously in the Shenandoah Valley of Virginia.

Evolutionary models can be conceived assuming (1) a stable base level, (2) a rise in base level, or (3) a fall in base level. In the first model, a gradual lowering of relief would be expected, greater effects occurring near base level than farther away, somewhat like a single cycle in the Davisian model. If base level rises, there would be a drowning of the lower valleys but very little effect upstream, and the topography would continue to lower. If base level falls, erosion would be accelerated near the new base level, and the acceleration would affect an increasingly large area. New adjustments of slope to rock resistance would be made simultaneously.

As has happened in the past, however, the model we adopt to explain landscapes must be related to prevailing thought in other fields of earth science.

INTRODUCTION

Many geomorphologists have abandoned the theory of the geographic cycle as a basis for landscape analysis, but no alternative has yet been generally accepted. In 1960, I advocated the use of the principle of dynamic equilibrium to explain the topography of erosional landscapes (Hack, 1960) and argued that the concept of multiple erosion cycles did not explain either the multilevel landscape of the Valley and Ridge province or the so-called dissected landscape of the Piedmont. Although I did not know it at the time, a criticism of the erosion-cycle concept very similar to mine had been made by an Australian geologist, E. O. Marks (1913), but Mark's paper received little notice outside Australia. The time was not then ready for such criticism.

The use of the dynamic-equilibrium principle was not new in geomorphology, for Gilbert (1877) had used it as the basis for his laws of erosion. Gilbert was quite familiar with thermodynamic principles, accepted them as physical realities, and commonly used them as a means of solving geologic problems.[1] He was not thinking in terms of models or theories of landscape evolution. He was explaining the origin of the landscape features observed in the Henry Mountains, but he realized that the explanation was of universal value, for he put it in terms of laws of nature. I thought in 1960 that I was advocating a similar approach.

In my opinion, the equilibrium concept should not be viewed as a model by itself. If one stated that the Appalachian Mountain system developed by downwasting that has been continuous since the latest orogeny, and that the variety of forms we now find are due to equilibrium of action in the erosional system — that would be a model. It must be recognized, however, that evolution is also a fact of nature and that the inheritance of form is always a possibility. The theory of the geographic cycle involving multiple peneplains, however, was inadequate as a model for the Appalachians and probably for all other similar terrains. It cannot be denied that the evolutionary changes during a cycle might look somewhat like those that Davis described, but this is quite different from the concept of multiple cycles and the inheritance of peneplain forms.

TEST OF THE CONTINUOUS DOWNWASTING MODEL

One way to test a theory or concept is to apply it systematically to explain field facts and relationships. This was the attempt in my study of the Shenandoah Valley published in 1965. Specific problems of the valley are explained by using the idea that topographic forms and processes are closely related to differences in the rocks, and the processes acting on them. The assumption was made that the erosional system had been downwasting a long time, and that it could be analyzed as though it were in a steady state. The analysis was carried as far as possible, and in my opinion it resolved many problems. Several examples are given below.

1. The specific geometries of drainage basins if examined quantitatively are related closely to the rock types, to exposure, and to other environmental factors throughout the Shenandoah Valley. The forms considered include size and shape of hollows, density of valleys, curvature of ridge tops, and

---

[1]Mr. Steve Pyne, University of Texas, who is studying Gilbert's life has been very helpful in advising me concerning Gilbert's approach to problem-solving.

channel slopes of streams of various sizes. Even the hypsometric curves of drainage basins differ in different rock types.

2. Major topographic features such as the great mountain ridges are closely related to geologic structure. For example, the height of the Blue Ridge is controlled by the thickness of the metabasalt of the Catoctin Formation where it is of low metamorphic grade and relatively unsheared. Some of the gaps in the mountain ridges where detailed geology has been done have been shown to be related to geologic structures. One example is Manassas Gap at Front Royal, where the Catoctin Formation is missing because of a thrust fault.

3. The major rivers avoid the resistant rocks, and the Potomac is a particularly good example.

4. The longitudinal profile of the Shenandoah Valley, including the general altitude of the hills, is related to lithology and structure. This point will be discussed in greater detail.

5. On the lowland surface of the Shenandoah Valley, the occurrence of as well as the thickness of unconsolidated residuum are determined by the rocks directly beneath.

6. The gravel terraces and aprons of the valley lowlands are distributed in close relation to large drainage areas in the resistant rocks that are sources of the gravels. The piedmont aprons can be shown to have formed like the "sheets of fanglomerate" on the pediments of the Henry Mountains (Hunt, Averitt, and Miller, 1953, p. 190).

7. Iron and manganese deposits occurring in a belt marginal to the west part of the Blue Ridge, which formerly were believed to be associated with the Harrisburg peneplain, are better explained as concentrates related to the residuum and alluvial cover over the Cambrian dolomites. The failure of the peneplain idea to explain these deposits is evidenced by the occurrence of ore in residuum derived from a few beds in the stratigraphic section. Furthermore, the ores vary in altitude from 100 feet below the Shenandoah Valley floor to 400 feet above the valley floor on spurs of the Blue Ridge front.

All of these features of the landscape show a remarkable dependence on geologic structure and on the distribution of rocks of different physical and chemical properties. If discordant surfaces of any extent had survived, it is almost beyond belief that the relation of the landscape elements to structures could be so orderly.

FLOOR OF THE SHENANDOAH VALLEY

Difficulties in explaining certain features of the Appalachian landscape using the geographic-cycle concept have been noted by others. In the Shenandoah Valley, an example is the sharp increase in gradient of the valley-floor surface, or Harrisburg peneplain, about 35 to 40 miles upstream from the Potomac River. This break in longitudinal valley profile occurs near Front Royal where the North and South Forks of Shenandoah River join at the north end of Massanutten Mountain. Keith (1894) noted the break in slope and postulated differential uplift of the peneplain. This idea and the possibility that more than one surface was preserved on the valley floor sufficed to explain its irregularity for many years.

As a result of detailed mapping of a part of the valley, King (1949) realized that the origin of the valley floor must have been much more complicated than was formerly supposed. South Fork is bordered by a series of extensive gravel terraces, the highest of which are much dissected. The high terrace gravels are unconformable in some places on still older gravels, which in turn are underlain by thick residuum. Although King did not deny the existence of a former peneplain such as the Harrisburg peneplain, he showed that it must be very ancient, if present at all.

When the valley floor is examined in the context of continuous downwasting and the adjustment of slopes toward equilibrium, the problem is simplified. In this explanation, the terraces or gravel deposits are present where the resistant rocks in the bordering mountains have large outcrop areas. The hard rocks shed debris at a higher rate than the secondary streams of the limestone valley can transport on the relatively low gradients. The gravel does not keep on accumulating but achieves a certain equilibrium of area. When spread in fans and terraces it is subject to solution and erosion at a more rapid rate. This argument is presented in detail elsewhere (Hack, 1965, p. 29-58).

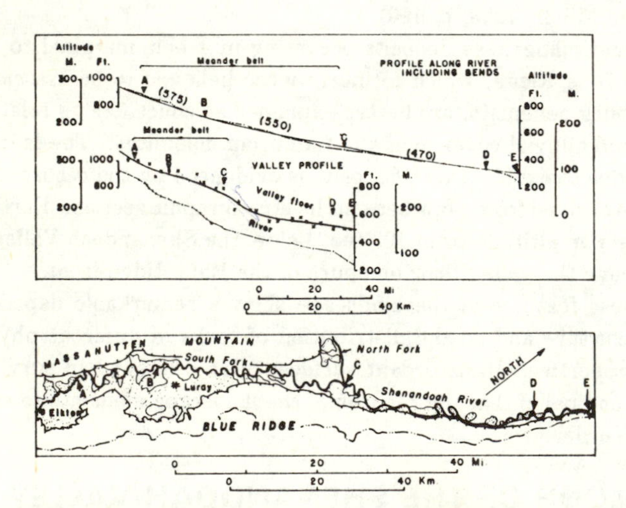

Figure 1. Profiles and sketch map of the South Fork of the Shenandoah River from Elkton to the junction with the Potomac River (at E). Actual river profile shown by short lines between dots. Profile constructed mathematically shown by fine smooth line. Numbers on the profile along river in parentheses are gradient index values. X's on the valley profile are altitudes of gravel terraces on the valley floor. The dotted pattern on the map represents gravel terraces and alluvium. Points A-E on map identify corresponding locations on the profiles.

The longitudinal profiles of the North and South Forks of the Shenandoah River provide a test of the use of the equilibrium concept as a solution. A map and profiles of the South Fork Valley and main Shenandoah Valley from Elkton to Harpers Ferry are given in Figure 1.

The river profile measured along the channel (upper curve) between localities A and E has a length of 140 miles (225 km) as compared with 80 miles (129 km) measured down the valley (lower curves), a ratio of 1.75 to 1. The difference in length of the two curves is due to the meandering character of the stream and the ratio between the lengths is called the sinuosity. The sinuosity is highest, however, in the reach between A and C where it averages 2.3. Downvalley from C to E it is only 1.4. The major break in the valley profile (lower curves) thus corresponds with the change in sinuosity.

The river-channel profile is a remarkably smooth curve, and in the figure, a logarithmic curve is plotted over it with the formula $H=3114-527 \ln L$, where H is altitude in feet and L is distance in miles from the head of the South Fork river system. The largest departures from the ideal mathematically constructed profile are in the reach above the junction of the North and South Forks (point C, Fig. 1) and the reach just above the junction with the Potomac (point E. Fig. 1), a river with much larger discharge. The gradient index values shown in parentheses permit a numerical comparison of the average gradients of the reaches. This index is a measure of the slope of a logarithmic curve defined by the equation

$$\frac{H_1 - H_2}{\ln L_2 - \ln L_1}$$

and, as shown elsewhere (Hack, 1973), correlates roughly with the power and competence of a stream.

The valley floor curve, which shows the elevation of gravel deposits on the valley floor, is irregular and flattens markedly below the junction of the North and South Forks at locality C as noted by Keith and others. It relates closely to the altitudes on the adjacent streambed, except that the depth of intrenchment increases downstream. The geometry of these curves does not support the idea of warping of a peneplain or other surface but indicates instead that the valley floor is in adjustment with the river and that the irregularity of the valley profile is a function of the sinuosity of the river.

A similar but even more pronounced change in the valley profile occurs on the northwest side of Massanutten Mountain and has been described by Hack and Young (1959). This part of the valley is drained by the North Fork, which has a sinuosity of 3; that is, the length of the meandering reach, if measured along the river, is three times the length measured downvalley. Thus, the meanders themselves impart to the valley floor a longitudinal gradient three times that upstream. In effect, both the North Fork and the South Fork flow down their valleys in series of switchbacks. The geometry of these gradient relations is wholly inconsistent with the concept of the Harrisburg as a valley floor peneplain, for the meanders having greater amplitude occur on the steepest parts of the valley floor. The coincidence between valley gradient and sinuosity is too close for the warping hypothesis to be tenable. It is really the bedrock geology of the valley that explains both the meandering and the steep slope. The high sinuosities occur where the river is in the Martinsburg Shale, a thin-bedded silty shale. The joints and cleavage in the shale are parallel to the meanders. Many other Appalachian rivers have similar exaggerated meander patterns in the Martinsburg Shale.

ACCORDANCE OF SUMMITS

Probably the kind of evidence that is most convincing to adherents of the theory of the erosion cycle is the belief in accordance of summits. The topography of the area north of Harrisburg, Pennsylvania, where much of the classic work on peneplains was done has a remarkable regularity, and from a good vantage point an observer can see many miles of mountain top that are at almost the same elevation. In Davis's time and until quite recently, knowledge of the geology was sketchy, and accurate topographic information was not easy to obtain. Now, modern geology and topographic maps at 1:250,000 scale are available, and it is a simple matter to compare topography and geology over large areas.

The mountain ridges near Harrisburg were believed to have been relics of the Schooley peneplain (Fig. 2). The ridges shown are produced by sandstones of the (1) combined Tuscarora and Juniata Formations (2) the Pocono Formation, and (3) the Pottsville Formation. Altitudes on the tops of the ridges are taken from the 1:250,000 scale Harrisburg topographic sheet of the U. S. Geological Survey. Differences in altitude are not great, but altitudes do differ significantly between the formations. Differences along each formation generally correlate with width of outcrop. The ridge formed by the Tuscarora sandstone just north of Harrisburg is 1,100 to 1,300 feet (335 to 396 m) in altitude. The ridge in the northwest corner of the figure, however, reaches 2,100 feet (640 m). This ridge includes a much greater outcrop of resistant rock.

The Pocono ridges average 50 to 100 feet (15 to 30 m) higher but vary with the outcrop area, reaching heights of more than 1,700 feet (518 m). The ridges on the Pottsville are systematically higher than the ridges on the narrow Tuscarora belt. The low altitudes [1,200 feet (365 m) or below] on the Pocono ridges are where the outcrop is narrow. A high degree of accordance does indeed seem to exist, but the details indicate that there are real differences in height systematically related to geologic structure and lithology. The explanation for the height of the ridges is that the sandstones, being

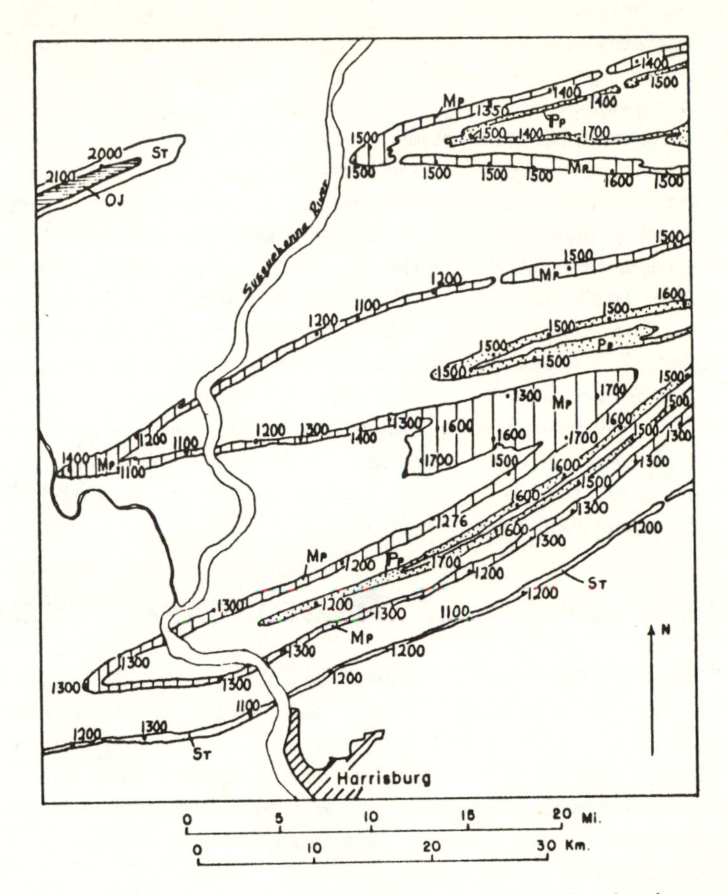

Figure 2. Map of the region north of Harrisburg, Pennsylvania, showing mountain ridges that were considered remnants of the Schooley peneplain. The figures next to dots are summit altitudes on these ridges.
Mp. Pocono Formation; Oj, Juniata Formation; Pp, Pottsville Formation; St, Tuscarora Formation.

more resistant, particularly to weathering, than the limestones and shales stand up at ridge heights in proportion to the area of the rock exposed at the present land surface. The physical properties of the sandstone, its thickness, and its altitude are all factors. When the sandstone bed flattens, as the Pocono does at the unbreached northeast end of the anticline in the figure, more resistant rock is exposed and the land surface is higher. The extreme regularity of the ridges one observes is partly a function of their height and width. A ridge that rises only a few hundred feet above the valley floor, is apt to give the impression of regularity because the distance from the crest to the base is not great enough for water to be gathered in hollows and channels. A wider and higher ridge, even though it may be in the same formation, can develop mountain hollows and first-order valleys on its flanks, and its crest is apt to be less smooth. The remarkable mountain pattern near Harrisburg, Pennsylvania, is thus due to the regularity of the folds, the consistently low thickness of the more resistant sandstones, and the spacing of the sandstones in the stratigraphic column.

The peneplain concept may be used in the explanation of the regularity and general accordance of these ridges. That is, their smooth tops are remnants of a nearly flat plain, preserved because the rocks of which they are composed are more resistant than the rocks between the ridges. However, if considered in detail, the correlation between the heights of the ridges, the specific rock formations, and their width of outcrop is still a fact and remains to be explained. The nearly flat plain, if it ever existed, must have had the same ridges on it as are there today, but they were lower. On the other hand, perhaps the peneplain was flat or had some other configuration, but during its dissection the rocks with greater resistance or wider outcrop were not lowered quite as far as the others. Either explanation is tenable, but in neither is the peneplain a necessary part.

OTHER ASPECTS OF THE GEOGRAPHIC CYCLE

Davis' model of landscape development included many ideas other than cyclicity or series of peneplains. Many of them can be used in the context of a continuous downwasting model, and they do not necessarily conflict with the tendency toward equilibrium. Even in the Appalachians, these other ideas are not necessarily invalidated by the rejection of multiple geographic cycles. Examples are the formation of wind gaps and water gaps, headward migration of divides, and stream piracies.

The adjustment of drainage to structure is truly remarkable in many places in the Appalachians, and this is perhaps the most striking feature of the region. It could have been accomplished only by piracies and divide migration on a grand scale. It is easier for some of us to believe that such a profound adjustment occurred during the erosion of many thousands of feet of rock over a long time span rather than during a short cycle in which a peneplain was dissected only a few thousand feet. Water gaps may be the result of adjustment to structures such as fault zones or fracture zones. On the other hand, they may be inherited from the past when superimposition occurred at a higher stratigraphic level and on a different structural pattern. Some wind gaps are the result of adjustment to structures. That is, they are a low place in a ridge where rock resistance has been overcome by a fault zone or other weakness. Some wind gaps may have formed, as Davis thought, by piracies.

Models Other Than the Steady State.

The continuous downwasting model in which a tendency for dynamic equilibrium is assumed does not necessarily imply a steady state. In fact, though a steady state is possible and is consistent with the idea of isostasy, it must be rare. One can speculate what would happen to a landscape under conditions of (1) a stable base level, (2) a rise of base level, and (3) a lowered base level.

If a landmass is uplifted and remains stable for a long period, it is eventually reduced by erosion and weathering to a level close to base level. During the reduction, an evolutionary sequence of changes occurs. If the rocks are homogeneous and of the same resistance, the topography that forms is determined by the drainage pattern and various erosional processes. Because discharge increases downstream and the ability to transport erosional waste is greatest in the larger streams, the lower end of the drainage system erodes more rapidly than the distal parts until an equilibrium of slope is achieved. This is the cause of the general concave-upward form of drainage systems.

As the interstream areas can rise only a certain distance above the streams, depending on the relative energy of the processes acting on them and the materials of which they are composed, a system of ridges develops. The relief in the system gradually becomes more gentle but is never eliminated.

Differences in rock resistance and the universal tendency for equilibrium of action cause differences of form to develop early in the evolutionary process. A differentiation of slopes, relief, drainage patterns, and other aspects of form takes place. These differences of form are never eliminated but do change if the erosion surface encounters different rocks at lower levels.

This kind of topographic evolution is not too different from that described by Davis, except that Davis believed that ultimately nearly all the topography would become bevelled. For example, he visualized the Piedmont of Virginia as an uplifted peneplain, a point of view clearly explained in his essay on the peneplain (Davis, 1909, p. 356-357). It is more likely that large parts of the Piedmont are close to the ultimate form that can be attained by a former mountainous area as a result of continued downwasting, except for parts near the coastal plain which have increased relief because of a drop in base level. In my opinion, the ultimate landscape in an area of stable base level is an orderly network of ridges and ravines that has a low relief. Such a landscape is well drained and is almost entirely in slope, though the average slope may be very low (Hack, 1960, p. 89). The more resistant rocks form a terrain of higher relief than the nonresistant rocks, and the differences are never completely erased.

Rise in Base Level.

Consider now a second case, that of a rise in base level and the effect of this change in a normal landscape that has already achieved a degree of adjustment. Imagine, for example, that sea level simply rises a thousand feet or so and floods the lower part of an area. Although the potential energy for erosion will be lowered, the area to be eroded will also be lowered, and upstream areas will probably be little affected by the rise. This conception is in harmony with the idea that river profiles and the general behavior of streams that are fashioning slopes are determined by upstream conditions rather than downstream conditions. Rubey (1952, p. 134) stated:

> "The slope at different points and the shape of the profile are controlled by duties imposed from upstream, but the elevations at each point and the actual position of the profile are determined by the base level downstream."

That this idea is now widely recognized is shown by the fact that the analysis of draining basins since Horton's (1945) work has been based on the ordering and measuring of streams and stream segments, proceeding from the head downstream.

The independence of streams from their base levels is also shown by the fact that tributaries do not enter master streams at grade. In fact, their slopes may increase as they approach the master stream, depending on the relative size of the two streams, the occurence of bordering terraces, and other factors.

The construction of a reservoir causes a rise in base level. The effect of one such reservoir has been discussed by Leopold, Wolman, and Miller (1964, p. 436-437). The obstruction of the Rio Grande River by the Elephant Butte Reservoir was thought by some to have caused increased sedimentation and a rise in river level 30 miles upstream from the reservoir. Analysis of sediment data show, however, that the gradual rise of the river began about 1901, antedating the reservoir by 14 years. The upper Rio Grande thus appears to be now depositing a greater load than it did before about 1901. Deposition is probably related to the cycle of gully erosion upstream that began about the turn of the century.

Lowered Base Level.

The third postulate is that of a lowered base level. The effects caused by this circumstance can be studied in the Appalachian region, for we know that the sea covered some of what is now the Piedmont during Tertiary time and withdrew in the Miocene. The amount of lowering can be fairly estimated in the vicinity of Washington, D. C., where Darton (1951) studied the overlap relations at the inner edge of the Coastal Plain. Figure 3 is a sketch map of this region which shows the extent of the Miocene outcrop as interpreted by Darton (1951, pl. 1).

The highest outcrops of probable marine origin shown on this map are at Tyson's Corner northwest of Alexandria, more than 10 miles (16 km) from the inner margin of the Coastal Plain. These outcrops underlie a group of hills 520 feet (150 m) in altitude that form the highest part of the Piedmont in the immediate vicinity. Ten to 20 feet (3-6 m) of fine sandy and silty clay that closely resemble the clay of the Calvert Formation of Miocene age are the lowest sediments exposed; they rest on crystalline rock at an altitude of

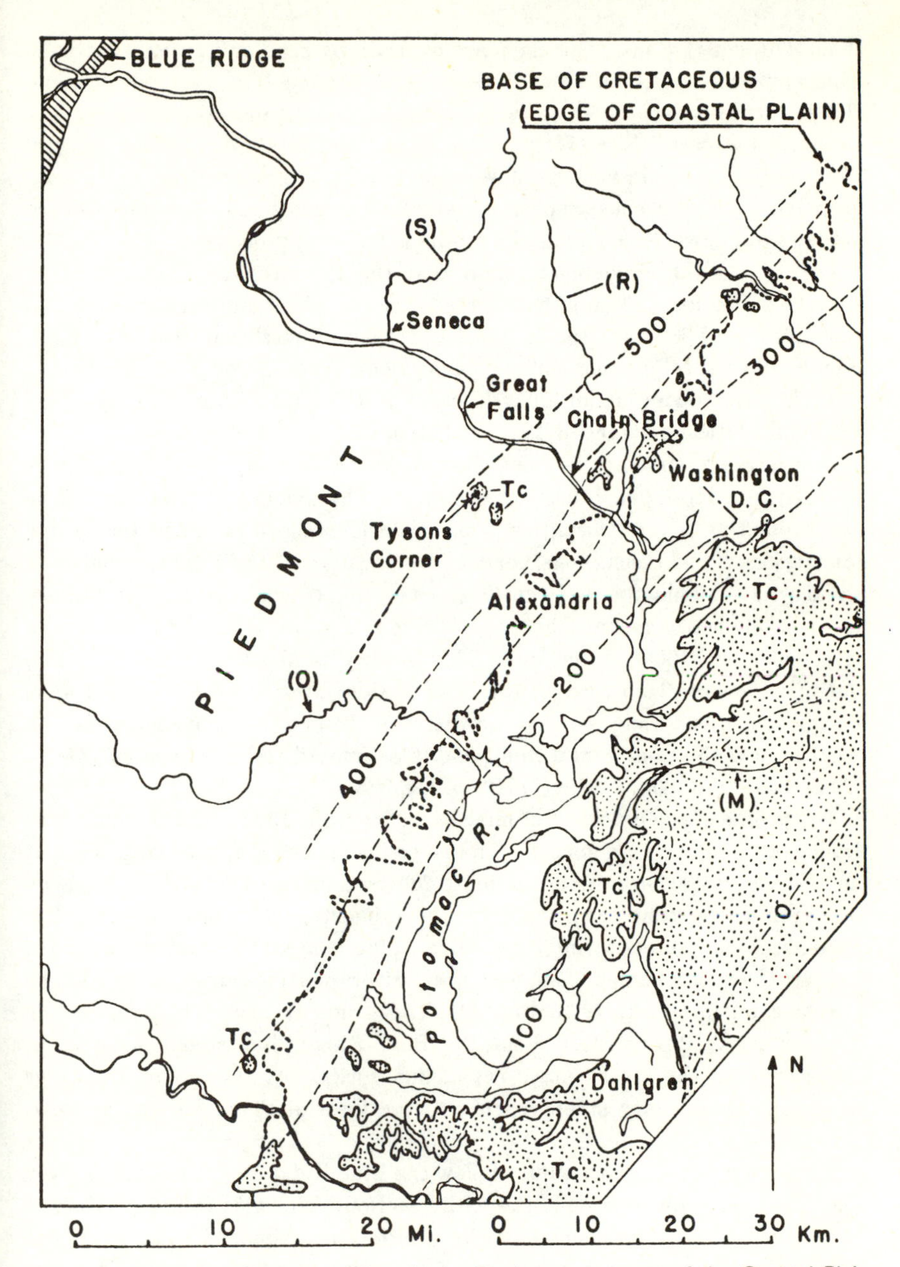

Figure 3. Map of the Potomac River in the Piedmont and part of the Coastal Plain near Washington, D.C. The fine stippled area (Tc) shows the outcrop of Miocene marine sediments and overlying gravels where present as mapped by Darton (1951). Contours are on the base of the Miocene. The heavy dotted line is the edge of the Coastal Plain as marked by the base of Cretaceous sediments. (S) Great Seneca Creek; (R) Rock Creek, (O) Occoquon Creek and its headwater, Cedar Creek; (M) Mattawoman Creek.

about 470 feet (143 m). The clays are overlain by sand and gravel that have generally been regarded as of Pliocene age, though they may be Miocene. These deposits are river gravels laid on the clayey sediments by the ancestral Potomac River (Wentworth, 1930; Schlee, 1957). Similar Coastal Plain outliers occur in northwest Washington, D. C., where they are unconformable on Cretaceous deposits and can be projected downdip toward deposits of known Miocene age in southwest Washington, D. C.

The Tysons Corner deposits show that the Potomac River flowed at an altitude of 500 feet (152 m) above present sea level in late Tertiary time. If Darton's (1951) 500-foot (152-m) level is projected northeastward (Fig. 3), it crosses over the Potomac Valley downstream from Great Falls where the river is now flowing at an altitude of 50 feet (15 m). Thus, the river has eroded a vertical distance of 450 feet (137 m).

Downstream, in what is now the tidewater section, the Potomac was lowered to a depth below present sea level. The amount of lowering can be determined at several localities from records of borings made for bridge foundations. The boring logs permit identification of the interface between the Tertiary sediments and the Quaternary fill (Hack, 1957). The locality farthest downstream is at Dahlgren, Virginia (Fig. 3), where the Pleistocene riverbed is at -160 feet (-49 m). This locality is about 70 miles (112 km) downriver from Chain Bridge, the head of tidewater. The lowest sea level during the Pleistocene was far downstream from this point, however, on what is now the Continental Shelf and is estimated to have been at -230 to -295 feet (-70 to -90 m) (Flint, 1970, p. 322-328).

The profile of the present Potomac River from the Blue Ridge Mountains to Dahlgren is shown in Figure 4. The river leaves the Appalachian Valley at an altitude of 300 feet (91 m) at least 200 feet (61 m) lower than sea level during Miocene time, as shown by the deposits at Tysons Corner. The Potomac River passes through the Blue Ridge at a steep gradient; gradient index values are higher than any the writer has measured in the upper Potomac Basin (Hack, 1973, Fig. 6). The gradient becomes lower in the Triassic rocks downstream from the Blue Ridge, but steepens again at Seneca where the river enters crystalline rocks. At this point, the river descends several very steep reaches to the Coastal Plain, where the now-submerged profile is quite gentle.

The high gradient index values below the Blue Ridge suggest that the river may have been rejuvenated in the entire section. The wide range in values is related to partial adjustment to rocks of different resistance. The Piedmont upland near the river averages about 150 to 200 feet (46 to 61 m) above river level as far downstream as Great Falls, where the river descends 100 feet (30 m) rather abruptly, entering a spectacular gorge that extends almost to Memorial Bridge on the Coastal Plain. The gorge indicates that in this reach the adjacent Piedmont surface has not achieved the same adjustment as the Piedmont above Great Falls.

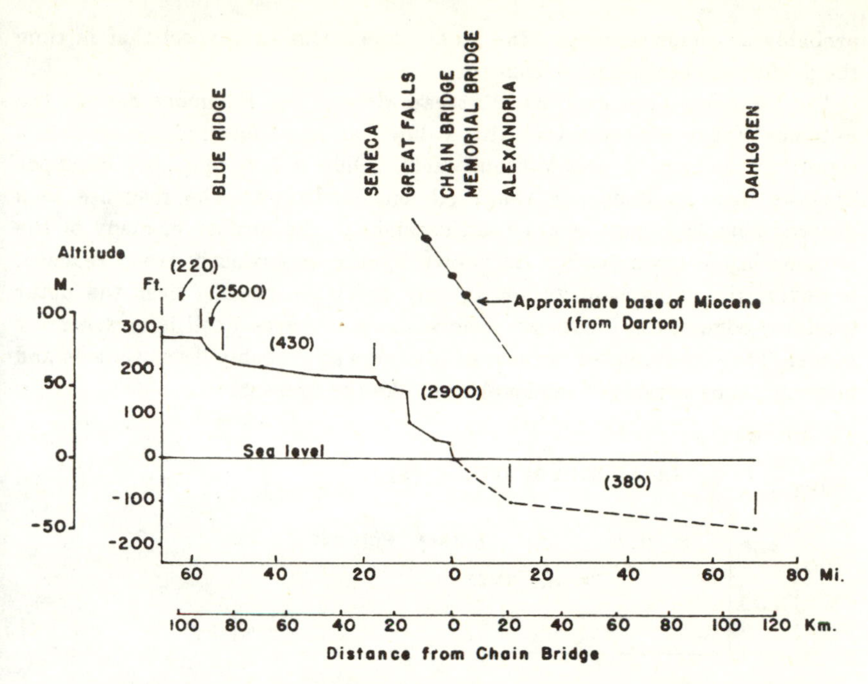

Figure 4. Profile of Potomac River from the Shenandoah Valley to Dahlgren, Virginia. Gradient index, or SL values, are shown in parentheses.

More can be learned from profiles of the tributaries, examples of which are shown in Figure 5. The tributaries above Seneca enter the Potomac at low slopes with gradient index values less than or the same as the Potomac itself, suggesting that they have achieved an adjustment to the main river. The profile of Great Seneca Creek (S, in Fig. 3) is an example. Below Great Falls, however, the profiles of the tributaries are not adjusted to the main river, as indicated by steep reaches near their mouths. Rock Creek (R) and Occoquan Creek (O) are examples.

Mattawoman Creek (M) is an example of a stream that is entirely of post-Miocene origin, for it enters the river from the east in an area almost entirely covered by Miocene as well as Pliocene and even younger sediments. The lower part of Mattawoman Valley is drowned. However the river is well graded and has a profile that is almost a straight line with few irregularities. It enters tidewater without changing its graded condition, but in order to join the Pleistocene Potomac at an altitude of -100 feet (-30 m) or lower, it would have to steepen its course considerably in the reach that is now drowned. Note that gradient index values increase downstream, an indication that the competence of the river increases. As the river flows in fine-grained sediments in its lower course, the increasing competence is not caused by the material transported by the stream. In its lower course, the stream is braided, probably as compensation for the steep slope. One can expect that in time the profile will become more concave.

The foregoing brief analysis of stream data in the Piedmont part of the Potomac Valley indicates that the valley has been lowered by erosion a significant amount in post-Miocene time. Only a few remnants of upper Tertiary deposits remain in the Piedmont landscape. The response to a lowered base level has been a readjustment of the profiles of many of the streams in the lower basin. As shown by the anomalously steep reaches, however, the adjustment as yet is only partial, especially near the outer (eastern) edge of the Piedmont, where the main river itself is in a narrow gorge. The withdrawal of the sea, of course, was probably discontinuous and interrupted by periods of sea-level rise, like the present.

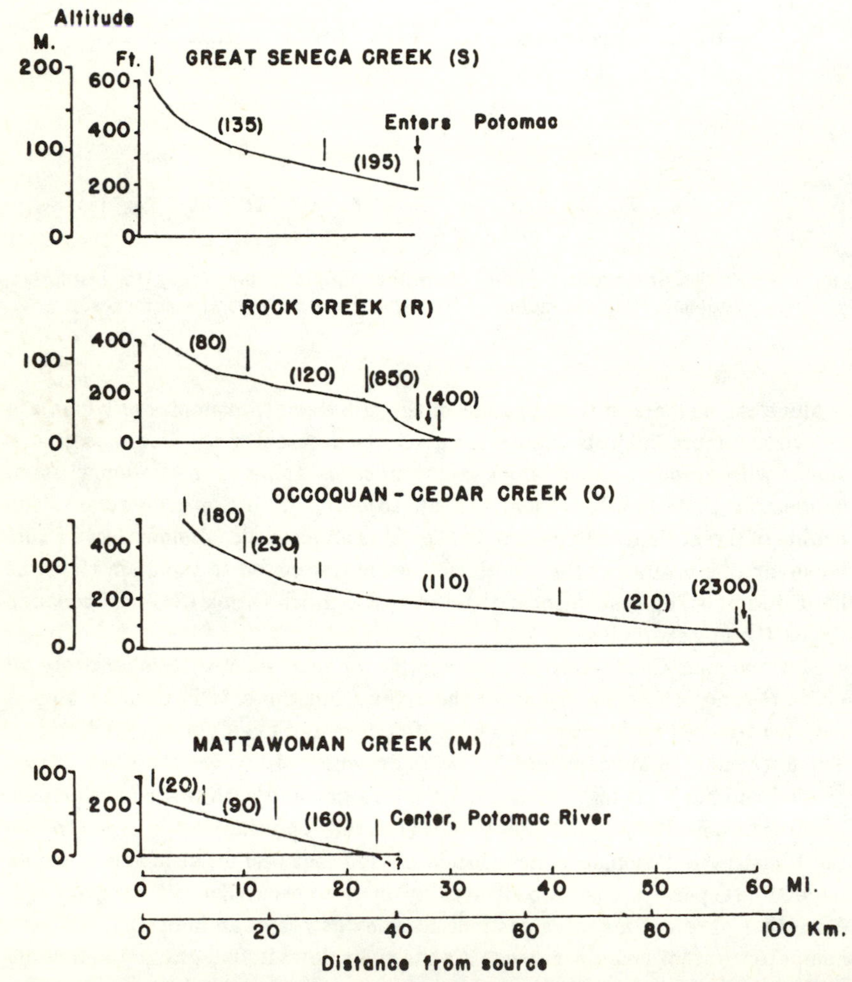

Figure 5. Profiles of four streams tributary to the Potomac River. SL or gradient index values are shown in parentheses.

SEARCHING FOR A MODEL

Chorley (1963) pointed out that the prevailing geomorphic theory in the past has been related to large systems of thought that embraced a major part of earth science. For example, in the 19th Century the ruling eustatic theory involved the ideas of continental stability and oceanic instability. Many interpretations in geomorphology were based on the eustatic theory. One example is the belief in the absolute horizontality of Coastal Plain terraces that were thought to be traceable from New Jersey to Florida. Another example is the former interest in seaward-facing terraces in the Piedmont and in New England.

Davis's geomorphology, or at least its almost universal acceptance, is closely related, according to Chorley (1963) to the concept of epeirogeny (Gilbert, 1890). Certainly the recognition of peneplains and their deformation was a concept ideally suited to define and study vertical movements of the continents. Although the recognition of peneplains may no longer be acceptable, the definition of true erosion surfaces can continue to be a valuable tool in a broader understanding of earth processes.

Plate tectonics is a new and profoundly important theory that though only 10 to 15 years old is now affecting all fields of geology. Geomorphologists are certain to become concerned, and the sooner they do, the better. I have read as well as heard the statement by devotees of the concept that topography is a "now thing." I am philosophically in complete sympathy with this idea, but analysis of Appalachian landforms and the perfection of their adjustment compels the belief that these mountains have been around for a long time. They may of course be now in motion and at a rapid rate — up or down, or in some other way.

Some simple but puzzling geomorphic facts concerning the Appalachian region can be cited as germane to this problem. The Appalachian Mountains can be divided into three parts that are quite distinct and different from each other.

The Southern Appalachians, extending to the Roanoke River at the north, have their highest area, by far, in the Blue Ridge, which here is a large part of the range. It is bordered on the east by an escarpment that appears to be retreating northwestward. The drainage is almost entirely to the Gulf of Mexico via the Mississippi and Alabama Rivers. The Central Appalachians extend from the Roanoke River to the Hudson River. Their drainage, except for the Plateau area, is entirely to the Atlantic, and the high parts of the range are found on the southeast as well as the northwest margins. The Blue Ridge belt is narrow and restricted to only the most resistant rocks; it ends in Pennsylvania south of the Susquehanna River, where the Valley and Ridge province widens and swings around to almost an eastward trend. Structural style is different here, as though this part of the mountain chain was exposed at a higher structural level in the crust or not lifted as high as the other parts. The Northern Appalachians are again different. Here, the areas of high mountains are in the zone of igneous and metamorphic rocks, and unmetamorphosed sediments form only a narrow zone bordering Lake Champlain. There is no plateau area, and the Adirondacks appear to be a vertical uplift that has no counterpart to the south. Here, as in the Southern Appalachians, the crystalline rocks form high mountains. Are these differences wholly related to the emplacement of the mountains and the break-up in Triassic time, or are they related to later events which we as geomorphologists might be able somehow to decipher?

REFERENCES CITED

Chorley, R. S., 1963, Diastrophic background to twentieth-century geomorphological thought: Geol. Soc. America Bull., v. 74, no. 8, p. 953-970.

Darton, N. H., 1951, Structural relations of Cretaceous and Tertiary formations in part of Maryland and Virginia: Geol. Soc. America Bull., v. 62, no. 7, p. 745-780.

Davis, W. M., 1909, Geographical Essays: Boston, Ginn and Co., 777 p.

Flint, R. F., 1970, Glacial and Quaternary Geology: New York, John Wiley and sons, 892 p.

Gilbert, G. K., 1877, Report on the geology of the Henry Mountains [Utah]: U. S. Geog. and Geol. Survey of the Rocky Mtn. Region (Powell), 160 p.

———————, 1890, Lake Bonneville: U. S. Geol. Survey Mon. 1, 438 p.

Hack, T. J., 1957, Submerged river system of Chesapeake Bay: Geol. Soc. America Bull., v. 68, no. 7, p. 817-830.

———————, 1960, Interpretation of erosional topography in humid temperate regions: Am. Jour. Sci., v. 258-A, p. 80-97.

———————, 1965, Geomorphology of the Shenandoah Valley, Virginia and West Virginia, and origin of the residual ore deposits: U. S. Geol. Survey Prof. Paper 484, 84 p.

———————, 1973, Stream-profile analysis and stream-gradient index: U. S. Geol. Survey Jour. Research, v. 1, no. 4, p. 421-429.

Hack, J. T., and Young, R. S., 1959, Intrenched meanders of the North Fork of the Shenandoah River, Virginia: U. S. Geol. Survey Prof. Paper 354-A, p. 1-10.

Horton, R. E., 1945, Erosional development of streams and their drainage basins: Geol. Soc. America Bull., v. 56, no. 3, p. 275-370.

Hunt, C. B., Averitt, Paul and Miller, R. L., 1953, Geology and geography of the Henry Mountains region, Utah: U. S. Geol. Survey Prof. Paper 228, 234 p.

Keith, Arthur, 1894, Geology of the Catoctin belt: U. S. Geol. Survey 14th Ann. Rept., pt. 2, p. 285-395.

King, P. B., 1949, The floor of the Shenandoah Valley: Am. Jour. Sci., v. 247, no. 2, p. 73-93.

Leopold, L. B., Wolman, M. G., and Miller, J. P., 1964, Fluvial Processes in Geomorphology: San Francisco, W. H. Freeman and Co., 522 p.

Marks, E. O., 1913, Notes on portion of the Burdekin Valley with some queries as to the universal applicability of certain physiographical theories: Royal Soc. Queensland, Proc., v. 24, p. 93-102.

Reed, J. C., Jr., Sigafoos, R. L., and Newman, W. L., 1970, The river and the rocks: Washington, D. C., U. S. Dept. Interior, 46 p.

Rubey, W. W., 1952, Geology and mineral resources of the Hardin and Brussels quadrangles (in Illinois): U. S. Geol. Survey Prof. Paper 218, 179 p.

Schlee, John, 1957, Upland gravels of southern Maryland: Geol. Soc. America Bull., v. 68, no. 10. p. 1371-1410.

Wentworth, C. K., 1930, Sand and gravel resources of the Coastal Plain of Virginia: Virginia Geol. Survey Bull. 32, 146 p.

CHAPTER 6

THE HYPOTHESIS OF UNEQUAL ACTIVITY

C. H. Crickmay

ABSTRACT

Today, two theories dominate the science of geomorphology: Davis', which visualizes universal, independent down-wasting of flat, near-level land as the rapid process; and Penck's, which holds the autonomous back-wasting of scarps as the dominant one.

Each of these theories fails to take account of most of the field evidence. The existence of flat, near-level plateaux contiguous to one another though at very different elevations is inconsistent with Davis' prime conclusion. The fact that some scarps have not wasted back at all during great lengths of geologic time is at variance with Penck's interpretation.

Numerous examples and combinations of pieces of evidence prove two conclusions: first, that among surface processes vastly different rates of action prevail; second, that no other surface process is as rapid as lateral erosion by a stream at grade. The neglected evidence includes: occasional antecedence among streams; the existence of monument-rocks and natural bridges; the predominance among natural features of flat, near-level land; the fact that many scarps without any corrading agent at their feet are much steeper than the normal angle of wasting; the general stair-step type of distribution of scarps and flat land; the existence, on top of some small, isolated plateaux, of fresh stream gravels. The field facts prove that discrepancies as great as a million to one subsist among the many and various kinds of surficial action on the face of the Earth.

INTRODUCTION

Today, the interpretation of scenery embraces (both openly and hidden) a number of theories. According to two of them, today's Earth-surface and the existing climates are very similar to those of most of geologic time. However false these notions may be, they are, plainly, an illegitimate outgrowth of sound uniformitarian doctrine. In view of the known facts, adherence to such theories demonstrates that belief and faith are the most unsound of all mental attitudes. On the other hand, some theories, notably those that emerged from the work of W. M. Davis, have shown great vitality — even in the face of adverse discoveries. With them, fault-finding must be done with circumspection.

REVIEW OF HYPOTHESIS

A notable Davis result was the conclusion that all scenery shows that it has gone through an erosional course of development and an even longer history of erosional repetition or, more exactly, a cyclical pattern of geomorphogeny. Stream erosion in the vertical dimension was well known long before Davis' time, but lateral erosion was very little appreciated, and oblique erosion was, as far as I can find, a Davis discovery. However, the great difficulty that that age of science faced (50 to 100 years ago) was the erosional reduction of the interfluves: what process was there to do it? Many observers had seen cultivated land washed into a river under violent storm conditions and, among them, the belief arose that the processes of wasting or, as Davis termed it, subaerial denudation, were competent to reduce the interfluves. This belief was a basic principle in Davis' mechanism of geomorphogenic process. Very unfortunately, it was no better than a guess: there is no real evidence for it. Furthermore, quite inconsistent with Davisian theories is the evidence of preservation on top of some small, isolated plateaux of ancient, but nevertheless fresh, stream deposits, which obviously had lain there unharmed by Davisian denudation during all the time required by stream erosion to carve deep valleys and thus to isolate the plateaux.

This and other defects in the Davis theory of universal down-wasting (equally defects in Equilibrium theory) became evident (*circa* 1920) to Walther Penck who made some discoveries among the facts of scenery toward which Davis and his disciples had been blind. Penck saw the stair-step pattern in which most scenery is arranged: the surface of the Earth being in great part flat, near-level lands separated by short slopes. This arrangement suggested to Penck that it had developed through the steady retreat of all the perceptible slopes that had ever come into existence — retreat that proceeded as a result of back-wasting, and which left behind it the flattish, very gently inclined surfaces that were supposed to bear testimony to this universal retreat.

There is, of course, no doubt that most isolated scarps have retreated: the perception of that by Penck was an outstanding advance. But there is no evidence that retreat is proceeding universally or all the time, or that it is brought about by the Penckian mechanism of unaided back-wasting. For there are a number of known scarps that during some length of time have stood in one location without retreat. The first examples that I discovered were remanent oxbow lakes perched on the gently sloping walls of certain river valleys of the western plains in North America; the valley side above the oxbow lake had not perceptibly retreated from the lake shore in all the time required to lower the valley floor the additional 20 meters or so. Some examples involving much longer time include such slopes as those of Quartz

Mountain, Oklahoma, which, still rising directly from the Permian re-entrant, have not retreated since the Permian. These exceptions demonstrate that there is no such process as the Penckian universal retreat of slopes. Universal retreat was an invention, not a discovery. A scarp retreats only when a local (as opposed to a universal) process forces it to. So far, only two such natural processes are known: corrasion at the base of the scarp by waves on standing water, and corrasion laterally by the meandering current of a stream at grade.

NEW HYPOTHESIS FORWARDED

When I had reached this point in my thinking, I saw that a completely different, not merely a modified, hypothesis was needed — one that admitted both geomorphogenic activity and geomorphogenic repose. In other words, there were different rates of action in surficial processes — ranging from highly vigorous to dead quiet, perhaps. At this point, it will be well to examine some evidence. For example, some streams have almost perfectly even gradients over a good part of their lengths, others have rapids here and there; from this one can conclude that the former have had time to make their gradients even, the latter have worked more slowly at smoothing out their courses than the uplift which thrust a steepened gradient upon them. Yet many streams have not fallen behind in competition with an uplift, for they have cut valleys into the uplifted land-mass, and in this process, wasting, too, has been rapid enough, proportionately, to widen the valleys to a V-shaped cross-section. Furthermore, there are such things as antecedent streams: plainly, these have worked as fast as, or even faster than, the uplift which took place across their courses. And finally, all the fluvial processes may go forward immeasurably more rapidly than any form of broad "subaerial denudation", or valleys could never come into being — for, if subaerial denudation lowered the surface of the land-masses (as Davis seemed to claim) as rapidly as drainage could work, no valley could ever be excavated.

Another important group of evidences that calls for careful interpretation comprises the monument-rocks and natural bridges. The former are widely scattered, the latter are located mainly alongside or across rivers and clearly represent a combination of lateral erosion at decreasing stream levels and survival, against the feeble powers of wasting, of the rock mass that was sufficiently elastic to stand above stream-level as the bridge. A disparity in accomplishment between wasting and stream corrasion is thus well illustrated: natural bridges are carved by the stream in a small fraction of the time that wasting would require to destroy them. Possibly, a similar inequality in erosional powers is exhibited in monument-rocks and natural arches — both of which have commonly been attributed entirely to wasting. Some indirect testimony is worth recording here: many of the monument-rocks and some of the arches rise above a rock floor or a terrace that bears some stream alluvium; and this points to the work of streams when they ran at higher geomorphic levels, though not, of course, at corresponding elevations.

As Penck observed, there is among natural scenery a predominance of flat, near-level land — at various elevations. To my thinking, the two prime, existing hypotheses of the origin of this flat land are both inadequate. But what was its origin? I began to find on some dry flat uplands remnants of alluvium that the geological reports seemed to have avoided reporting; and I asked myself, can these broad areas, far from running water, be a form of long-abandoned flood plain? And if so, can they have been made, one after another, at different levels, by successive episodes of lateral corrasion? I said little, but listened carefully to Sidney Paige (1912), E. B. Knopf (1924), Eliot Blackwelder (1931), and Douglas Johnson (1931). But no one seemed able to prove convincingly the strength and adequacy of the lateral action of streams that I, following these pioneers, suspected was the only possible mechanism. Then at last (1960) I established an example in which a rate of progress in lateral corrasion could be measured: I had found a small river that in recent years had pushed one of its meanders through a wall of bed-rock of more than 9 meters thickness in 30 to 35 years, thereby establishing a rate of about 0.3 m a year — incidentally, a much higher rate than that of any normal vertical erosion. In this case, the river had planed in 30 years or so at this locality about 360 square meters of new valley floor — the near-horizontal erosional progress having been measured from several successively-made maps. The rate of lateral erosion established thereby means that a plain 1000 km wide could be carved out by this small stream in one third of a million years — geologically, not a long period. This result is interesting to compare with the accomplishments of the great Mississippi, which for two centuries has been cutting through flood-plain alluvium at an average rate of 13 m a year, and since the beginning of the Pleistocene has cut away bed-rock, as it widened successive flood plains, at the rate of 0.6 m a year. These measures establish lateral erosion as a more powerful process and in some cases a more rapid action than any other in the surficial realm.

Most of all expressed opinion opposes this conclusion; and with good reason, for what the observer most readily grasps is that running water on the face of the Earth can be seen only, where it is eroding, to be working vertically and, because of that, can make no perceptible progress in any other direction. This adverse view reaches an extreme in the words of W. B. Scott (1932, vol. I, p. 251):

> "The destructive work of rivers is, in the aggregate, far less important and extensive than that accomplished by the atmosphere, but it is much more striking because it is concentrated along narrow lines, not spread over the entire land surface."

Very few sayings by competent geologists have been as far from the truth. In this matter, Professor Scott's failure to perceive stems from a difficulty that began to arise in the late Cenozoic when renewed diastrophism started to hoist the lands to higher and higher elevations; and this raising up of land surfaces, exceptional with respect to most of geologic history, has continued and increased ever since. At present, the trend is extreme. The result is that, today, most streams are confined to a narrow, valley-bottom position, in which they can not act normally and wander far and wide over flood plains of tens to hundreds of kilometers width.

As for Professor Scott's comparison with the work of the atmosphere, one may well say at this point that the atmosphere can do almost nothing until running water has given it a slope on which to work. The relative weakness of atmospheric influence is shown up strongly in localities that were glaciated in the Pleistocene but have been free of glacial ice for fifteen to twenty thousand years: in that length of time, the atmosphere has done very little toward erasing the glacial striae, though many a stream has cut a canyon 100 meters or more deep. Furthermore, despite all the exertions of the atmosphere, many a flat, level-topped plateau has preserved its surface perfectly from the Cenozoic to the present.

Among flat, near-level lands, stand the scarps that, characteristically, separate the several levels typical of such lands. And many of these scarps (excluding recent fault-scarps) are much steeper than any sloping surfaces that have long been exposed to wasting. Though not recognized as such, this is an anomalous occurrence; for, whether one regards the scarp (with Davis) as down-wasted, or (with Penck) as back-wasted, one would expect these old scarps to exhibit the typical wasted-surface angle, the gradient of repose. The fact is, most of the scarps (whether near a stream or not) show the same oversteepened slopes that are currently being undercut (and thereby steepened) by rivers. In my view, the answer is that the isolated, oversteep scarp was pushed to its present position by a long-since departed stream and, subsequently, has been little affected by the much feebler processes of wasting.

Attention must be called once more to the general stair-step arrangement of most of the Earth's flat lands — in Davis' terms, one peneplain below another, which expression is modified by modern Penckians as one pediplain above another. The first of these two views is readily disposed of: no lower peneplain could well be formed by Davisian down-wasting without destroying all the higher ones. Davis seems never to have thought of that difficulty. The Penckian interpretation is somewhat more plausible, though it takes no account of such inconsistencies as the fact that most of the pediments are carpeted with fluvial alluvium rather than the expectable angular rock waste. Nor of the additional difficulty that no pediment above river levels can be proved to be growing — as Penckian doctrine claims they all are. If the investigator can verify the evidence found by Paige (1912), Blackwelder (1931) or Johnson (1931), and follow the thinking of these pathfinders, he sees pediplains as products of lateral corrasion by rivers; though, surprisingly, not even Blackwelder saw this process as a discontinuous one — a process not necessarily happening on all pediplains at the present time. An existing pediplain may well be almost absolutely inactive: it may have been completed in the past by streams long since gone from the scene; the water now running down its sloping surface having had no part in making it, nor yet having a part in destroying it.

DISCUSSION

This interpretation seems the more likely for at least four reasons. First, the stream gravels on pediplains usually include a greater variety of rock types than occur locally *in situ*; in other words, the river that did the work was not a local one. Second, most pediplains slope at an angle too steep for the condition of grade to prevail in a water current running down-slope: thus, with the only condition under which lateral corrasion takes place eliminated, it is unlikely that pediments were formed by the small side-streams now running down their slopes. Third, the surface which now accommodates these side-streams had to be made before a down-slope current could run down it. Fourth, the lateral continuity of both the sloping pediplain and its upward margin is so smooth as to suggest that it came into being, not everywhere in its surface at once, but part by part, and through the work of a current that ran, not down its present slope, but across it and nearly parallel with the existing contour lines. I visualize the pediplain's being carved either near-level as a terrace is or obliquely downward as a slip-off slope. Its making was accomplished by streams that ran at higher geomorphic levels than today's drainage, but probably at much lower elevations; its shaping may well have been entirely ended before any of the small-sized, present-day side-streams began to run as they now do.

One remaining group of evidences must be examined: the occurrence on top of isolated, highly elevated, flat uplands of fresh though ancient stream gravels and other forms of remanent river alluvium. Quite obviously, these deposits were laid down when what is now upland was part of a broad plain at a much lower elevation and was traversed by streams that are no longer there. The illustration (Fig. 1) shows the essentials of my best example; the case in point is from South Africa (the locality being 65 km northwest of Middelburg, C. P.) The essentials are: a small, isolated, elevated, flat-topped upland, about 0.5 km wide, in which lies a dry, alluvium-filled stream channel with, in its small surviving length, neither source nor destination. Besides the few million, small, ellipsoidal pebbles that constitute this evidence of preservation in the upland, there is a loose scattering of well-rounded cobblestones over the whole flat surface. Both these types of gravels have been remarkably well preserved on the isolated upland while deep valleys were excavated by stream erosion on all sides into the ancient, broad plain of which a small part survives as the diminutive gravel-capped plateau.

Reasoning from this example, let us say that the erosion of the valleys round this plateau has proceeded for perhaps half a million years at a rate of

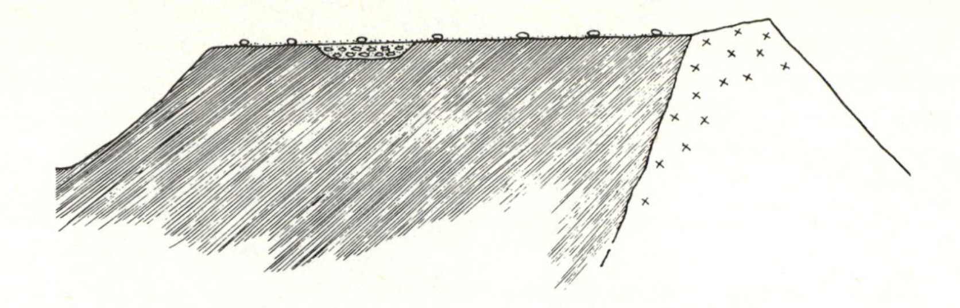

Figure 1. Cross-section of a small plateau.

(again perhaps) 0.00015 m of depth a year. During that half million years, not even the original stream-induced smoothness has gone from the little pebbles on the upland. Here, then, is a perfect case of a discrepancy in rates of wear-and-tear that exceeds a million to one.

Similar, though perhaps less beautiful, examples may be found in many places: no continent is entirely without them. They all tell, some more perfectly than others, the same true story — the story that exceedingly different rates of surficial activity prevail. In my "Preliminary Inquiry" article (1959), I called this mode of understanding the evidence the Hypothesis of Unequal Activity. I continue to use this designation, my most recent and most complete exposition of it having appeared earlier this year in my book, *The Work Of The River* (1974), a usable volume but an unfortunate one in that it was given many errors and blemishes by incompetent editors (who knew nothing of geology and very little of common English) — after I had had my last opportunity to correct the page proofs.

REFERENCES CITED

Blackwelder, Eliot, 1931, Desert plains: Jour. Geol., v. 39, p. 133-140.

Crickmay, C. H., 1959, A preliminary inquiry into the formulation and applicability of the geological principal of uniformity: published by the author, Calgary, Alberta, 55 p.

———————, 1960, Lateral activity in a river of northwestern Canada: Jour. Geol., v. 68, p. 377-381.

———————, 1974, The Work of the River: The MacMillan Press, London, 271 p.

Johnson, Douglas, 1931, Planes of lateral corrasion: Science, n.s., v. 73, p. 174-177.

Knopf, E. B., 1924, Correlation of residual erosion surfaces in the eastern Appalachian highlands: Geol. Soc. American Bull., v. 35, p. 633-668.

Paige, Sidney, 1912, Rock-cut surfaces in the desert ranges: Jour. Geol., v. 20, pp. 442-450.

Scott, W. B., 1932, An Introduction to Geology, 3rd Edition: MacMillan, New York, 604 p.

CHAPTER 7

LANDFORMS THAT DO NOT TEND TOWARD A STEADY STATE

William B. Bull

ABSTRACT

Although most aspects of river systems tend toward a steady state, many elements of hillslope and depositional landscape systems do not. The height, volume, or other dimensions of a landform may change with time as a result of differential erosion or net deposition. Examples include topographic inversion where a lava filled valley in metamorphic rocks becomes a ridge, stepped topography formed by different weathering micro-environments of massive granitic rocks, mass movement and cliff retreat processes, and aggrading landforms such as alluvial fans and glacial moraines.

For landscape elements and processes that are changing at different rates, the model of allometric change is useful for analysing the tendency for adjustment between interdependent materials, processes, and landforms that do not tend toward a steady state. The power function

$$y = a\ x^{b}$$

may be used to analyse the dynamic interrelations of geomorphic variables at different times during the history of a landform and the static interrelations at one time.

INTRODUCTION

One of the major contributions of geomorphologists such as G. K. Gilbert, Arthur Strahler, Luna Leopold, John Hack, Richard Chorley, Stanley Schumm, and Walter Langbein has been the concept that drainage basins are open systems. Insight into some of the complexities of fluvial systems may be gained if one adopts the viewpoint that landforms and geomorphic processes tend towards an equilibrium configuration — steady state. "The most probable distribution of energy in certain geomorphic systems could be derived by considering the geomorphic system as an open system in steady state." (Langbein and Leopold, 1964, p. 784). "The principle of dynamic equilibrium states that when in equilibrium a landscape may be considered a part of an open system in a steady state of balance in which every slope and every form is adjusted to every other." (Hack, 1965, p. 5).

Leopold and Maddock (1953) found that a tendency toward steady state existed in streams, and that adjustments between changing streamflow variables could be defined by power-function equations. They termed these adjustments "quasi-equilibrium" because of the substantial scatter about their regressions, and because they could not be certain that steady state had been attained. Bull (1975) expanded on the concept of landscape elements and geomorphic processes that are changing at different rates — the concept of allometric change — by analysing streams, hillslopes, and depositional environments that tend toward a steady state.

Although most aspects of river systems tend toward a steady state, many elements of hillslopes and depositional environments do not. For many landforms, height, volume, or other dimensions change progressively with time instead of tending toward a time-independent size or configuration. For those landforms and processes that do not tend toward a steady state, a model that focuses on the interdependent changes between variables in a changing open system is more useful than the steady-state model.

The purposes of this article are: 1) to discuss examples of erosional and depositional landforms that do not tend toward a steady state, and 2) to use the allometric approach. For each example an argument will be presented to show why the landform does not tend toward a steady state and, for some examples, the allometric adjustment of part of the system will be discussed.

ALLOMETRIC CHANGE

Allometric change is the tendency for adjustment between interdependent materials, processes, and landforms in a geomorphic open system. The allometric concept has been used by biologists, paleontologists, geomorphologists, and social scientists. Although many mathematical expressions apply to allometry, one approach has been to correlate changes of an element of a system with changes of a different measure of the same element or with other elements of the system. The correlation commonly may be expressed by multivariate equations or by the bivariate power-function

$$y = a x^b, \tag{1}$$

where a and b are constants, and a is always positive. The power-function is widely used in geomorphic studies because of its statistically significant fit to most data, simplicity, and relative ease of interpretation. The use of the allometric approach encourages recognition of the tendency for adjustment between interdependent materials, processes, and landforms, even where steady state does not exist. Furthermore, it permits assignment of statistical degrees of adjustment between different pairs or suites of variables of the system. The reader is referred to Bull (1975) for a review of the allometric concept, limitations of allometry as applied to geomorphology, comparison of the allometric model with other geomorphic models, and allometric analyses of processes and landforms that tend toward steady state.

Stephen Gould has recognized two modes of allometric analysis (Gould, written commun.; p. 3670 of Mosley and Parker, 1972). He considered allometry as "a general term for all relationships, dynamic or static, . . . and involving change of shape correlated with size increase." Geomorphic dynamic allometry is a time-dependent relation that refers to the interrelations of measurements made of a landform or process at different times during its history. Static allometry refers to the interrelations of measurements of a landform or process at one time in its history. The hydraulic geometry of streams provides examples of allometric change that tend toward steady state. Analysis of streamflow in the downstream direction at a given flow frequency is static allometry, and the interrelations between variables at a station is dynamic allometry. A stream that is rising at a station after an intense rain is not in equilibrium but tends toward a steady state (equilibrium) through an orderly adjustment between the changing variables. Steady state would be a condition of zero rate of change of variables and would be represented by a single point on a graph portraying dynamic allometry. Steady state does not apply to static allometry because only a single point in time is considered.

An example of dynamic allometry that does not tend toward a steady state would be the relation between the cumulative volume of lateral moraines, and the distance that the glacier had moved downvalley. In this case the pairs of values of the two variables represent many points in time, even if the data were collected at a single point in time after recession of the glacier. A power-function describing the curvature of an end moraine being pushed down the valley would be based on measurements for a point in time and would be an example of static allometry. The relation between mean width and length of the active part of an earthflow at different times as the earthflow moves down a hillslope would be an example of dynamic allometry. The interrelations of slope fall and slope length from a ridgecrest in the earthflow would be an example of static allometry.

Examples of Allometric Change of Landforms that Do Not Tend Toward a Steady State.

Many landforms would not tend toward steady state, even if independent variables of the open system such as climate, relief, and base-level remained constant. In some erosional environments self-enhancing feedback mechanisms cause progressive changes in the relative dimensions of hillslope elements. Depositional landforms commonly represent the endpoint of processes operating in geomorphic systems, and because deposits increase in volume with time, they do not tend toward a steady state.

DIFFERENTIAL EROSION OF HILLSLOPES

Contrasts in the erodibility of some rocks and their weathering products may be eliminated by changes in slope steepness, because the rate of erosion of a given material increases with increasing slope. For certain moderately contrasting materials the erosion rates of resistant materials may equal those of the less resistant materials when the slopes of the resistant material become sufficiently steep to balance the contrasts in erodibility (Fig. 1). G. K. Gilbert (1879, p. 93-150) and John Hack (1965, p. 5-10) used the above concept to help develop the principle of dynamic equilibrium.

> "When the ratio of erosive action as dependent on declivities becomes equal to the ratio of resistance as dependent on rock character, there is an equality of action" (Gilbert, 1880, p. 100).

Under ideal conditions an equilibrium (steady state) is approached in which all parts of a landscape downwaste at about the same rate.

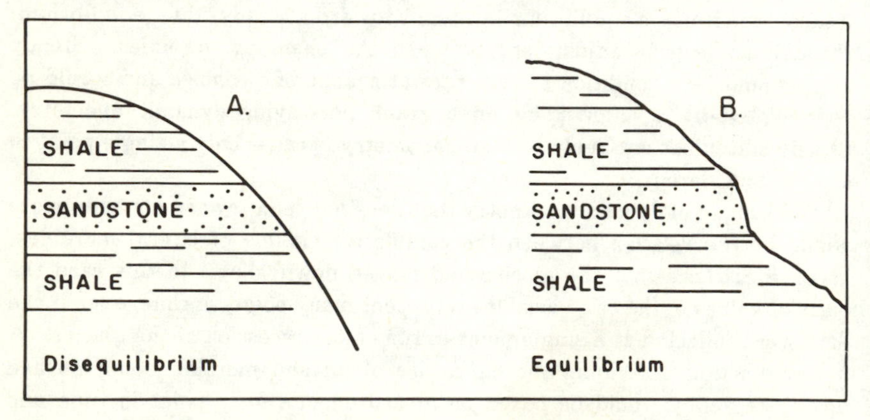

Figure 1. Diagrammatic sketches illustrating Gilbert's equality of action on hillslopes. A. Disequilibrium, B. Equilibrium.

The interrelations between the variables of slope steepness, slope materials and geomorphic processes commonly result in hillslopes that do not tend toward a steady state. There is a limit to the steepness that a given slope can attain, and most lithologies undergo changes in weathering characteristics with change in slope and soil cover. Where erodibility is more important than slope steepness, resistant lithologies will become progressively higher relative to the adjacent softer lithologies.

Two classic localities for differential erosion rates are the Table Mountains of the Toulumne (Bateman and Wahrhaftig, 1966, p. 137) and San Joaquin Rivers (Wahrhaftig, 1965, p. 1170) in the foothills of the Sierra Nevada in California. Geographic data about this and other sites are summarized in Table 1. About 9.4 m.y. (million years) ago (Dalrymple, 1963) a basaltic lava

Table 1. General information about the study sites

Site	Location	Altitude (meters)	Climate (Koppen-Geiger, 1954)	Lithology	Vegetation
San Joaquin Valley	Fresno County, California	75-280	Mediterranean, hot	alluvium	sparse grass and bushes
Sierra Nevada	Fresno County, California	100 to more than 2,500	Highland	granitic	conifer forests, oak-grass woodland
Threatening Rock	San Juan County, New Mexico	1,900	Steppe, cool	massive sandstone, shale	sparse grass, scattered bushes, and trees.
Downpour Gulch	Cochise County, Arizona	1,200	Tropical desert	silty clay	barren to sparse grass and trees

flow filled the gorge of the San Joaquin River (Fig. 2A). The basalt was more resistant to weathering and erosion than the adjacent granite, and differential erosion occurred. The resulting topographic inversion has left the basalt flow and its underlying river gravels as a sinuous ridge that is as much as 400 m above the San Joaquin River (Fig. 2B).

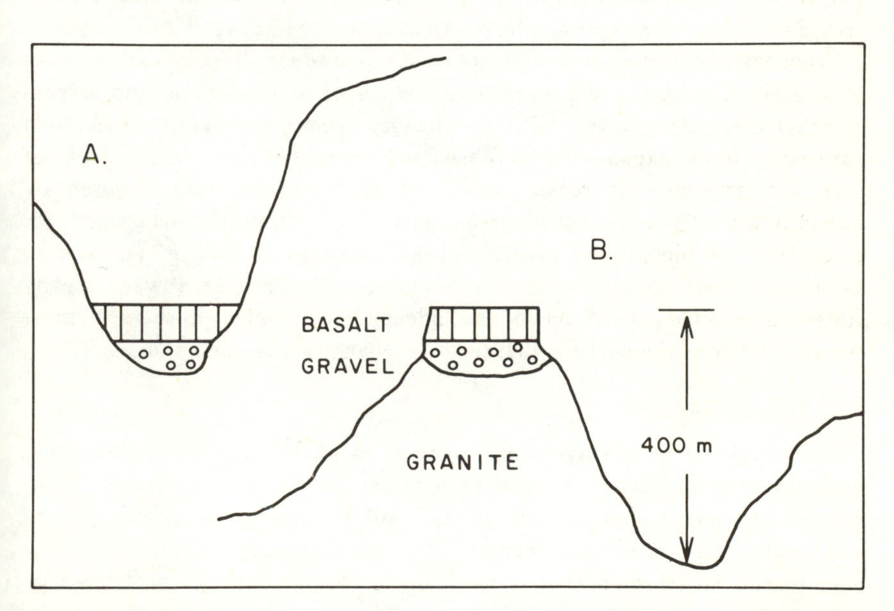

Figure 2. Diagrammatic sketches showing the topographic inversion of the San Joaquin River, California. A. About 9.4 million years ago. B. At Present.

Even the Henry Mountains, studied by Gilbert (1880), and the Appalachians appear to be examples of differential erosion. Hunt (1969, p. 127) points out that the laccolithic rocks of the Henry Mountains are becoming progressively higher relative to the more erodible rocks of the adjacent Colorado Plateau. Examples of differential erosion in the humid terrain of the Appalachian Mountains have been noted by Godfrey and others (1971) as a result of their studies of the rates of chemical denudation of mica schist, quartzite, limestone, and serpentine. In each of the above areas the interdependent variables have undergone gradual long-term adjustments as a result of the lack of a tendency toward a steady state.

Wahrhaftig (1965) proposed that a progressive increase of the areas of exposed granitic rocks has occurred in the western Sierra Nevada of California. Because granitic rocks weather more rapidly in moist micro-environments than when exposed, an accidental exposure of rock will result in enlargement of the exposure. As grus is eroded from the edges of the outcrop, the outcrop will increase in area and height relative to the surrounding rock that is still mantled with grus. Random exposures of granitic rocks coalesce with time to form steep fronts that are separated by gently sloping treads still mantled with grus.

The progressive increase of areas of bare rock over time spans of several million years may be considered an irreversible change because the feedback mechanism of rapid runoff from massive rock outcrops tends to enhance the erosion of grus at the margins of the outcrop. Thus, Wahrhaftig's study provides a nice example of landforms that do not tend toward steady state.

From a dynamic allometric viewpoint, stepped topography landscapes may be described in terms of progressive change of both hillslope and stream morphologies. As a given hillslope changes from grus-mantled rock to a bare steep front, exponential increases and decreases occur in the areas of bare and grus-mantled rocks. Areas of fresh granitic rock exposed by streamflow may become the ridgecrest part of the terrain (Wahrhaftig, 1965, p. 1181). Such topographic inversion requires that the drainage net as well as the hillslopes undergo progressive change. A tendency toward steady state is impossible because one of the independent variables in the system — the erodibility of the surficial materials — changes with time and space.

Cliff Retreat.

Some types of mass movements, such as rockfalls, may accelerate exponentially once the initial threshold resisting stress has been exceeded. Cliff retreat by the process of rock or soil fall provides a nice example of non-steady state allometric change. A combination of lack of horizontal support and partial removal of vertical support results in tensional stresses in massive materials ranging from clay to sandstone. The locus of cracking is commonly along desiccation or pressure-relief cracks. Separation along the tension fracture permits the top of the block or monolith to rotate toward the adjacent valley at an exponentially increasing rate. There are no steady states, and the endpoint of the process is collapse of the block into a pile of rubble in the adjacent valley. The time needed for the entire process ranges from a few minutes, in the case of undermining of the sandy banks of a stream by raging flood waters, to more than 2,000 years for high cliffs of massive sandstone (Schumm and Chorley, 1964, p. 1051).

The exponential increase in the rate of movement of a massive sandstone block away from the top of a cliff can be demonstrated with data from Schumm and Chorley (1964). Equation 2

$$Y = 1.47X^{1.74}, \quad 1 \leqq X \leqq 5 \tag{2}$$

indicates that the rate increased rapidly during the five years preceding the fall of Threatening Rock in Chaco Canyon National Monument. Figure 3 is not an allometric plot because time is plotted as a variable (see the derivation of equation in Bull, 1975, p. 1490). In order to make an allometric plot, some other aspect of the monolith-cliff system (such as overhang) would have to be plotted against the crack separation.

The cliffs that form the sides of an arroyo provide a setting where the same processes of cliff retreat can be studied on a smaller scale than at Threatening Rock. The silty clay banks of an arroyo at a site named Downpour Gulch (Table 1) provide data about bank overhang — a function of the amounts of bank undercutting and block rotation. Data from other sites show that crack and block widths are also functions of cliff height and lithology as well as the amount of overhang.

Maximum overhang and crack and block width were measured at the top of a 4.2 m high cliff. The data for the two blocks of Figure 4 were obtained by making measurements from plumb lines spaced at 30 cm intervals. Block width remained constant despite additional undercutting and crack widening.

Crack width and maximum overhang were measured at the top of a block that had been undermined intermittently for more than a year. Sufficient time had passed that crack width varied as a function of the stresses induced by variations in maximum overhang. Equation 3

$$Y = 1.14X^{0.67}, \quad 21 \leqq X \leqq 71 \tag{3}$$

shows that crack width increases exponentially with increase of maximum overhang, but that negative allometry exists (the exponent of less than 1.0 shows that crack width increases less rapidly than does maximum overhang). The ideal data-collection procedure for the relation in Figure 4A would be to measure change in crack width and overhang at a single point after every streamflow, or at regular intervals, for two years. Such a procedure would provide an example of dynamic allometry. Figure 4A can also be considered

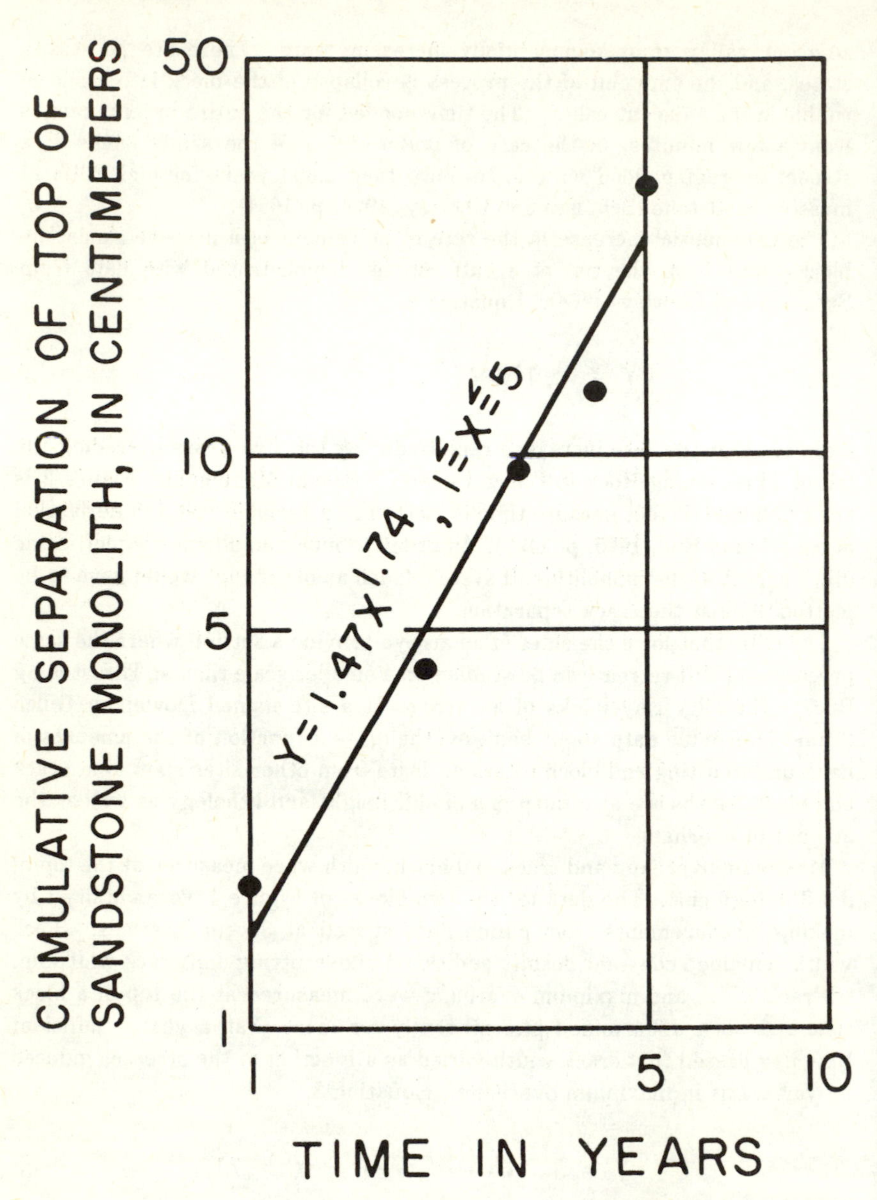

Figure 3. Increase in rate of cumulative separation of the top of Threatening Rock sandstone monolith from a cliff in Chaco Canyon National Monument, New Mexico. Data from Schumm and Chorley, 1964.

an example of dynamic allometry because space data (many points along the block) have been substituted for time data.

At a second location, block width was compared with overhang for a block that was only three days old. The narrow crack width was the same for

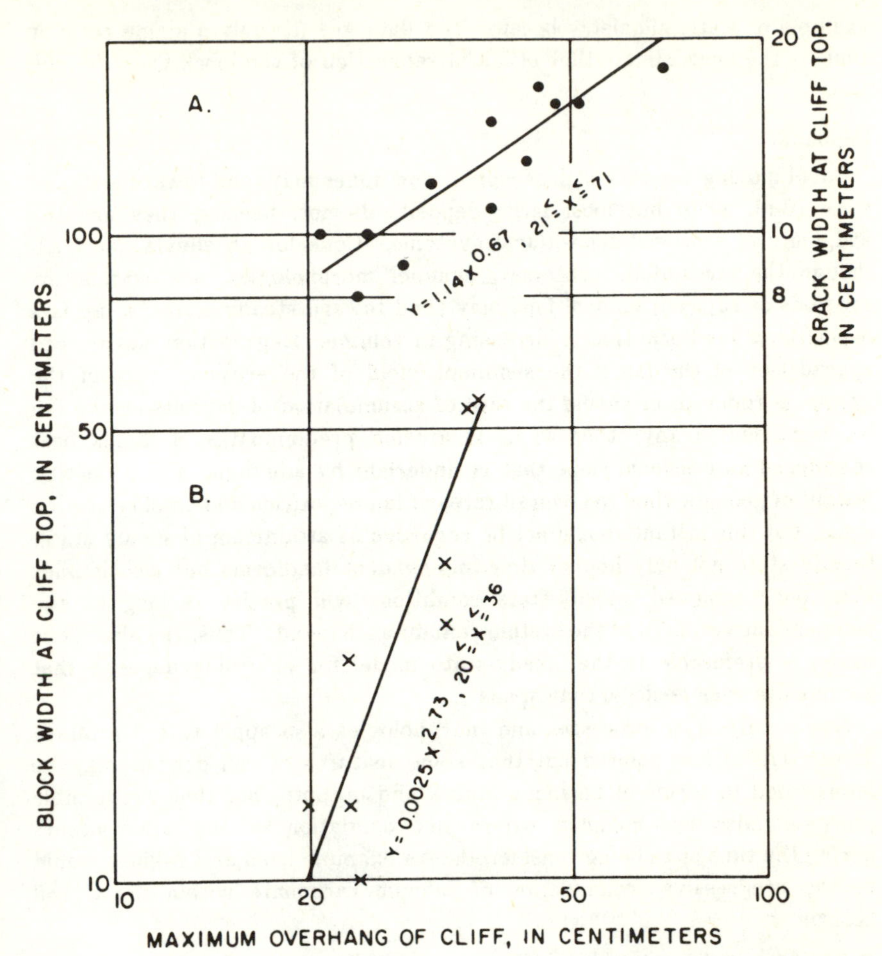

Figure 4. Cliff retreat along an arroyo at the Downpour Gulch site, Arizona. A. Relation of maximum overhang of the cliff to crack width at the top of the block. B. Relation of maximum overhang of the cliff to the block width at the time of undercutting.

almost the entire length of the block. Apparently insufficient time had passed for variations in overhang to cause variations in crack width. The intent of the measurements was to compare block width with an index of undercutting at the time of initial undercutting. In equation 4

$$Y = 0.0025X^{2.73}, \quad 20 \leq X \leq 36 \qquad (4)$$

block width increases exponentially with increase of maximum overhang, but positive allometry pertains to this case (the exponent of more than 1.0 shows that block width increases more rapidly than overhang). Figure 4B is an example of static allometry because the data are for only a single point in time — the approximate time of initial separation of the block from the cliff face.

Deposits.

Small moving deposits such as ripples and dunes may tend toward a steady state (Bull, 1975) but most large deposits do not, because they are the endpoints of erosional-depositional systems. Consider an alluvial fan. Although the streamflow processes, channel morphologies, and area interrelations of adjacent alluvial fans may tend toward steady states; a fan is a depositional landform that is increasing in volume. Degradation may exceed aggradation of the fan if the sediment yield of the erosional part of the system is reduced, or should the rate of accumulation of deposits on the fan be decreased in any other way. If erosion predominates, a fan is best considered an erosional slope that is underlain by alluvium. For a fleeting instant of geologic time the overall rates of fan deposition and erosion may be equal, but this instant should not be regarded as attainment of steady state. Steady state not only implies time-independent landforms but also implies that, once achieved, steady-state conditions will persist as long as the independent variables of the system remain unchanged. Thus, the allometric model is preferable to the steady-state model for analysing deposits that accumulate over geologic time spans.

Non-steady state processes and morphologies also apply to soil profiles. Yaalon (1971) has pointed out that some features of soil genesis may be interpreted in terms of balanced inputs and outputs, but that irreversible processes also are common where net eluviation or illuviation occurs during the time span being considered. An example from arid regions would be the progressive accumulation of calcium carbonate within the C soil horizon.

Deposits in Sedimentary Basins.

Allometry can be used to analyse depositional relations in sedimentary basins such as those where flood-plain, lacustrine, and alluvial-fan deposits are competing for space. Roger Hooke (1968) proposed that the areas of alluvial fan and playa deposits in a closed basin tend toward a steady state that is defined by their relative rates of increase in thickness. Hooke's concept pertained to dynamic allometric adjustments such as the changes in areas of adjacent depositional environments. In his model a hypothetical increase in the rate of deposition on one fan would tend to increase the rates of deposition on all the other fans and the playa. For example, his model (Hooke, 1968, p. 616) required that the area of each fan, A_f, be proportional to the volume, V, of debris contributed to it per unit time. The constant of proportionality, dT/dt, is the rate of increase of fan thickness, T. Thus:

$$V = \frac{dT}{dt} A_f \tag{5}$$

Hooke considered equilibrium (steady state) to exist when the rates of deposition in all parts of the closed basin were the same.

If dynamic allometric analysis is used, the lack of change in area of adjacent alluvial fans through time would plot as a single point on a graph and would suggest a steady-state condition. However, if change in fan area were compared to change in fan thickness, the plot would consist of a line instead of a point — indicating a lack of steady state. Landforms that are undergoing progressive increases or decreases in volume can not tend toward a steady state that includes all the variables of the system for the time span being considered. This example illustrates the need for analysing more than one pair of variables in order to demonstrate steady-state conditions for a landform.

The allometric model can also be used to analyse adjacent depositional environments that are accumulating at different rates. In an aggrading valley with a through-flowing stream, the relative rates of accumulation of detritus derived from the two bordering mountain ranges will determine the type of stratigraphic contact between the deposits from the two source terrains. In Figure 5, as long as the relative rates of accumulation of detritus from the granitic and sedimentary source terrains remain the same, the contact between the deposits will remain vertical, as is the case below horizon A. The relative rate of accumulation of deposits from the sedimentary source has doubled at horizon A. As a result, the depositional area supplied by the sedimentary source becomes larger, and the depositional area supplied by the granitic source becomes smaller as the two areas tend toward equal rates of aggradation. The curving contact above horizon A indicates a rapid shift of the stratigraphic contact immediately after the change in sediment yield. Equal rates of aggradation are approached asymptotically and are therefore unlikely to be attained (Hooke, 1968, p. 615, 616).

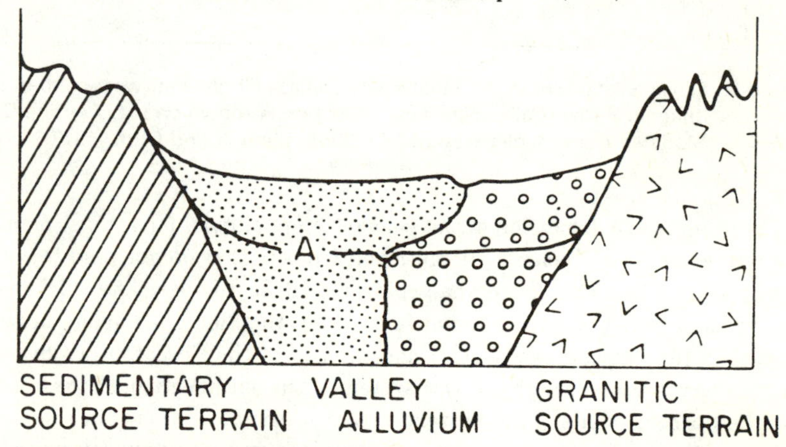

Figure 5. Diagrammatic sketch depicting depositional relations in an aggrading valley with two sources of sediment, where the sedimentary source terrain undergoes a rapid increase in sediment yield.

A cross section of late Quaternary valley fill in the San Joaquin Valley, California is shown in Figure 6. The time-transgressive contact between the clayey sand derived from the sedimentary rocks of the Coast Ranges and the micaceous arkose derived from the Sierra Nevada is easily discerned in numerous core holes and electric logs. As in Figure 5, the deposits from the two mountain ranges are competing for space. The marked encroachment of the Diablo Range alluvial fan deposits over the Sierran alluvium during the 0.6 m.y. (Janda, 1965, p. 131) since the deposition of the Corcoran Clay Member of the Tulare Formation was largely the result of a relative three- to six-fold increase in sediment yield (Bull, 1970). The sediment yield increase was caused chiefly by increased hillslope steepness of the Diablo Range. Comparison of mean deposition rates above the Corcoran and "A" lake clays (Fig. 6) shows that the rates of deposition have increased from about 0.30 m per thousand years for the 0.6 m.y. period to about 0.55 m per thousand years for the 27,000 years (Croft, 1972) since the deposition of the "A" clay.

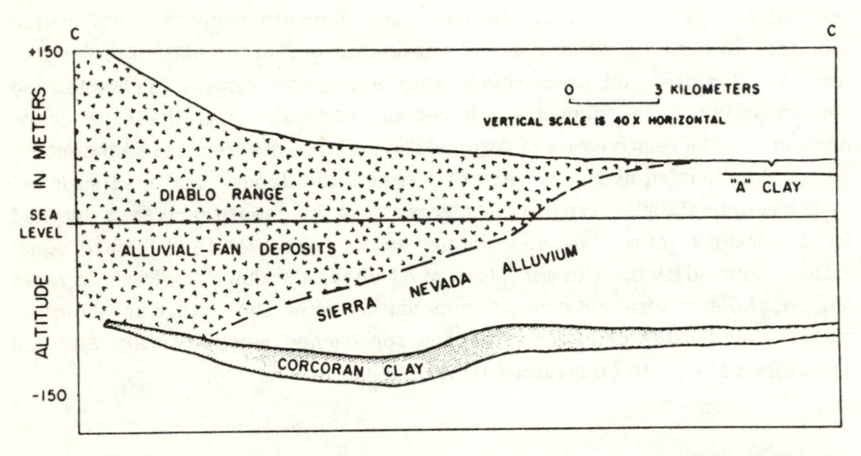

Figure 6. Cross section of Late Quaternary valley fill in the west-central San Joaquin Valley, California. Line of section is shown on Fig. 7 as line C-C'. Modified from Magleby and Klein (1965, plate 4) and Croft (1972).

The progressive expansion of the Diablo Range alluvial fans (Figs. 6 and 7) is a nice example of dynamic allometry of deposits that do not tend toward a steady state. One allometric analysis would be to compare the thicknesses and areas of post-Corcoran alluvial fans, using the Sierran-Diablan alluvium contact (a time-transgressive line) and the top of the Corcoran (a time line) for measuring both changes of the system. The intersection of the contact and the Corcoran defines the eastern extent of the alluvial fans at the end of Corcoran time. The fan piedmont area was planimetered between lines C-C' and D-D' (Fig. 7) for thicknesses of 20, 40, 70 and 110 m of accumulation of fan deposits.

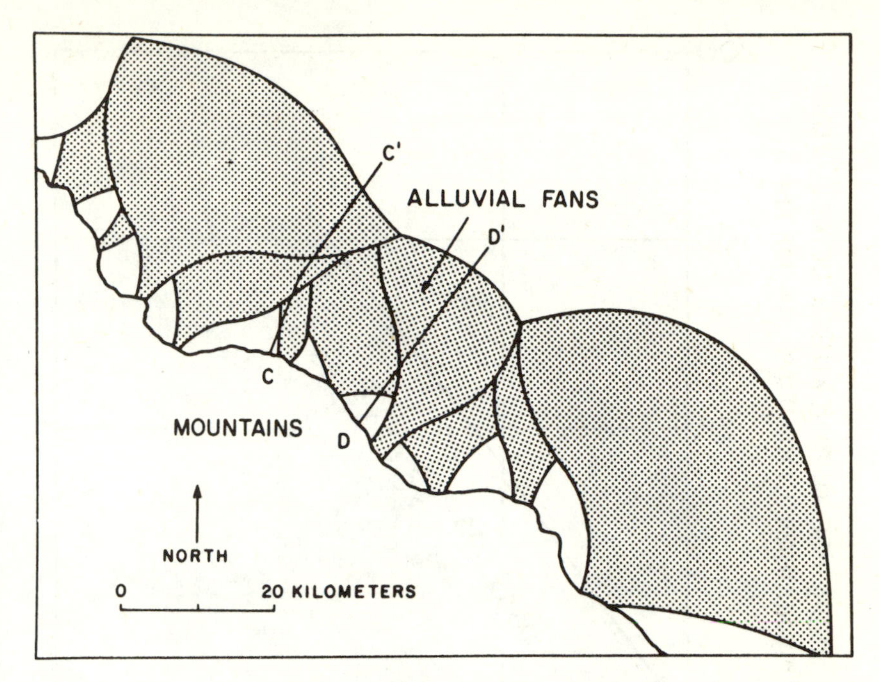

Figure 7. Aerial interrelations of the larger alluvial fans of western Fresno County, California. Section C-C′ is shown on Fig. 6.

The results of the analysis are shown in Figure 8. If increase in fan thickness is Y, and increase in fan area is X, then

$$Y=0.085\ X^{1.35},\quad 56.5 \leqq X \leqq 199. \tag{6}$$

Equation 6 describes a markedly different type of allometric adjustment than would be expected if the fans were not expanding. If the fans were not expanding the regression would be vertical, the exponent infinity, and the coefficient zero. A narrow band of piedmont was used in this analysis so as to include only a single mountain-front structural unit. Analyses of the vairation of Equation 6 along 100 km of the eastern margin of the Diablo Range should reveal some interesting effects of variations of sediment yield and the times and amounts of post-Corcoran tectonic activity.

The large thickness of fan deposits adjacent to the mountains (Fig. 6) and the general lack of weathering profiles near the apexes of the present-day fans indicate rapid uplift of the mountain front. If du/dt is the rate of vertical differential uplift of the mountain front, dw/dt the rate of stream downcutting in the mountains upstream from the mountain front, and ds/dt the rate of accumulation of fan deposits downslope from the mountain front, then the following relation has been characteristic of the mountain-fan

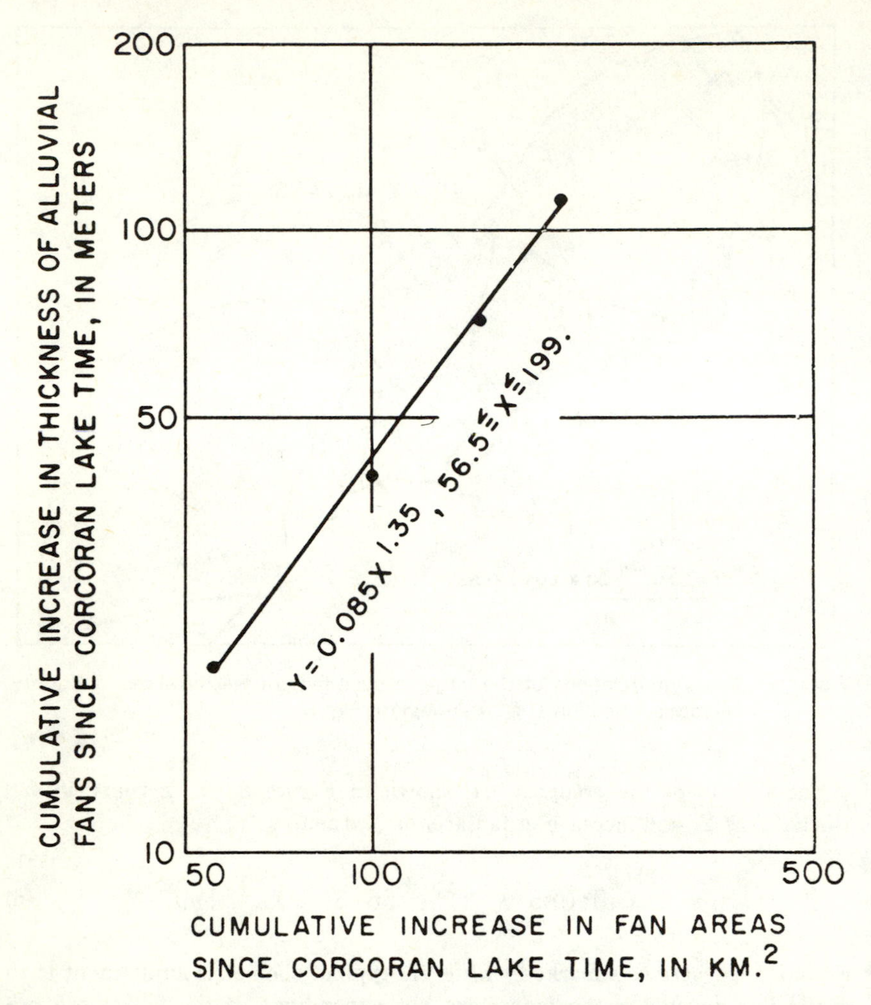

Figure 8. Interrelations of increases of alluvial-fan thickness and fan area during the last 0.6 million years, western Fresno County, California.

boundary since Corcoran time:

$$\frac{du}{dt} \geqq \frac{dw}{dt} + \frac{ds}{dt} \tag{7}$$

Uplift tends to keep the stream channel at the mountain front higher than the fan apex, and in Equation 7 is sufficient to offset both channel downcutting and fan deposition which tend to cause fanhead trenching conducive for the development of weathering profiles.

The map of alluvial fans in western Fresno County (Fig. 7) suggests that the fans competed with each other for space in the depositional basin as they increased their areas by encroachment to the east (Fig. 6). Each fan has an area that is, in part, a function of the sediment yield of its source area. Large drainage basins or easily eroded source materials produce larger fans than do small drainage basins or resistant source materials.

The areas of adjacent alluvial fans and their source areas can be analysed to determine if allometric adjustment exists for fans that have undergone increases in area and rates of aggradation during the last 0.6 m.y. The points for each plot in Figure 9 are for fans that have source areas with similar rock types, climate, and vegetation. The small amount of scatter (which is largely a result of variations of drainage basin slope) indicates good allometric adjustment between the erosional and depositional parts of this large multi-basin fluvial system. The coefficients of the equations are measures of the fan areas for drainage basins of 1 km^2. The lithology of the source area affects the static allometric relation between fan area, A_f, and drainage-basin area, A_d.

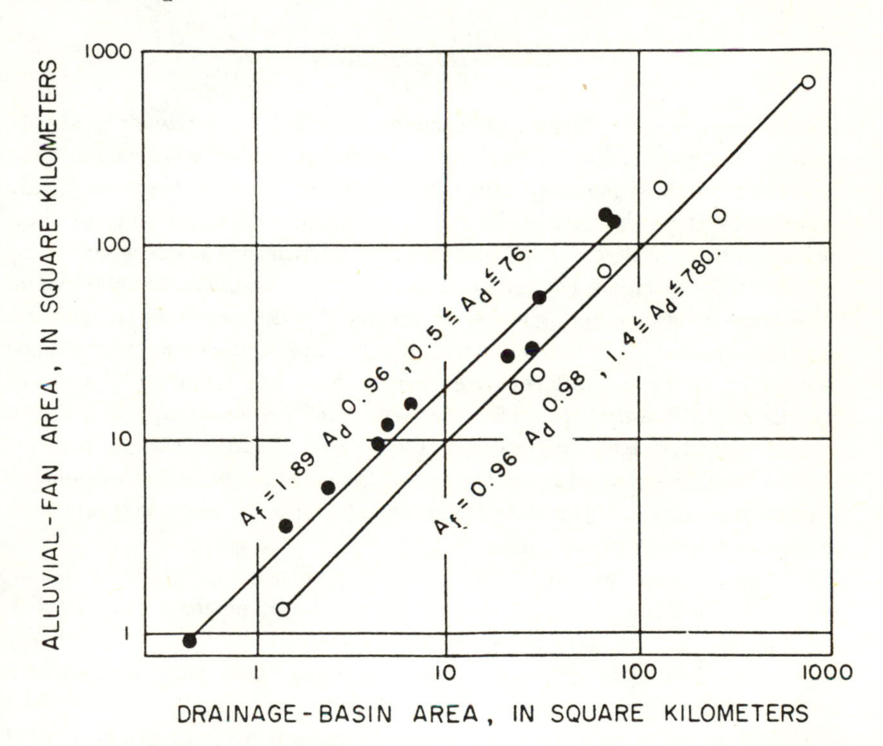

Figure 9. Relations of alluvial-fan areas to drainage-basin areas and lithologies in western Fresno County, California; for drainage basins that are underlain by 48-86 percent mudstone and shale (●), and 56-68 percent sandstone (o). Modified from Bull, (1962).

For fans derived from mudstone source areas,

$$A_f = 1.89\ A_d^{0.96}, \quad 0.5 \leqq A_d \leqq 76. \tag{8}$$

For fans derived from sandstone source areas,

$$A_f = 0.96\ A_d^{0.98}, \quad 1.4 \leqq A_d \leqq 780. \tag{9}$$

The lower coefficient in Equation 9 suggests that roughly half as much sediment is yielded from the drainage basins underlain by sandstone than from the drainage basins of the same size that are underlain by mudstone and shale (Bull, 1962). The factors affecting the exponents for the types of relation described by Equations 8 and 9 are discussed by Hooke (1968).

CONCLUSIONS

Many elements of hillslope environments do not tend toward a steady state. Erosional landforms that are undergoing progressive increase or decrease in height, volume, or other dimensions do not tend toward a steady state. Lack of steady state is the result of changes in independent variables such as climate, base level, and erodibility of surficial materials which cause complex and continuing changes in the dependent variables of the system. Differences in the weathering and erosion rates of rocks may cause topographic inversion. Examples include a lava-filled valley in granitic rocks, and stepped topography formed by different weathering micro-environments of massive granitic rocks (Wahrhaftig, 1965). Self-enhancing processes also do not tend toward a steady state. For example, the process of cliff retreat by rock fall starts with a tension fracture at the top of the cliff that permits the monolith to rotate toward the adjacent valley at an exponentially increasing rate until collapse produces a pile of rubble.

Most depositional landforms may be considered as endpoints of erosional-depositional systems. In order to exist as a three-dimensional body, depositional landforms must have grown in volume instead of maintaining a steady state size. Individual deposits (such as alluvial fans) may have smooth surfaces that can be described mathematically, and that are statistically related to variables such as source area and lithology in the erosional parts of the systems. On a larger scale, individual deposits compete with another for space in the depositional environment, and have stratigraphic contacts that provide a history of the relative inputs of sediment and growth of adjacent deposits.

The model of allometric change is useful where landscape and/or geomorphic processes are changing at different rates. The model can be used to analyse the tendency for adjustment between interdependent materials, processes, and landforms even where a tendency toward a steady state does not exist. The power-function

$$y = a\, x^{b},$$

may be used to analyse the dynamic interrelations of geomorphic variables at a sequence of times during the history of a landform and to analyse the static interrelations at one time.

The emphasis of future geomorphic work may change. Even if the existence of non-steady state landforms and processes should become generally recognized, the tendency toward steady state will continue to be a useful concept for some processes and landforms. However, where geomorphologists are concerned about the effects of changes in the independent variables — such as climate, tectonics, man, and the erodibility of surficial materials — they may place added emphasis on the complexity of interactions between variables (Schumm, 1973), self-enhancing feedback mechanisms, and thresholds between different modes of operation of fluvial open systems.

REFERENCES CITED

Bateman, Paul and Wahrhaftig, Clyde, 1966, Geology of the Sierra Nevada, *in* Geology of Northern California: Calif. Div. Mines Bull. 190, p. 107-172.

Bull, W. B., 1962, Relations of alluvial-fan size and slope to drainage-basin size and lithology in western Fresno County, Calif., Article 19 *in* Short papers in geology and hydrology: U. S. Geol. Survey Prof. Paper 450B, p. 51-53.

———————, 1970, Effect of climatic and tectonic changes on denudation rates in part of the Diablo Range, California: Geol. Soc. America Abstracts with Programs, v. 2, p. 77.

———————, 1975, Allometric change of landforms: Geol. Soc. America Bull., v. 86, p. 1489-1498.

Croft, M. G., 1972, Subsurface geology of the late Tertiary and Quaternary water-bearing deposits of the southern part of the San Joaquin Valley, California: U. S. Geol. Survey Water Supply Paper 1999-H, 29 p.

Dalrymple, D. B., 1963, Potassium-argon dates of some Cenozoic volcanic rocks of the Sierra Nevada, California: Geol. Soc. America Bull., v. 74, p. 379-390.

Gilbert, G. K., 1879, Report on the Geology of the Henry Mountains, 2nd ed., U. S. Geog. and Geol. Survey of the Rocky Mountain Region, Washington, D. C.: U. S. Govt. Printing Office, 170 p.

Godfrey, A. A., Reesman, A. L., and Cleaves, E. T., 1971, The importance of chemical erosion: dynamic disequilibrium: Geol. Soc. America Abstracts with Programs, v. 3, p. 767, 768.

Hack, J. T., 1965, Geomorphology of the Shenandoah Valley, Virginia and West Virginia, and the origin of residual ore deposits: U. S. Geol. Survey Prof. Paper 484, 84 p.

Hooke, Roger LeB., 1968, Steady state relationships on arid-region alluvial fans in closed basins: Am. Jour. Sci., v. 266, p. 609-629.

Hunt, C. B., 1969, Geologic history of the Colorado River, *in* The Colorado River region and John Wesley Powell: U. S. Geol. Survey Prof. Paper 669, p. 59-130.

Janda, R. J., 1965, Quaternary alluvium near Friant, California: Internat. Assoc. Quaternary Research, 8th Cong., U. S. A., 1965, Guidebook for Field Conf. I, p. 128-133.

Koppen-Geiger, 1954, Klima der Erde (map): Justus Perthese, Darmstadt, Germany. (American distributor, A. J. Nystrom and Co., Chicago.)

Langbein, W. B., and Leopold, L. B., 1964, Quasi-equilibrium states in channel morphology: Am. Jour. Sci., v. 262, p. 782-794.

Leopold, L. B., and Maddock, Thomas, 1953, The hydraulic geometry of stream channels and some physiographic implications: U. S. Geol. Surv. Prof. Paper 252, 57 p.

Magleby, D. C., and Klein, I. E., 1965, Ground-water conditions and potential pumping resources above the Corcoran Clay — an addendum to the ground-water geology and resources definite plan appendix, 1963: U. S. Bur. Reclamation Open-File Report, 21 plates.

Mosley, M. P., and Parker, R. S., 1972, Allometric growth: a useful concept in geomorphology?: Geol. Soc. America Bull., v. 83, p. 3669-3674.

Schumm, S. A., 1973, Geomorphic thresholds and complex response of drainage systems: *in* Fluvial Geomorphology Marie Morisawa, ed., Binghamton, New York, SUNY Pub. in Geomorphology, p. 299-310.

Schumm, S. A., and Chorley, R. J., 1964, The fall of Threatening Rock: Am. Jour. Sci., v. 262, p. 1041-1054.

Wahrhaftig, Clyde, 1965, Stepped topography of the southern Sierra Nevada: Geol. Soc. Am. Bull., v. 76, p. 1165-1190.

Yaalon, D. H., 1971, Soil-forming processes in time and space: *in* Paleopedology; Origin, Nature, and Dating of Paleosols, D. H. Yaalon, ed., Jerusalem, Keter Press, p. 29-40.

CHAPTER 8

THE CONCEPT OF CLIMATIC GEOMORPHOLOGY

Louis C. Peltier

ABSTRACT

It is argued that because climates in different parts of the world, or at different times in a single place, differ, that the combination of geomorphic processes also differs. As a result denudation proceeds at varying rates and landforms vary systematically in detail. The theoretical basis for this argument lies in a combination of the writings of Davis and Penck whereby landform is expressed in terms of geologic structure, rate of uplift, rate of erosion and time. Emphasis is here placed upon the rate of erosion.

Climatic geomorphology is considered to be a component of regional geomorphology, along with tectonic geomorphology and continental stratigraphy. Useful geomorphic components are described. Landform, in the gross sense, is presented as the product of relief, lithology, rate of weathering and erosion, and time. A general formula for landforms is postulated.

Maps illustrating a method of predicting morphogenetic regions are presented. A test of this method suggests that because climate has varied rapidly during and since the last glacial episodes, it is likely that most landforms are polygenetic.

INTRODUCTION

Conceptual Structure of Regional Geomorphology.

Climatic geomorphology is one part of a broader structure of regional geomorphology which includes both tectonic geomorphology and the stratigraphy of continental deposits. In combination the argument is made that the geomorphology of the world varies in a reasonable, explainable and even predictable way.

Tectonic geomorphology argues that the major geomorphic features are geographically distributed in a rational manner as the result of the operation of a relatively small number of forces including gravity, centrifugal force, inertia, radioactive decay, and the resistance of crustal material to tangential stress. Thereby the gross relief pattern of the world comes into being concurrently with denudational processes and it is upon this gross pattern that the more subtle operations of climate exert their influence.

Continental stratigraphy considers the Earth's surface as an unconformity which in places may be buried by deposits. If an episode of erosion is postulated it should be possible to account for the volume of material removed in adjacent deposits. The sequence of tectonic or eustatic fluctuations should be reflected in both the geomorphology and the sedimentary deposits and in both chronologic and volumetric agreement. So far, the time scale, within which geomorphic interpretations are constrained, has been provided primarily by stratigraphy and leads to two important conclusions: 1) that the climatic changes of the Pleistocene must have been global in their influence and 2) that the time which has elapsed since the last major glacial episode and its associated influences was so brief that the landforms over much of the world must be polygenetic in nature.

Climatic geomorphology, within this context, presents the argument that the rate of geomorphic process, and therefore the duration of the stages of denudation, varies geographically and further that the detailed configuration of landforms varies in accordance with the process mix of the region. Emphasis is placed upon the relative importance of the different geomorphic processes in the production of the landform and the effects of regional variation in this relative importance.

Definitions.

It is suggested that, in first approximation, geomorphology may be expressed in terms of relief, slope, landform, texture and lineation. When combined with measurements of geologic and organic materials, including plants, which comprise and cover them and with the processes which do or did operate upon them, one may proceed to geomorphic analysis in quantitative terms. Climatic geomorphology depends for its arguments upon such a quantitative analysis.

For analytical purposes, the following definitions are suggested:

1. Relief is the maximum difference in elevation within a unit area. Operationally a one-mile square unit area has proven useful.
2. Slope is the rate of change in elevation with respect to distance.
3. Landform, in this context, is the curvature of slope or the rate of change of slope with horizontal distance.
4. Texture is the frequency of reversal of slope direction per unit distance.
5. Lineation is the variation in texture with respect to compass direction. It may be operationally expressed as the ratio of minimum to maximum

values.

It is useful, in the pursuit of this type of analysis to identify a basic topographic unit, such as a hill "similar to that illustrated in Figure 1", composed of an assemblage of slopes and landforms which can be analyzed by transects extending from the valley bottom upward over and across the top or crest. On such a transect one may classify the slopes or surfaces into bottom slopes, footslopes, foreslopes, cliffs and the round. These are defined respectively as:

1. Bottom slope — the distal surface of deposition including playa, floodplain and alluvial terrace surfaces.
2. Footslope — the proximal surface of erosion or alluvial deposition including pediment, alluvial apron and alluvial fan surfaces.
3. Foreslope — the surface of slopewash or mass movement including alluvial cones.
4. Cliff — the gravitational surface from which materials have fallen including scarps formed by landsliding, glacial action, stream erosion or wave erosion.
5. Round — the residual surface or surface of down-wastage generally above the brow of the hill or negative landform.

The slope of these components and their relative importance in the composition of the hill are important for they tend to vary with the lithology, stage of denudation and the process-mix characteristic of the climate. On any hill this sequence may be repeated; all components need not be present in every hill.

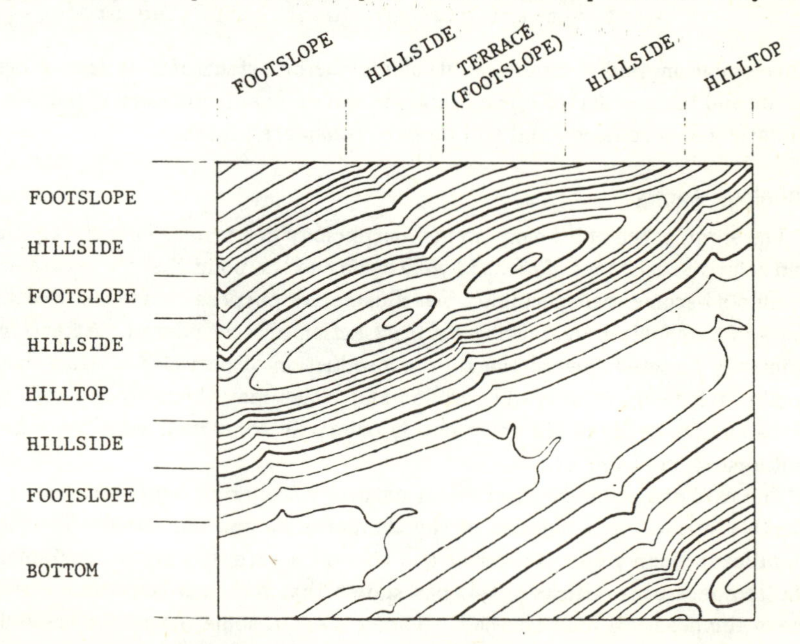

Figure 1. Diagram showing significant parts of landforms.

A THEORY OF QUANTITATIVE GEOMORPHOLOGY

General Landform Equation.

The basic equations for the understanding of landforms presented by Davis (1899); landform is a function of structure + process + stage and by Penck (1924); landform is a function of the rate of uplift with respect to the rate of degradation, may be taken as a point of departure for more detailed analysis. They may be combined as shown in Table 1.

Table 1

$$LF = f(m, dm/dl, de/dt, du/dt, t)$$

where:

LF = landform

m = geologic material

dm/dl = rate of change of geologic material with distance (a structural factor)

de/dt = rate of erosion

du/dt = rate of uplift

t = total time of duration of the process

This equation thus contains a lithologic factor, structural factor, erosion factor and tectonic factor. The lithologic and erosion factors are important to climatic geomorphology and will be here considered further.

Lithologic Factor.

The significance of lithology in geomorphology lies primarily in the absolute and relative resistance of rocks to weathering to the point that the weathering products become transportable. For illustrative purposes, chemical weathering and mechanical weathering by frost action are considered. Attention is primarily directed toward the rate of production of particles of sand size or smaller which are most readily eroded and which may themselves comprise a transporting medium. In short one is primarily concerned with the rate of soil formation (Table 2).

The rate of a chemical reaction, at naturally occurring temperatures on the Earth's surface, is a function of the temperature and ionization. Thus it is significant when water is present and then at a rate directly proportional to the temperatures. Hicks (1944) has shown that for such considerations the mean temperature may be used. It may also, in some places, be related to the absorption of atmospheric carbon dioxide in the ground water and in

Table 2

Lithologic Factor

$m = f(Wc, Wm, Dp, Fc)$

where:
- m = geologic material (in a geomorphic sense)
- Wc = rate of chemical decomposition of rock
- Wm = rate of mechanical disintegrating of rock
- Dp = diameter of soil particles produced
- Fc = cohesive force between particles

$Wc = k(Tm \cdot Rd)$

where:
- Wc = rate of chemical weathering
- k = constant
- Tm = mean absolute temperature
- Rd = days of precipitation expressed as a percentage of the period for which Wc is calculated.

$Wm = K_f(ntc/t \cdot Rd) + K_h(nSt.H)$

where:
- Wm = rate of mechanical weathering
- K_f = constant reflecting the relative importance of frost action
- ntc/t = number of passages of temperature across the freezing point per unit of time
- Rd = days of precipitation expressed as a percentage of the period of time for which Wm is calculated
- K_h = constant reflecting the relative importance of salt action
- nSt/t = number of storms in which the rate of precipitation is less than the rate of evaporation per unit of time
- H = concentration of soluable salts in the soil

$$m = k(TmRd) + K[K_f(ntc/t \cdot Rd) + K_h(nSt/t \cdot H)] + f(dp, Fc)$$

where: K is a constant of lithology dependent on permeability, solubility, void ratio and vulnerability to chemical reaction.

other places to the concentration of organic decomposition products. Because of the requirement for water and ions and because there is generally a slight tendency for vertical motion in ground water the horizon of greatest chemical weathering is most likely to occur at or near the surface of the ground water table.

Mechanical disintegration is dominated by the pressures of crystallization created by the interstitial crystallization of ice or deliquescent salts. In both instances a continuous supply of feedstock for crystal growth through capillary action is important. In a broader sense the rate of the process is related to its frequency of repetition, of freezing and thawing or wetting and drying.

Thus the equation for mechanical weathering has a slightly different form from that for chemical weathering.

Because of the importance of water in all of these processes, the lithologic factor must be further modified in recognition of varying rock permeabilities both in joints and in pore spaces, as shown in a detailed examination of the slope of Salem Knob in eastern Missouri (Fig. 2). This component is considered as a constant for the rock whereby the ability of water, when present, to penetrate the rock is measured.

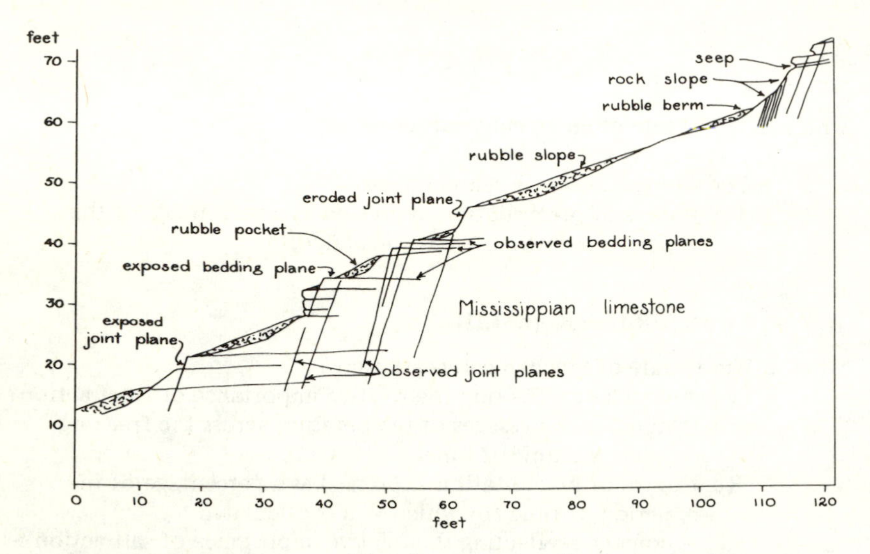

Figure 2. Salient features of the slope of Salem Knob, eastern Missouri.

Erosion Factor.

The erosion factor expresses the rate at which the weathering products are eroded and transported under the action of gravity in the presence of fluid media of varying density and viscosity: air, water, mud, ice. The rate of erosion over an area is dependent on the exposure of materials of a certain size which is critical for each medium and the probability of occurrence of fluid media of significant or critical velocities. In this context an outcrop is interpreted as a point at which the rate of erosion exceeds the rate of weathering. It is convenient to separate channel erosion from non-channel erosion, the former comprising the action of rivers, mountain glaciers, rock avalanches, aerodynamic weirs and tidal weirs. The latter comprises slope wash, slump, mass wastage, congeliturbation, soil flowage, continental glaciation, beach erosion and deflation.

Thus, in very general terms, for pluvial-fluvial erosion, two sets of relationships (Table 3) may be postulated.

Table 3

Erosion Factor

$de/dt = f(\rho_1 - \rho_2, \eta, v, \phi, G)$

Where: de/dt = rate of erosion (removal and transportation)

$\rho_1 - \rho_2$ = difference in density between the transporting medium and the material transported.

η = viscosity of the medium

v = velocity of the medium

ϕ = frequency of occurrence of the transporting process with respect to space and time.

G = gravitative constant

$ep/dt = k_s \cdot k_v\ [^{P}m(30)\ (ns/yr)\ Am]$

Where: ep = rate of pluvial, non-stream erosion

$^{P}m(30)$ = mean quantity of rain falling from an excessive storm in 30 minutes.

ns/yr = average number of excessive storms per year

Am = percentage of area existing as interfluves

k_s = a constant related to the average length of slope

k_v = a constant expressing the protective effect of vegetation.

Summary.

In summary then, the relationships of erosional geomorphology may be expanded as shown in Table 5. From this it is argued that landform varies according to:

1. Relief.
2. The geographical variation in lithology particularly with respect to its geographic variation in vulnerability to weathering.
3. The rate of weathering and erosion.
4. The duration of the assemblage of geomorphic processes.

The erosion factor, as it applies to stream channel erosion, may be expressed in similar terms as shown in Table 4. Thus, the rate of erosion, de/dt, is dependent upon both the rate of surface erosion and the rate of channel erosion. The relative rates at which the two proceed should determine the curvature of the lower part of the profile or landform. It is further argued that because the rate of weathering and erosion — and their special characteristics — vary with the climate, the detailed characteristics of the landforms must also vary with the climate. Because climatic characteristics vary

Table 4

Channel Erosion

$$ec/dt = A_c\,(v^2/d)\,(nF/yr)$$

Where: ec/dt = rate of channel erosion
v = mean flood velocity
d = mean flood height
nF/yr = number of floods per year
A_c = percentage of area in channels

Table 5

General Landform Equation

$$LF = f(m, dm/dl, de/dt, du/dt, t)$$

AND

$$LF = f(t[k\,(Tm \cdot Rd)] + K[Kf(ntc/t \cdot Rd) + K_h(nSt/t \cdot H)]\; k_s - k_v\,[{}^{P}m(30)(ns/yr)Am],\; A_c(v^2/d)(nF/yr),\; dm/dl,\; du/dt)$$

LF therefore varies according to
(1) the relief
(2) the geographical variance in lithology
(3) the rate of weathering and erosion and
(4) the duration of the process.

systematically throughout the world so that they may be grouped into climatic regions, the detailed geomorphic characteristics of the landscape may also be grouped into morphogenetic regions.

GEOGRAPHIC VARIANCE IN GEOMORPHIC FACTORS

The geographic variance in weathering and erosion within conterminous United States serves to illustrate further the need to recognize differences in certain aspects of geomorphology.

Weathering.

Chemical weathering has been postulated on a map (Fig. 3) on the assumption that the relative rate of a chemical reaction in nature was, in first approximation, equal to the mean annual temperature in degrees Celsius multiplied by the frequency of days of rain in percent. It shows a low rate in the arid and seimarid areas of the southwest and a high along the Gulf Coast and northern Appalachians and Pacific Northwest.

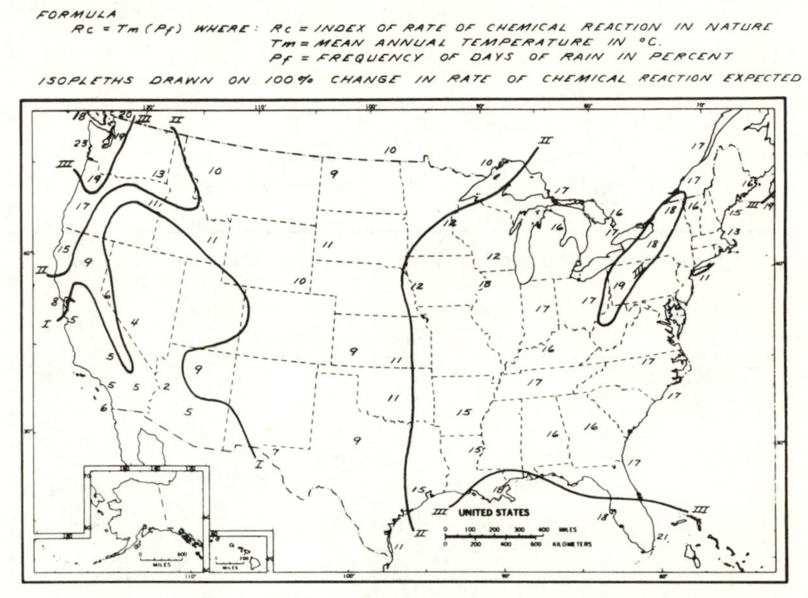

Figure 3. Expectable rate of chemical reaction in the United States.

Mechanical weathering is represented by the average depth of frost penetration (Fig. 4) which is here shown to be significant in the northern and mountain areas.

Erosion.

Pluvial erosion is presumably related to heavy storms in which the rate of rainfall exceeds the rate of infiltration into the soil. The geographic distribution of clearly heavy storms, the average number of storms over one inch per hour within a year (Fig. 5), shows a concentration of potential erosion in the southeastern quadrant of the country. Because of low relief, highly permeable ground and vegetative protection this potential is not always reached.

The country has been further divided into five categories which express the relativeness of vegetation as an element in retarding pluvial or wind erosion (Fig. 6). Least effective are the areas of sage, mesquite and creosote bush, moderately effective are the grasslands and most effective are the forests with their surface mantle of duff or litter.

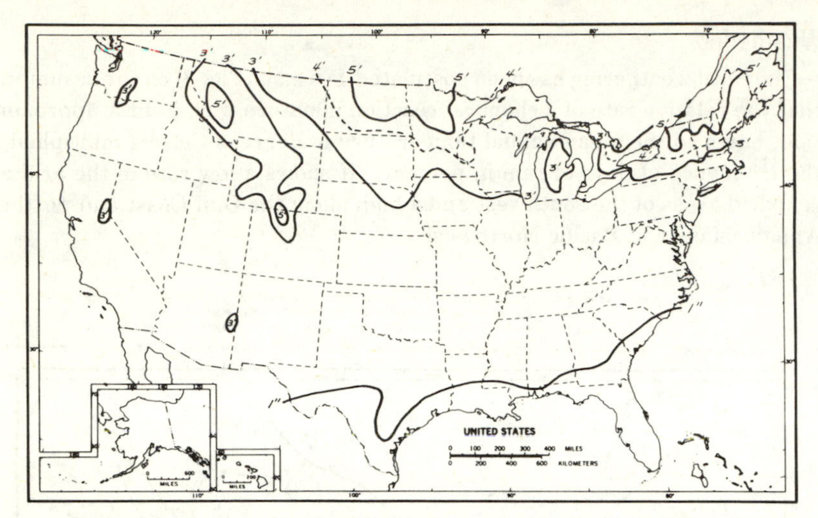

Figure 4. Average depth of frost penetration in the United States.

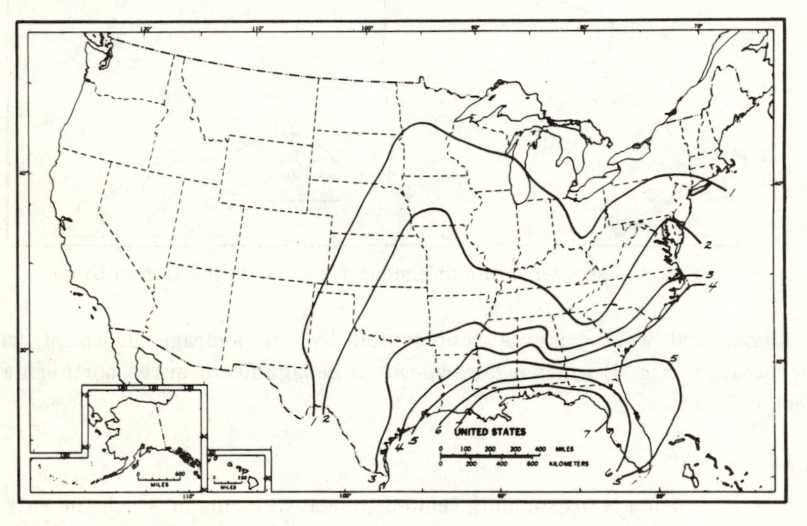

Figure 5. Average number of storms of one inch per hour per year of precipitation.

The geographical patterns of weathering and pluvial erosion have been combined to show the general geography of geomorphic process in the United States (Fig. 7). Further modifications for soil permeability and vegetation should be added. This map does, however, serve to emphasize the geographical variation in geomorphic process and lead one to expect differences in geomorphic detail, such as slope and texture, in different parts of the country.

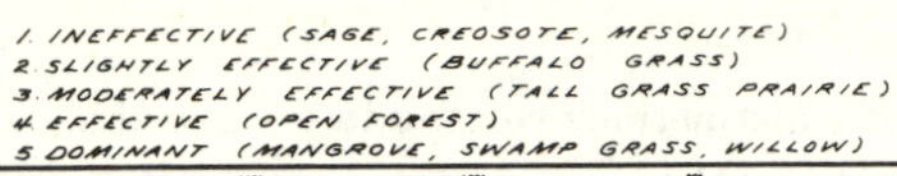

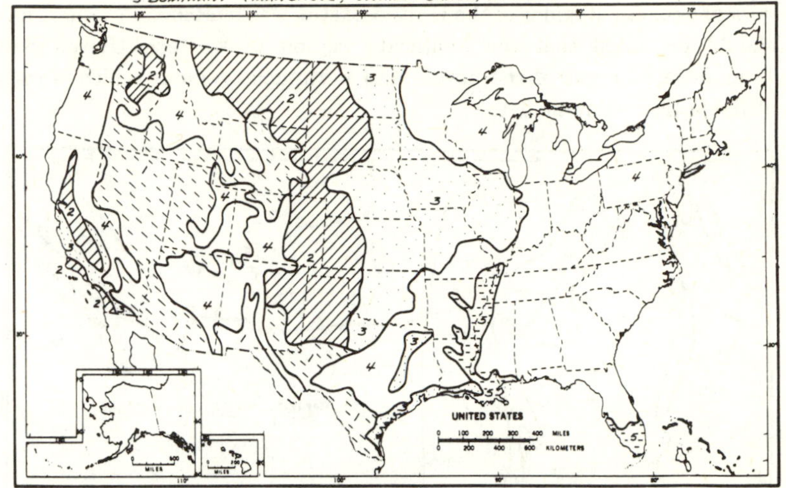

Figure 6. Relative effectiveness of vegetation against erosion.

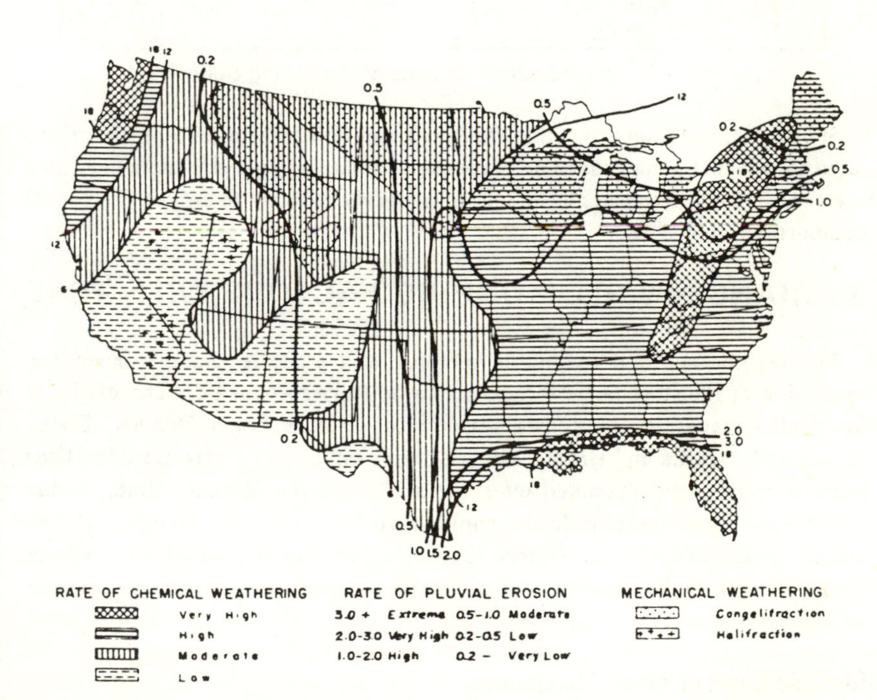

Figure 7. Geography of geomorphic processes in the United States.

Morphogenetic Regions.

On the basis of climatic distinctions related to geomorphic process a map of the Morphogenetic Regions of the United States was sketched in 1950 (Fig. 8). It should be noted that the moderate region of western United States contains isolated areas of the boreal and periglacial regions which were not identifiable from climatic data.

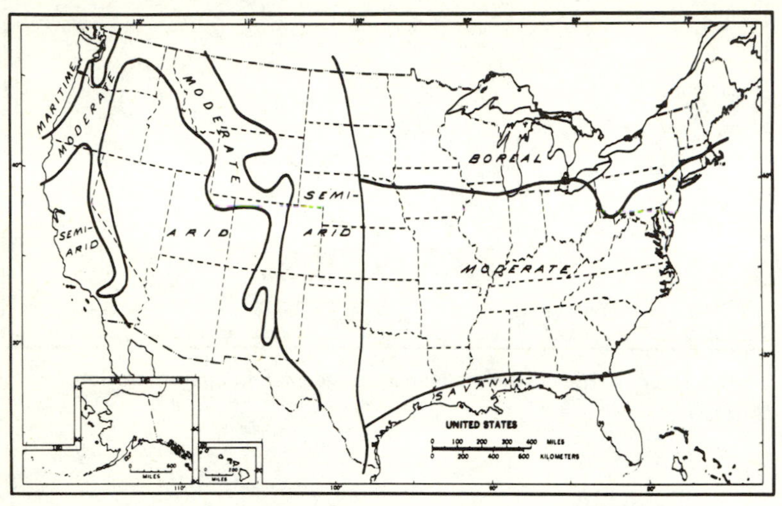

Figure 8. - Morphogenetic regions of the United States.

Subsequent attempts to refine this map and adjust it to field observations have shown that a map based on regions of similar polygenetic sequence since the time of the Wisconsin glacial maximum is more likely to reflect the actual geomorphic conditions.

VARIANCE IN GEOMORPHIC FACTORS WITH TIME

The significance and extent of tectonic and eustatic changes since the beginning of Pleistocene time remains to be fully recognized. The evidence for such changes in the coastal areas, the Pacific Coast Ranges, Rocky Mountains, Ozarks and Great Lakes is impressive. As a stratigraphic time scale is refined and extended over the continent the absolute time, during which specific geomorphic forms could have been created, becomes limited and it becomes possible to explore the rate of geomorphic process in greater detail. Under these circumstances it becomes easier to relate paleoclimatology to landforms.

Climatic Changes of the Pleistocene.

The Pleistocene is commonly associated with the "Ice Age" and with areas of glaciation. It must be emphasized that it was a period of time characterized by conditions of continentality, diastrophism and significant climatic fluctuations of global extent. Because of changes within air-mass source regions the climatic changes were not mere displacements of climatic zones. In geomorphic terms environments in one region might vary between glacial and moderate, in another between boreal and savanna and still elsewhere between arid and savanna or even moderate. A few observations in the tropics suggest that there may have been a variation in parts of the selva region between selva and savanna. The time involved in such shifts must have been geologically brief and required not more than 30,000 to 50,000 years. Within these episodes of not more than 3000 to 5000 years duration may be recognized. Thus, one can no longer think in terms of an infinitesimally slow process acting over infinite time to produce a landform, for time is clearly limited. It is important, instead, to think of a process-mix operating over a short period of time followed by a somewhat different process-mix operating over another brief episode, the net products of which produce the landform as it is observed.

Evidences of Climatic Change.

Insofar as the soil group and soil profile are products of the climatic environment they must change as the climate changes. In those places where sedimentation has been present one may find completely buried soils overlain by another complete soil profile. In some places, where only a thin sediment overlies the older soil subsequent soil development causes the newer profile to penetrate and modify the older soil. In those instances in which no sedimentation is present or even in which surface erosion has removed part of the original soil profile the new soil, developed in accord with the new climatic environment, develops on, and penetrates the old soil creating a transformed soil profile.

Similarly, insofar as the floristic limits of plants reflect the climatic conditions changes in the environment may result in a retreat of these limits leaving relic floral associations in favored sites or their stems, leaves and pollen may remain as fossil evidence of their former presence. Conversely the floristic limits may advance but here it is important to recognize the significance of time. Floras may disperse from a source point or points at a finite rate depending upon the characteristics of the species. All suitable places need not be occupied if time has been insufficient for optimum distribution. For these reasons floristic limits alone are not particularly reliable indicators of climate.

Polygenetic landforms, themselves, provide evidence of climatic change. The argument is based on the notion of environmentally stable profiles. According to this notion each assemblage of geomorphic processes, unique to a climatic environment, tends to produce its own unique hill profile as distinguished by the radius of curvature of the round, the frequency or probability of cliff development, the length and gradient of the foreslope and footslope. Periglacial hills (Fig. 9) tend to have a large radius of curvature at the top, an extensive low-gradient foreslope and broad bottomlands; semiarid hills tend to have a small radius of curvature at the top, important cliff and bottomland development and relatively small but moderately sloping foreslope. Tropical hills tend to have a moderate to small radius of curvature at the top, significant but steep foreslopes and a moderate to extensive bottomland. It is further argued that a change in climate leads to a change in the stable profile toward adjustment to the new slope-forming processes. In the interim both may coexist on the same hill.

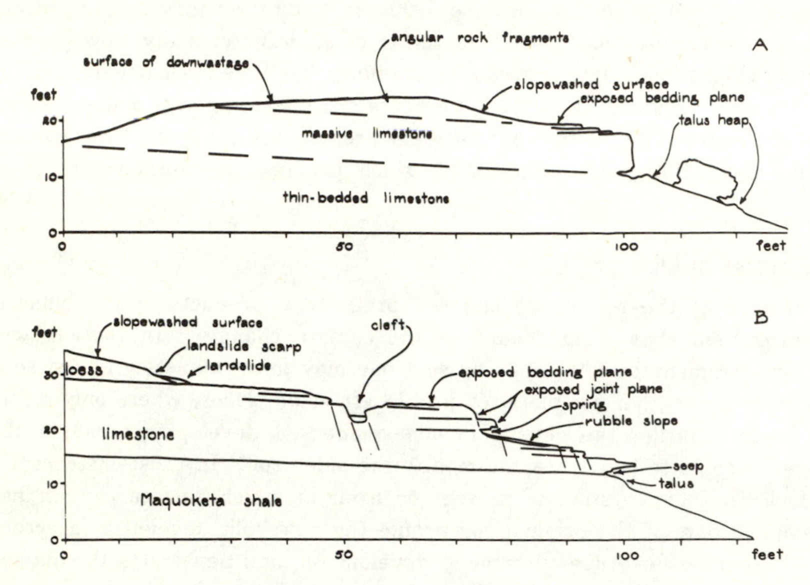

Figure 9. Comparison of typical periglacial hillslope (A) with hillslope of a former periglacial region (B).

Under some circumstances such a change may cause climatically accelerated soil erosion on the slopes and alluviation of the bottomlands. Such appears to be the case in the change from moderate to periglacial conditions or arid to savannah conditions. A climatic change which reduces slope erosion may result in a down-cutting in the bottomlands and the creation of climatic terraces.

SUMMARY

Most if not all of the land surfaces which one sees were formed, or at least strongly modified, during the last 50,000 years. Some of these surfaces show a strong influence of characteristics inherited from the past, but truly old surfaces which have remained unchanged since the more distant past must be extremely rare. The actual surfaces can be interpreted in terms of the succession of various process-mixes which have operated upon them. For this reason maps which describe the polygenetic sequence are more realistic in expressing landforms than those which merely describe the operation of current process.

REFERENCES CITED

Hicks, E. W., 1944. Note on the estimation of the effect of diurnal temperature fluctuations on reaction rates in stored foodstuffs and other materials: Jour. Council Sci. and Ind. Research Comm. Australia, Vol. 17, p. 111-114.

CHAPTER 9

THE COMPATIBILITY OF STRUCTURE, LITHOLOGY AND GEOMORPHIC MODELS

Robert C. Palmquist

ABSTRACT

The reference systems of Davis, Penck and Hack are not as incompatible as may be inferred from recent reviews of their writings. Davis develops a reference system of landscape evolution which includes models based upon varying rates of uplift; he recognizes the control lithology exerts on hillslopes. Hack utilizes a reference system based upon the open system to develop a model of land forms adjusted to lithology; he allows that landform evolution occurs under stable conditions.

These three reference systems and portions of the models are compatible. The recognition of lithologic adjustments in the models of Davis and Penck and the land form evolution allowed by Hack's reference system facilitates the merger of their systems. A composite model of landscape evolution can be derived from a systems oriented reference system and the basic assumption that the rock mass above base level is a disequilibrium factor. In an open system, some factors may achieve equilibrium before others; thus land form evolution — mainly relief reduction — may occur while hillslope and stream gradients are in a state of equilibrium with lithology. Landscapes thus evolve in a manner similar to Davis' models while maintaining the equilibrium features of Hack's model. The end product of such a landscape evolution is a structurally - lithologically controlled peneplain — a phenomenon implicit in Davis' own writings.

A relict peneplain should not be recognizable in a maturely dissected landscape. "Statistical peneplains", those preserved by accordancy of summits, are as well explained by dynamic equilibrium as by peneplanation. Peneplains preserved on even-crested interfluves are more easily recognized but much of the published evidence is incompatible with both the concept of dynamic equilibrium and peneplanation.

INTRODUCTION

The title of this paper refers to the evolutionary geomorphic models of W. M. Davis and W. Penck and the lithologic-structural equilibrium model for landforms of J. T. Hack. These three models are the most commonly utilized reference frames for viewing the land surface.

The abbreviated manner in which the concepts of Davis, Penck and Hack are presented in recent textbooks and compendia distorts their views. Nowhere in the discussions of the landscape models of Davis and Penck, which appear in Thornbury (1969), Garner (1974), Ruhe (1975), Young (1972), Carson and Kirkby (1972), Leopold, Wolman and Miller (1964) and Harris and Twidale (1968) does mention occur of the adjustments of the landscape to lithology and structure which are found in their writings. Likewise, these same authors fail to mention the time-dependent models which appear in Hack's (1960, 1965) writings. The uninformed reader would leave these texts with the idea that, in Davis' opinion, landscapes owe their shape to time and initial conditions; that Penck believes that the shape of the land surface is related to the rate and direction of tectonic movement; and that Hack postulates landscapes to be open systems in a state of dynamic equilibrium and unchanging in form. These impressions are incomplete and may cause readers to find a greater difference between the ideas of Davis, Penck and Hack than actually exist.

In many of the comparisons of the ideas of Davis, Penck and Hack a separation of the model from the reference system does not occur. As pointed out by Carson and Kirkby (1972, p. 11), the cycle of erosion postulated by Davis is a system of reference in the sense that it maintains that the landscape evolves through time. The mode of landscape development envisaged by Davis constitutes one possible model within this evolutionary system. This distinction is important because the invalidity of a particular model, be it Davis', Penck's or Hack's, does not invalidate the reference system from which it is drawn. Likewise, the invalidity of detail within a model does not necessarily invalidate that model, though it demands its modification.

Since Strahler and Hack introduced geomorphologists to general systems in the 1950's and 1960's, several papers on systems analysis have been published which provide another reference system for viewing the landscape. Chorley (1962) analyzed general systems and recognized that Davis' geographic cycle is a closed system and that Gilbert's treatment of landscapes as well as that of other "dynamic equilibriumists" was essentially that of open systems. Although Chorley's analysis emphasized the differences between dynamic equilibrium and the geographic cycle, he did note that the historical element of relief reduction does not fit into the equilibrium model just as the idea of steady state equilibrium does not fit into the cyclic concept. Further considerations of systems by Howard (1965), of equilibria by Tanner (1968), of allometric growth by Woldenberg (1968) and Bull (this volume) and of the effects of time and space upon equilibria by Schumm and Lichty (1965) have gone far to clarify and to reconcile the problem noted by Chorley.

This paper is not a comprehensive review of the ideas of Davis, Penck and Hack. A brief review of their ideas is included to point out the similarities between their models. It will concentrate on the concept of lithologic adjustment in Davis' and Penck's models and the evolutionary aspects found in Hack's writings in an attempt to show that all three geomorphic models have both evolutionary aspects and aspects of structural-lithologic adjustments. The paper will focus on the ideas of Davis and Hack, and give some consideration to the ideas of Penck.

The paper will conclude that the concepts of Davis, Penck and Hack are compatible. Hack's concept explains the spatial variations in landforms during Davis' cycle, and Davis' cycle postulates changes in landforms after achievement of stable environmental conditions. The product of such an evolution is a structurally and lithologically controlled peneplain. Dissected remnants of such an adjusted peneplain should be difficult to identify once the youthful stage is past. Thus geomorphic models of landscape evolution are compatible with the lithologic-structural adjustments postulated by Hack.

The format of the paper will be to first present the ideas of Davis, Penck and Hack. Their ideas will be summarized under the categories of goals, basic assumptions, and either time-dependent model or adjustment model and then briefly assessed. The basic scheme of a composite model will then be proposed, and will end with a discussion of the peneplain problem.

DAVIS' MODEL

The "geographical cycle" of W. M. Davis is presented in a collection of his articles entitled "Geographical Essays" (Davis, 1954). A more extensive review of Davis is presented in Flemal (1971).

Davis' goal in formulating the geographical cycle was to provide the basis for a systematic, genetic classification of landforms. The possession of such a classification should, in his opinion, promote the collection and description of observable facts, and its understanding will assist the trained reader in appreciating the descriptions of the trained explorer (1954, p. 295). Davis' strong emphasis upon the importance of the descriptive power of the geographical cycle is displayed in his 1909 article, "The systematic description of landforms."

The reference system developed by Davis is more universal than the geographical cycle for which he is remembered. The reference system is that landforms change in an orderly manner as processes operate through time such that under uniform external environmental conditions an orderly sequence of landforms develops. This is the cyclic concept which Penck (1953, p. 7) considered to be Davis' contribution to landform analysis. Based upon this reference system, several models of landscape development, known as cycles of erosion were formulated such as those for arid regions (Davis, 1954, p. 296), and for humid-temperate climates; the last cycle, the normal geographical cycle, is of concern here. Davis developed two variants of the normal cycle, one for rapid uplift followed by stability and the other for slow uplift (Fig. 1A) though he recognized that numerous variations occur (Davis, 1954, p. 283). As Penck (1953, p. 12) noted, it is unfortunate that Davis, his followers and his detractors have emphasized the simple, normal geographical cycle in their discussions for this concentration on the model has slighted the reference system.

Six premises are necessary for Davis' cyclic system and for his model of landscape evolution. The assumptions required for the cyclic reference system are (Davis, 1954, p. 279): 1) landforms are the evolved product of the interactions of internal and external agencies and, 2) the evolution of landforms is orderly such that a systematic sequence of surface forms develops in response to an environmental change. Three of the remaining four assumptions are related to the general model of a geographical cycle: 1) streams erode rapidly downward until the graded condition is achieved, at which time lateral erosion becomes dominant (Davis, 1954, p. 258), 2) hillslopes become progressively graded from the base upward and have a gradient controlled by the texture of the soil (Davis, 1954, p. 267-269), and 3) climatic changes do not occur. The sixth assumption varies with the particular model. For the ideal model, rapid uplift of the landmass followed by a prolonged standstill is assumed. For the other variant, a slower rate of uplift is assumed (Davis, 1954, p. 283).

Davis' best developed time-dependent model is his ideal geographical cycle (1954, p. 249-272). In the scheme of the ideal geographical cycle a complete sequence of landforms is developed; other variants with different rates of uplift are permissible but do not yield all the landforms of the simple, ideal model (Fig. 1).

Davis does not develop a model of lithologic adjustment, but his writings are filled with references to the adjustments of landforms to lithology and structure. The adjustment model which can be inferred from Davis' writing is as follows: In a mature landscape the major streams follow lines of structural and/or lithologic weakness, and the interfluves are localized along resistant rock units (1954, p. 262). The slopes of the graded streams will be adjusted to the discharge and to the volume and texture of load, probably more to the discharge of water and load than to the load texture (1954, p. 257). Even though most major gradient changes related to lithology are minimized, irregularities persist related to tributaries, floods and "varigradations" (1954, p. 396). The hillslopes are graded with respect to the stream at their base (1954, p. 396), and the slopes rise above them with a gradient related to the coarseness of their soil. Variations in rock resistance cause upper portions of the slope to be graded to lower resistant rocks (1954, p. 268-269). The graded streams within the basin control the erosion of the

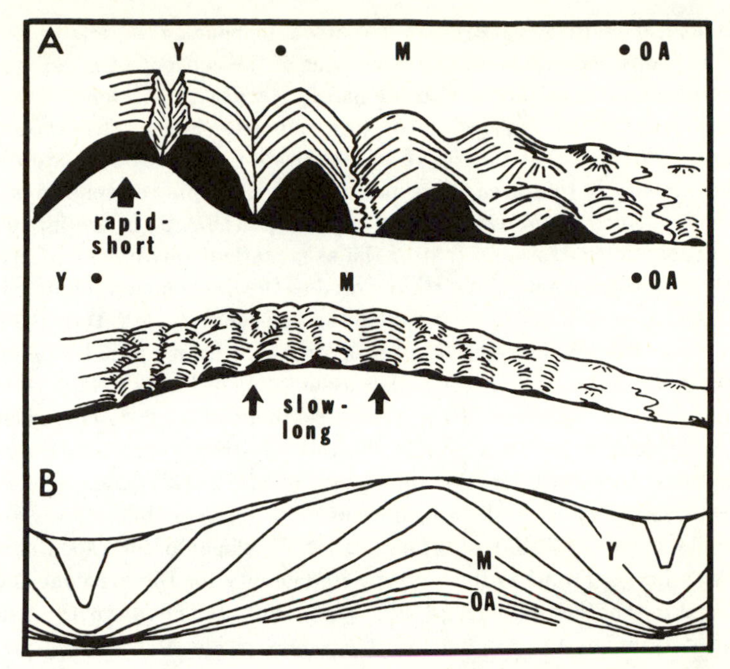

Figure 1. Models of landscape evolution according to Davis, (A) The simple geographical cycle traces landscape evolution through the stages of youth (Y), maturity (M) and old age (OA) after a landmass has been rapidly uplifted during a short interval of time followed by protracted stability. Another model has slow uplift of long duration which shortens the youthful stage and extends the mature stage (redrawn from Lobeck, 1939, p. 162). B. Changes in slope profiles from youth to old age during the geographical cycle (redrawn from Young, 1972, Fig. 8, based upon Davis, 1912).

basin as a whole. Graded streams in neighboring basins, alike in stage and volume, may differ in gradient and the basins differ in relief if the rocks are different in resistance (1954, p. 396). The adjustment of drainage, which begins in the early stages of dissection, is most advanced during the mature stage and may degenerate during old age; however, the degeneration is not extensive because the adjustments carry over from one cycle to the next and become progressively more pronounced (1954, p. 343-344).

PENCK'S MODEL

Penck's ideas on landscape development are presented in an incomplete manuscript published in German after his death. The following remarks are based upon the 1953 Czech and Boswell translation of that work. Penck's work is reviewed elsewhere by Davis (1932) and Simons (1962).

The goal of Penck's studies and landform scheme was to deduce the cause and development of crustal movements from exogenetic processes and morphological features, a process he called morphological analysis (1953, p. 6). This goal led him to the development of the concept of relief types or "landform associations" and not to the naive view of parallel slope retreat and of slope shape reflecting tectonism (see Simons, 1962). The appropriate reference system is that the shape of landforms in an area is related to the tectonic activity of that area. The basic premise of this reference system is that the modelling of the Earth's surface reflects the ratio of the intensity of the endogenetic processes to that of the exogenetic displacement of material (1953, p. 11). Penck's writings are devoted to the development of a model for slope development which is based upon the reference system that the shape of a hillslope reflects the relative rates of stream incision and hillslope erosion (Penck, 1953, Chapters 3-6). The basic assumption of Penck's slope reference system is that the gradient of the lowest segment of a hillslope is such that the rate of denudation on it equals the rate of stream incision. Secondary assumptions necessary for the development of Penck's slope model are: (1) The weathering of rock, which renders it more mobile, tends toward uniformity (1953, p. 51-55). (2) Denudation on slopes occurs spontaneously when weathering produces the appropriate mobility for the gradient and at a rate equal to that of weathering (1953, p. 73-74). (3) The intensity of stream erosion determines the gradient of the slopes rising above them and the details of slope form depend upon the character of the rocks (1953, p. 178). (4) Breaks of gradient are base levels which allow the development of the up-gradient segment independent of lower base levels (1953, p. 125-128). (5) A slope unit once developed will recede from the drainage at a constant angle and at a rate determined by the intensity of denudation (1953, p. 135-136).

Penck's temporal-response model to changes in the intensity of stream erosion may be summarized as follows (Fig. 2A): Along the lines of a drainage net fresh slopes of uniform gradient arise to the extent that the streams are incising. The denudation of the slopes is in equilibrium with the intensity of stream erosion such that if the latter changes a new slope necessarily appears with a gradient adjusted to maintain the equilibrium. With increasing intensity of erosion, steeper and steeper slope units arise thus producing convex slope profiles and an increase in relative height (waxing development); with decreasing intensity of erosion, the slope units become flatter and flatter thus producing concave slope profiles and a decrease in relative height (waning development); with a constant intensity of erosion, the slopes maintain the same gradient thus producing straight slope profiles and a constant relative height (uniform development). Between the successively produced sections of slope, breaks of gradient occur which denote the levels to which the surface above each is related. These surfaces and related local base levels migrate away from the drainage at a rate determined by the intensity of reduction and progressively consume the uphill slope units. The succession, one above the other, of slope units with different gradients thus provides a sensitive means of deducing the temporal

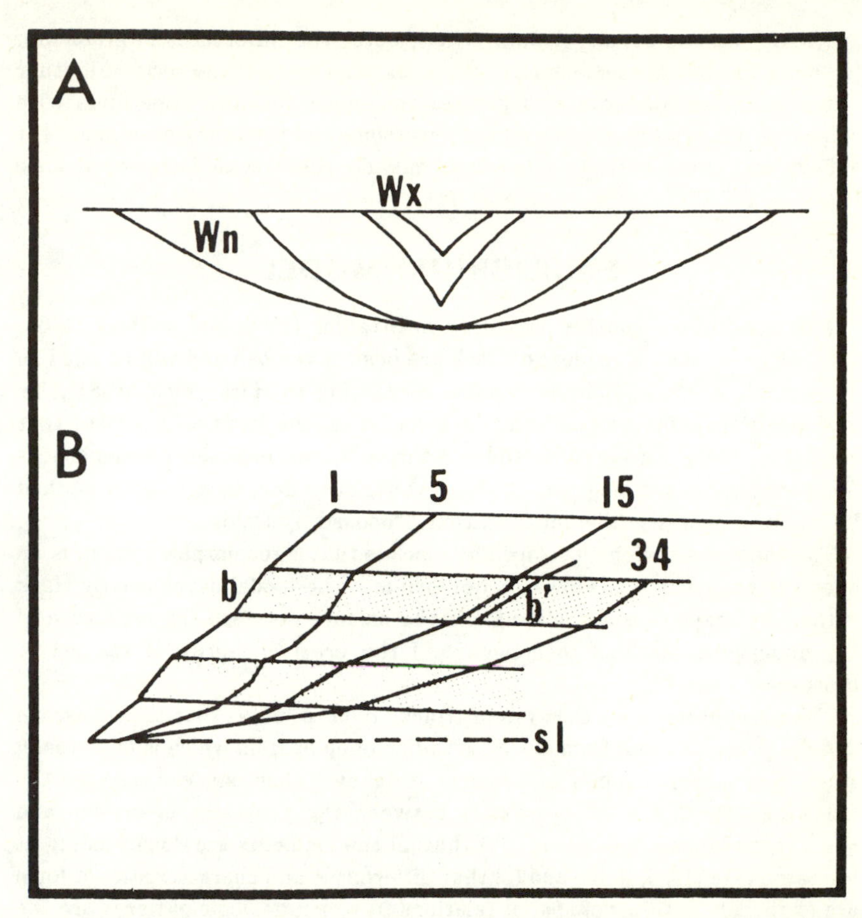

Figure 2. Models of hillslope and landscape development according to Penck. (A) Changes in slope profile as a landscape passes through waxing (Wx) to waning (Wn) development. (B) Development of concave slopes in an area underlain by heterogeneous rocks. Note the effect of the resistant bed (stippled) upon slope gradient (redrawn from Simons, 1962, Figs. 1 and 2).

variations in erosional intensity at a definite place (1953, p. 130-180).

Penck developed his lithologic-adjustment model to a higher degree than did Davis. His model may be summarized as: Rocks of different character have different degrees of resistance to chemical and mechanical destruction. The more resistant rocks cause breaks of gradient (structural base levels) that do not recede upslope and vanish as do the normal slope units, but rather follow the shifting of the rock boundary (Fig. 2B). At the top of the more resistant unit a convex break of slope always occurs and a concave break of gradient occurs at the downhill boundary. As the resistant unit is slowly lowered, slope units above it are produced with flatter and flatter gradients (waning developments). The greater the differences in gradients, the more apparent become differences in the character of the rock. All other things being equal, rocks with greater resistance produce slope units with steeper gradients and a lesser rate of recession, and thus they have narrower valleys and greater relative heights than do rocks with lower resistance (1953, p. 167-177).

EQUILIBRIUM MODEL

The equilibrium model is presented in Strahler (1950) and in Hack (1960, 1965, this volume). The ideas of Hack are best developed and will be used as an example of the equilibrium model. According to Hack (1960, p. 81), his goal was to explain the landforms in a region on the basis of processes that are acting today through the study of the relations between phenomena as they are distributed in space. It was a conscious effort to develop a method of landform analysis different from that proposed by Davis.

The reference system developed by Hack is that a geomorphic system is an open system which tends toward equilibrium. The model developed by Hack is that the shape of landforms reflects the balance between the resistance of the underlying material to erosion and the erosive energy of the active processes.

The basic premise for the system (Hack, 1960, p. 81) is that the landscape and the processes that form it are part of an open system which is in a steady state of balance. From this premise it follows that one can assume the following: (1) that a balance exists between the processes of erosion and resistance of rocks (1960, p. 86), (2) that all the elements are downwasting at the same rate (1960, p. 85) and (3) that differences and characteristics of form are explicable in terms of spatial relations in which geologic patterns are the primary consideration (1960, p. 85). Another assumption, that processes operating today are the processes which shaped the landscape (1960, p. 80), is the test to recognize historical features, which are deposits or landforms out of harmony with the processes of the present (Hack, 1965, p. 64).

Hack (1960, p. 81) does not develop fully a model of change through time in response to changing environmental conditions. He recognizes that changes occur as equilibrium conditions vary but maintains that it is not necessary to assume the kind of evolutionary change envisaged by Davis. However, he does consider variations in relief for several cases of balance between rates of uplift and of erosion: (1) rate of uplift is balanced by rate of erosion — if rates are high, a high relief topography is formed and maintained as long as rates remain constant (1960, p. 86); (2) rate of uplift decreases to zero — relief is decreased and even though the ridge and ravine topography is maintained the ridges may become more rounded (1960, p. 95); (3) rate of uplift increases — relief is increased to maintain rate of erosion (1960, p. 86). Hack maintains that as long as the diastrophic movements are gradual, so that they are balanced by erosive activity, the topography will maintain a state of balance as it evolves from one form to another. He admits that if the diastrophic movement is sudden, relict landforms may be preserved in the landscape until a new equilibrium is achieved (1960, p. 86). Hack (1965, p. 8) paraphrases Davis' ideal geographical cycle in terms of the equilibrium concept and develops a similar evolutionary scheme. An initial disequilibrium stage (youth) of rapid stream incision is followed by an equilibrium stage (mature) wherein the rounded interfluves are lowered as potential energy decreases though they do not change in form.

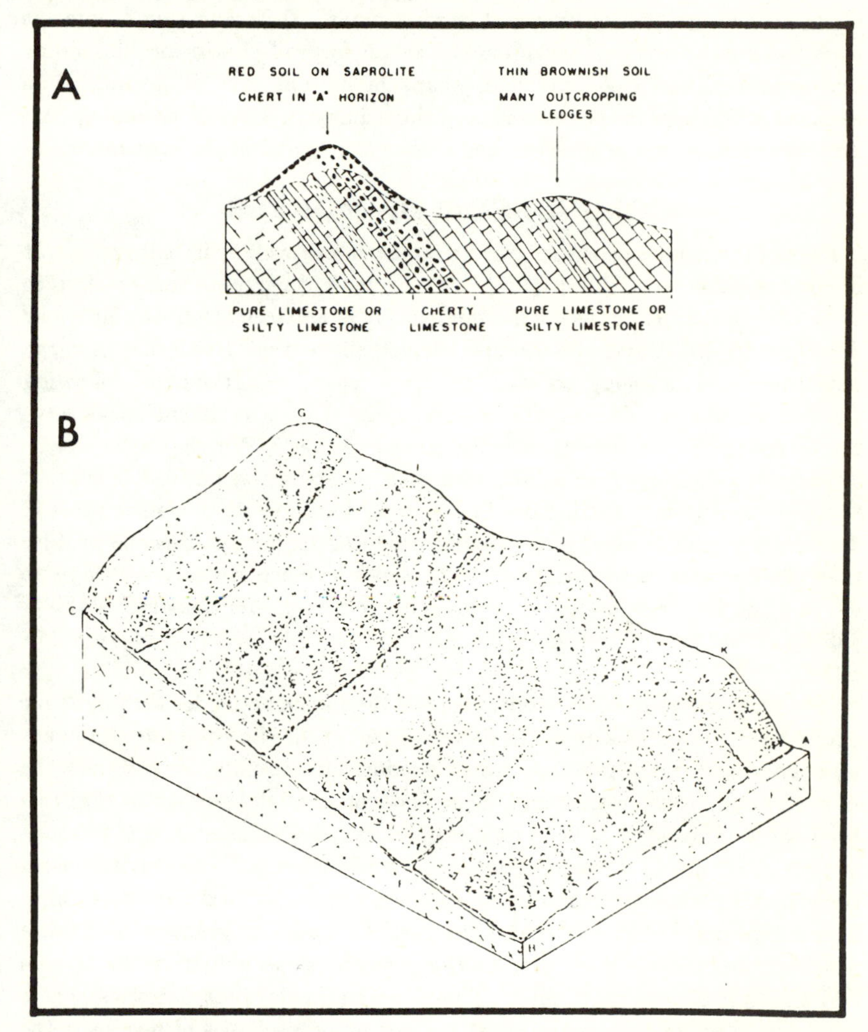

Figure 3. Landform adjustments resulting from dynamic equilibrium within the landscape according to Hack. (A) Relationship between erosional resistance and slope profiles (from Hack, 1960, Fig. 1). (B) Adjustment of interfluve profile and relief to drainage net (from Hack, 1965, Fig. 4).

The adjustment model in Hack (1960, p. 86-91) is well developed. He assumes that a balance exists between the processes of erosion and the resistance of the rocks as they are uplifted or tilted by diastrophism. In part the landforms are adjusted to the rates of diastrophism such that the necessary potential energy is available for erosion to balance uplift. Within this framework, differences in the landscape are related to differences in geology; on resistant rocks, stream and slope gradients as well as relief will be greater than those on nonresistant rocks (Fig. 3); likewise, landforms differ, sharper crests being required on resistant rocks than on nonresistant rocks for the removal of debris at similar rates. Differences in landforms from one area to another, including the relief, form of stream profile, valley cross-sections, width of floodplain, shape of hilltops, are all explicable in terms of differences in the bedrock and the manner in which it breaks up into different components as it is handled on the slopes and in the streams.

COMPARISON OF MODELS

Each of the models described has its opponents as well as its adherents. It is not the purpose of this paper to review their diverse arguments but to show the similarities between the models. It is apparent from the previous descriptions that Davis, Penck and Hack believe that landforms undergo adjustments of lithology as well as time-dependent variations following environmental changes. All the models allow that 1) resistant rocks have landforms developed upon them with steeper slopes and higher relief; 2) all or part of the hillslope is in equilibrium with the stream at its base, such that 3) rapid rates of downcutting produce steeper hillslopes than slower rates of downcutting; and 4) rapid downcutting produces straighter hillslope profiles than does slower downcutting, which allows a more concave hillslope to develop. In all three models, a prolonged interval of environmental stability will allow the progressive lowering of relief to yield a region of very low relief.

It is not possible to completely reconcile the Penckian model with dynamic equilibrium. The Penckian model demands that at the time of its development each slope unit has a gradient which produces an equilibrium between the intensities of stream erosion and hillslope denudation and postulates that this gradient is adjusted to soil and rock properties. The slope unit at the time it is formed is thus in a state of dynamic equilibrium with its environment, because of its constant form during an energy-material flux. However, once a new slope unit begins to develop the resulting break of gradient becomes a functional base level which isolates the previous slope unit from the stream and allows it to become a relict. The Penckian model thus demands that a hillslope consist of a sequence of relict slope units, each of which reflects the dynamic equilibrium conditions extant at its time of origin. According to the Penckian model, the entire landscape can not be in a state of dynamic equilibrium under most conditions but the lowermost slope unit must be in such equilibrium. However, each slope unit is adjusted to lithology and spatial patterns in shape should exist which reflect variations in lithology. They would not, however, reflect the existent energy environment as in the dynamic equilibrium model.

Equilibrium of the landscape in its entirety can be achieved in the Penckian model under two conditions. When the rate of hillslope denudation equals the rate of stream erosion and a straight slope unit extends from stream to divide, a state of dynamic equilibrium exists as the elevation of the region is decreased; but at the same time slope form and relative relief remain constant. A state of static equilibrium is achieved within the landscape when the slope of minimum gradient, established during a prolonged interval of waning development, extends from stream to divide and denudation essentially ceases. The Penckian model merges with the equilibrium model during uniform development and with the Davisian model during waning development.

In this writer's opinion, once the adjustment of landforms to lithology is recognized in the Davisian model, many of the differences between it and the equilibrium model as expressed by Hack (1960, 1965) disappear. The Davisian and equilibrium models can be reconciled in theory but perhaps not in application. The reconciliation is related to the extent of time and space under consideration as suggested by Schumm and Lichty (1965) and Schumm (this volume). No conflict between the models exists when dynamic equilibrium is limited to short durations and small areas such as a hillslope during "present time" and the Davisian model is considered to describe the landform evolution through long intervals of time characterized by stable external conditions and progressively changing internal conditions. The Davisian model is thus a series of dynamic equilibrium states which develop in response to a progressive shift of the equilibrium constant through time. Howard (1965) points out that the dynamic equilibrium model demands numerous changes in the landforms as climatic fluctuations occur and as the distribution of rock units varies during lowering of the land surface.

The merging of the Davisian and equilibrium systems results in part from the recognition that an open system is a complex multivariate phenomenon which responds at different rates to different external variables such that it need not be in equilibrium with all variables at once. As Carson and Kirkby (1972, p. 2-3) point out, it is trivially true that no geomorphic system is completely closed and that Davis' model is an example of "closed system thinking." However, it should be pointed out that when tectonic stability exists, a geomorphic system is partially closed to materials; that is, the initial quantity of material above base level is fixed. Thus the behavior of the system during denudation becomes predictable; a continued decrease in relief must occur. Therefore, a dynamic equilibrium within a drainage basin or a hillslope as a whole cannot exist for material flux except when the rate of uplift equals the rate of denudation as will be demonstrated later in this paper. The imbalance in the material flux means that changes in the size, elevation or form of the system must occur. Even though change occurs in the open system because of an imbalance in the material flux, dynamic equilibrium conditions may exist between the other variables. Thus a landscape may exhibit spatial variations in landforms which are related to geologic patterns and which reflect dynamic equilibrium between process and rock resistance and yet the landscape may be evolving because of a base level constraint. The concept of an open system thus encompasses both Hack and Davis.

Application of the Davisian model to landscape interpretation produces conflict with the Hackian model. Hack (1965, p. 4) views the conflict as one of whether or not the topography of a region can be viewed as having undergone one or more partial geographic cycles which are indicated by dissected peneplains or whether the same topographic relationships can be interpreted in terms of dynamic equilibrium. Bretz (1962) apparently agrees and cites topographic and stratigraphic evidence for the existence of dissected peneplains in the Ozarks which he contends the dynamic equilibrium concept cannot explain.

GEOMORPHIC MODELS AND REALITY

The work of Hack (1957, 1965), Hack and Goodlett (1969), Strahler (1950, 1952), and Melton (1957) among others indicates that stream gradients and densities, hillslope angles, local relief and hypsometric relations can be related to lithology. This body of work suggests that the lithologic adjustments recognized by Davis, Penck and Hack are reasonable.

Studies on developing drainage basins by Schumm (1956) and Carter and Chorley (1961) indicate a landform evolution very similar to that postulated by Davis, and slope behavior similar to that attributed to waxing development by Penck. It should be pointed out that the drainage basins were in the youthful and early mature stages. As the basins evolved into the late youth and late mature stages, such phenomena as parallel slope retreat and a constancy of hypsometric integral suggest that portions of the basin may be approaching a state of dynamic equilibrium. The studies are thus not incompatible with any of the models.

Studies by Ruhe (1950, 1969) of drainage densities and slope angles on Pleistocene surfaces of different ages in Iowa can be interpreted to indicate the achievement of equilibrium in drainage density prior to its attainment for mean slope angle (Fig. 4). Other studies by Ruhe (Ruhe, 1969, p. 88) indicate that the hillslopes developed during this interval contain three partially dissected erosional surfaces and that consequently much of the landscape preserves history; that is, it is not in a state of dynamic equilibrium. These studies by Ruhe suggest that a landscape evolution does occur and that some variables such as drainage density may rapidly approach an equilibrium state while other variables such as hillslopes with relict features remain in a state of disequilibrium.

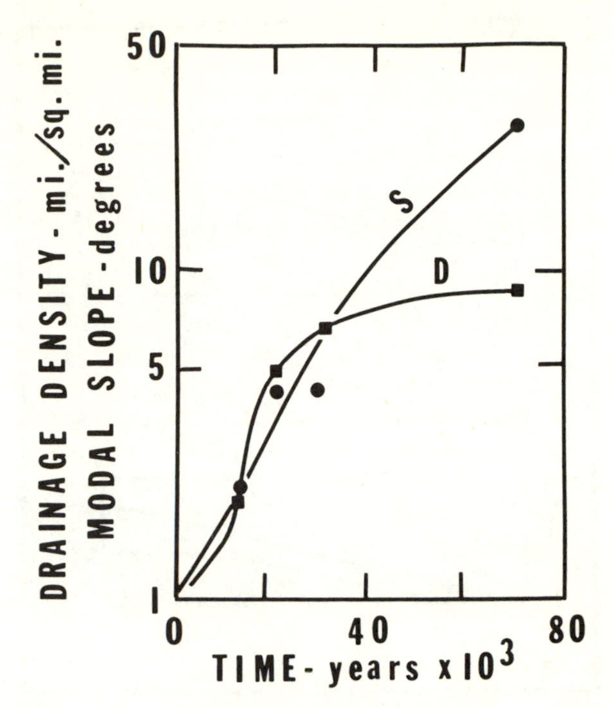

Figure 4. Changes in drainage density and modal slope on drift sheets of different ages in Iowa. Note that drainage density (D) approaches equilibrium conditions after 30,000 years (based on Ruhe, 1969, Fig. 3.13). Modal slope angle (s) does not approach equilibrium (based on Ruhe, 1950).

Process-response modelling by Carson and Kirkby (1972) indicates that the manner of slope evolution is dependent upon climate and lithology (Fig. 5). Note that in Figure 5 stages dc through msm are similar in gross aspects to Penck's waxing development and to the early stages of his waning development; the details of slope development differ; stages dc through cc are similar to the profiles developed by Davis for his youthful and early mature stages (Figure 1B). Stages cc and fs are similar in behavior and in form to Hack's dynamic equilibrium and in form to Davis' mature stage. In its early stages, the model of Carson and Kirkby (1972) does not differ from the descriptions of slopes in the Perth Amboy Badlands (Schumm, 1956). The slope model of Carson and Kirkby is similar to the forms postulated by Davis and by Penck to develop under stable conditions though the postulated mechanics of development may differ.

The analysis of modern rates of orogeny and denudation by Schumm (1963) indicates that a disparity in rates exists. Maximum average rates of uplift exceed maximum average rates of denudation by a factor of 8 (Schumm, 1963, p. 7). He concludes that the disparity in rates makes it unlikely that hillslope form can be used to decipher the Earth's recent diastrophic history;

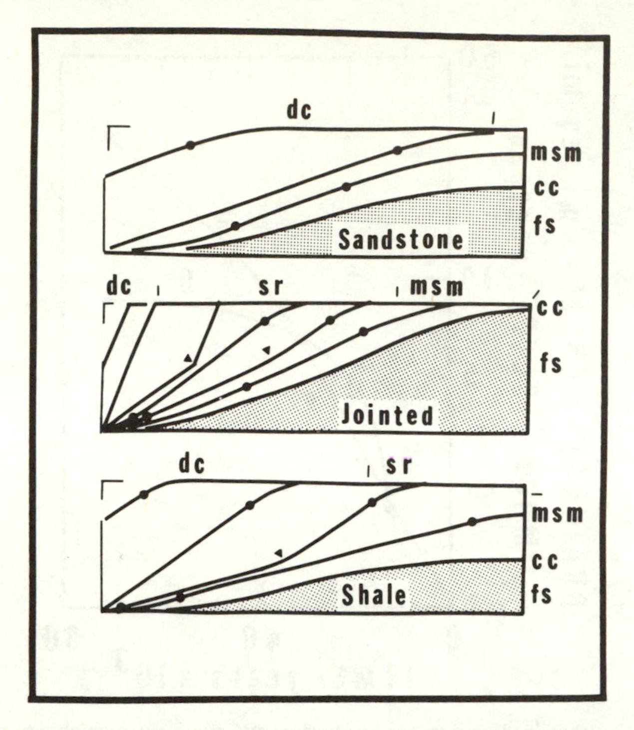

Figure 5. Summary of slope evolution models for various kinds of rocks in a humid temperate environment. Active stream incision (dc) causes development of straight slopes. Following stream stability the initial slopes are replaced by more stable, mantled slopes (sr). The stable main slopes are lost as erosion causes encroachment of concave and convex slopes (msm). Stable slope profile is convexo-concave (cc). This profile is maintained as relief is reduced (fs and stippled). (Modified from Carson and Kirkby, 1972, Figs. 15.7, 15.8, 15.9).

a balance between rates of uplift and denudation will not yield time-independent landforms; and the rapid rates of denudation makes peneplanation a very likely event under past conditions (Schumm, 1963, p. 12).

The cited studies indicate that landscapes do evolve through time at least until the mature stage of Davis and that the younger landscapes contain historical, relict features. Evidence suggests that in the mature stage a constancy of landforms develops which may indicate an approach to dynamic equilibrium. Slope evolution models indicate that Davis' postulated sequence of landforms should develop under conditions of stability and that slope forms do become constant after development of a convexo-concave profile. The observations are compatible with Penck's postulates that straight and convex slopes indicate active stream downcutting and that concave slope profiles indicate stream stability, though his mechanisms of slope unit development and recession are not verified.

A COMPOSITE MODEL

The introduction of systems theory into geomorphology since the formulation of the theories of Davis and Penck requires that they be modified as indicated by Hack's arguments. However, aspects of Davis' and Penck's reference systems as well as models appear valid today. A merger of the three systems thus seems desirable.

Only two premises are necessary to produce a reference system which allows both for landform evolution and dynamic equilibrium. These are: 1) geomorphic systems are multivariate open systems which tend toward a steady state equilibrium, and 2) the mass of rock existing above base level constitutes an external variable to which the system is in constant disequilibrium. From the first premise the following can be deduced about the system: 1) the internal form and processes are controlled by external factors such as diastrophism and climate as modified by secondary and feedback responses such as microclimate and weathering; 2) each internal variable has a characteristic response to disequilibrium conditions as reflected in its initial resistance to change and its rate of adjustment toward equilibrium; 3) each internal variable will tend to approach equilibrium at a rate proportional to its distance from the equilibrium value; 4) most internal variables in a system will be in a near state of equilibrium to slow changes of minor magnitude in the external variables, and the system will contain few relict features; likewise, many variables in a system will be in disequilibrium to rapid changes of great magnitude in the external variables, and the system will contain many relict features.

The concept that the mass of rock above base level constitutes a disequilibrium condition may be derived from considerations of potential energy and material flux. Base level beneath land may be conceived of as a surface gently rising from sea level which everywhere has a gradient just sufficient for the stream removal of load. Beneath interfluves base level is warped upward toward the divides with a gradient enabling slope processes to just remove the load. At base level the potential energy of position is no longer convertable into kinetic energy for erosion. Given this concept of base level, any mass of rock rising above base level is in a zone wherein erosion can occur. Thus a constant tendency toward relief reduction exists in areas above base level.

In an open system both energy and materials may pass through the system. The system will be defined as a drainage basin, bounded laterally by a vertical surface through the divides, at the bottom by base level and at the top by the land surface (Fig. 6). The discussion will be restricted to the flux of rock materials through this drainage basin and will ignore the flux of water, ice, air and dust. The inflow of rock material to the basin can only be through the diastrophic uplift of material across base level. The outflow of rock from the basin is by fluvial erosion and transport through the basin

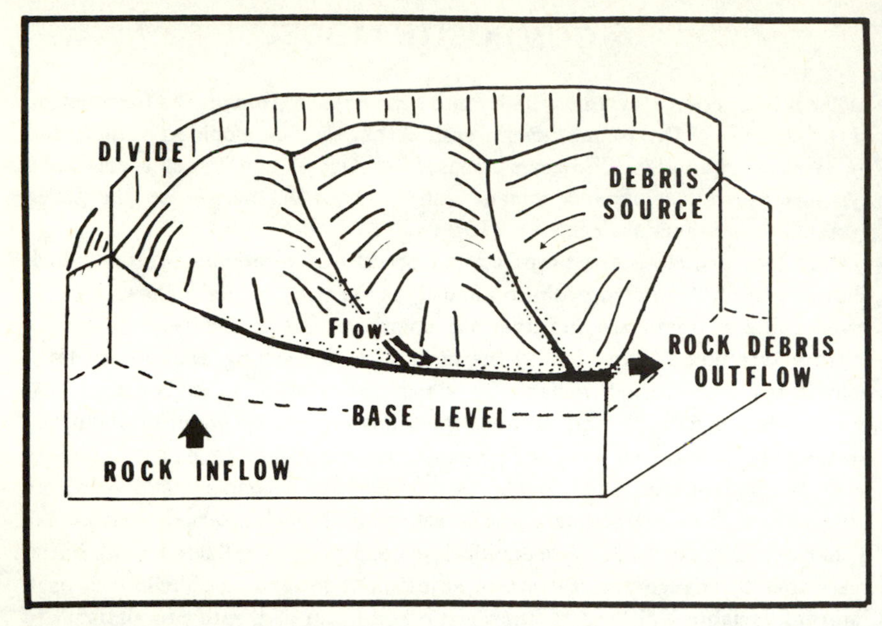

Figure 6. Definition of the boundaries of a drainage basin when considered as part of an open system.

mouth. For a dynamic equilibrium to exist such that a constant geometry is maintained within the basin, inflow must equal outflow. The only way to balance the flux is for the mass of rock uplifted across base level to equal the mass of rock leaving the basin through its mouth. Whenever an imbalance in this flux exists, the relief within the drainage basin must change. Thus the material budget of a drainage basin controls the evolution of landforms within that basin. A positive budget constantly renews or increases potential energy which results in stream incision and a consequent increase in relief and slope gradient, whereas a negative budget has the opposite effects (Fig. 7).

The basic premises cover many of the important basic assumptions in the reference systems of Davis, Penck and Hack. The assumption that geomorphic systems are open systems includes all the premises of Hack's reference system and many of the assumptions of Davis and Penck. The assumption that the mass of rock above base level constitutes a disequilibrium condition which necessitates a continuous change in landforms allows for the evolution of landscapes as proposed by Davis and allow consideration of Penck's postulate that the intensity of stream erosion determines the gradient of the slopes rising above it as modified by rock characteristics. The inclusion of climate as an external variable accommodates the objection of Garner (1974, p. 17). The acknowledgement that rates of adjustment of the different internal variables are not uniform allows for the preservation of relict

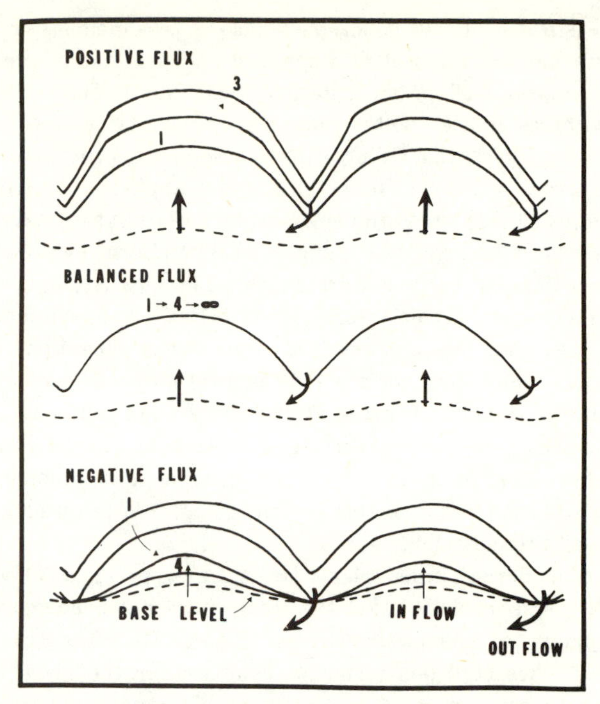

Figure 7. Hypothetical cross-section showing changes in relief and form within a drainage basin under different mass budgets wherein net flux = inflow - outflow.

features for varying lengths of time.

The generation of a landscape model requires assumptions as to the character of the external variables. An evolutionary scheme very similar to that of Davis' ideal cycle can be generated in systems terms of rapid uplift followed by diastrophic and climatic stability are assumed; Hack (1965, p. 8) has developed this model. Other evolutionary models will result from different assumptions as to the character and rates of change of the external properties and the kinds of internal secondary and feedback responses. In any model the change in the relief of the system will be related to the flux of rock material through the system. The shape of the landscape as it evolves in response to the assumed disequilibrium conditions will be a function of the kind of processes operating and the characteristics of the rock within the system. The landscape will reflect climate to the extent that it influences the effectiveness of the various exogenetic processes and thus influences the mass budget as well as slope form.

THE PENEPLAIN PROBLEM

Conceptually, no reason exists why, during a sufficiently long period of tectonic stability, the landforms in a stable region should not evolve to produce the surface of low relief known as a peneplain. Schumm (1963) indicates that modern rates of diastrophism and denudation are compatible with the assumptions of the ideal geographical cycle. The concept of plate tectonics with its mobile belts separating large stable areas is likewise compatible and suggests that Gilluly's (1949) assertion of continuous diastrophic activity is correct worldwide but certainly not applicable to any particular region. The previously cited studies on expanding drainage basins (Schumm, 1961; Carter and Chorley, 1961) suggest that landform evolution occurs at least as far as Davis' early mature stage. The discussion of systems by Wilson (in Harris and Twidale, 1968), Howard (1965), and Schumm and Lichty (1965) indicates that the existence of present-day conditions of dynamic equilibrium in a region does not preclude long-term evolution of the landforms as long as the environment remains stable. As Wilson points out, with stable external conditions (tectonism, climate) a continuous change in the internal conditions (base level restraint) should produce a negative allometric growth, or landform evolution (see Bull, this volume). A proper question is, therefore, where have all the peneplains gone?

The lack of modern day peneplains is not a new problem; Davis faced the same question and concluded that the Quaternary has been too unstable for the development of peneplains (Davis, 1954). He cites the widespread distribution of dissected peneplains as evidence for the geographical cycle having gone to completion in the more stable environments of the past (Davis, 1954, p. 363-374). Most recently, Harris and Twidale (1968) repeat the same argument. It is difficult to refute that the Quaternary is an environmentally unstable period. Garner (1974, p. 37) calculates that 87 to 111 climatic changes occurred during the Pleistocene. Fairbridge (1961) describes the repeated sea-level fluctuations of several hundred meters during the Pleistocene. King (1965) reports on the Pleistocene tectonism in the United States. The important question then becomes, how does one recognize a dissected peneplain?

Davis, in his 1899 response to Tarr's assertion that there exists in the highlands of New Jersey ". . . a very evident general sympathy between the present topography and the rock texture", reaffirms that the best evidence for a dissected peneplain is the indifference of the peneplain to the various structures it systematically truncates (Davis, 1954, p. 270, 360). A more detailed set of criteria is presented in Thornbury (1969, 182-185), Trowbridge (1921) and in Bretz (1962). The lack of structural and lithologic influences on the peneplain appears in Davis' writings many times. Is this a valid assertion?

The contention by Davis and many of his adherents that a peneplain is indifferent to structure is not valid. A peneplain must be a structurally controlled surface. Both the reasoning presented in Davis' writings and the merger of the dynamic equilibrium concept with Davis' cycle demand that it be so. The following quote from an 1896 article of Davis (1954, p. 343-355) indicates why a peneplain must be structurally controlled.

> "... the processes of spontaneous adjustment of streams to structures ... involves the adjustment of many subsequent streams to the weaker structures and the shifting of many divides to the stronger structures. Adjustment begins in the early stages of dissection, advances greatly in the mature stages, and continues very slowly toward old age, while the relief is fading away. Indeed, when the region is well worn down, some of the adjustments of maturity may be lost in the wanderings of decrepitude, but this will seldom cause significant loss of adjustment except in the larger rivers. Now, if a region thus base-leveled, or nearly base-leveled, is raised by broad and even elevation into a new cycle of geographical life, the rivers will carry the adjustments acquired in the first cycle over to the second cycle. Still further adjustments may then be accomplished. The master streams will increase their drainage area in such a way that the minor streams will seldom head behind a hard stratum. In a word, the drainage will become more and more longitudinal, and fewer and fewer small streams will persist in transverse courses ... It should be further noted that in the early stages of the second cycle the residual reliefs of the first cycle will still be preserved on the uplands and that they will be systematically related to the streams by which the dissection of the upland is in progress ..." (Davis, 1954, p. 343-344).

Note that the streams and divides are structurally (lithologically) controlled and that most of his control is not lost during peneplanation. Some of the larger streams may lose their structural control but the divides do not. Later, in the 1899 response to Tarr previously mentioned, Davis (1954, p. 359-360) states, after noting that the weaker rocks will be worn down below the inferred level of the dissected peneplain:

> "... It is a matter of necessity that the present topography of an uplifted and dissected peneplain should exhibit sympathy between form and structure, for where should better accordance of form and structure be expected than in such a region of adjusted drainage ...".

It is evident from these two quotes that Davis expected that the preserved portion of a dissected peneplain should be lithologically controlled, and that these remnants would be the former interfluves. Why then the assertion by Davis that a peneplain "is indifferent to structure"? Penck (1953, p. 169) comes to the same conclusion; it is not unique to Davis. Perhaps Davis means that structural differences become less apparent to the eye; or in other words, as Penck points out (Penck, 1953, p. 15), the relief variations due to lithologic differences become less as the local relief and slope gradients decrease. If this is the case, the inferred peneplain surface could bevel structures; that is, be indifferent to structures, and yet show higher relief on

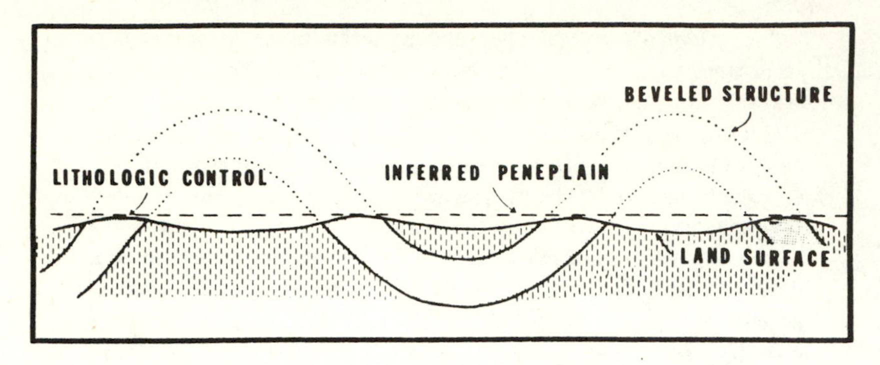

Figure 8. A hypothetical lithologically controlled peneplain beveling structure as required by Davis.

the more resistant rocks than on the less resistant rocks and have the divides on more resistant rocks and the valleys on less resistant rocks (Fig. 8).

The larger streams on a peneplain are postulated to lose their structural control (Davis, 1954, p. 269), which is a postulate consistent with Davis' concept of a graded stream in which lateral planation dominates. It is also a convenient phenomenon to use as an explanation for streams crossing resistant ridges by way of superposition. Davis (1954, p. 443-445) pictures local superposition from the thick alluvium deposited by the stream. However, Davis is not consistent in his picture of a partially planated peneplain surface, for in his 1912 textbook he shows development of a peneplain whereon the valleys are narrow and separated by broad, low convex divides (Fig. 1B). However, the 1912 version is not compatible with his later 1930 version. It is not illogical to assume that even though the channels of the larger streams may locally lose detailed structural control, their valleys are for the most part so controlled. This must be true if the adjacent divides are structurally controlled.

Given the view of a structurally-lithologically controlled peneplain what evidence does one have to identify a dissected peneplain? Two kinds of dissected peneplain can be recognized; individual remnants preserved on level-crested interfluves and those inferred "statistical" surfaces preserved by summit accordance. The statistical peneplains are easily explained by dynamic equilibrium. The landscape preserving a statistical peneplain is usually in the mature stage of the current erosion cycle. It is during this stage that structural adjustments are best developed. Therefore, there should be no relict surfaces to identify, and the accordancy should be explicable by the lithologic adjustments of the current cycle. Whatever accordancy was inherited from the previous cycle should be similar to that developed during the current cycle. Thus any peneplain surface preserved by the statistical accordancy of summits should be unrecognizable. A statistically preserved peneplain is thus an example of the equifinality inherent in an open system. Hack (1965) concludes that accordancy of summits, which has been used as evidence for the Schooley and Harrisburg Peneplains, is accounted for by dynamic equilibrium and that no evidence thus exists for these surfaces.

Peneplains preserved on even-crested interfluves should be more recognizable. The evidence cited by Bretz (1962) for the Ozarks pertains in large part to this type of peneplain preservation. If we consider Bretz's evidence as typical, several problems of peneplain recognition exist. Some of the evidence is compatible with the assumption of a current condition of dynamic equilibrium, some is incompatible with the concept of a lithologically adjusted peneplain, and some is inconsistent with the concept of a peneplain.

The recognition that a peneplain, as described in Davis' writings and as required by the merging of dynamic equilibrium and the geographical cycle, is a lithologically and structurally controlled surface invalidates most of the evidence cited for dissected peneplains. Evidence for a lack of structural and lithological control on upland surfaces and for unadjusted drainage is as incompatible for relict peneplain remnants as it is for modern surfaces in dynamic equilibrium.

SUMMARY

This paper has shown that the concepts of Davis, Penck and Hack are more compatible than is apparent in many comparisons of their work. The paper concludes that the reference systems of Davis, Penck and Hack are compatible and may be merged to developed models of landscape evolution using, as a reference system, the concept of an open system. The end product of landscape evolution under stable external conditions, the surface of low relief known as a peneplain, is a structurally-lithologically controlled surface which upon rejuvenation of the area becomes difficult to recognize after the Davisian stage of youth is passed.

The merging of the reference systems is facilitated by the recognition that the models of Davis and Penck include the lithologic and structural control of landforms and that the model of Hack allows for evolution of landscapes during changes of equilibrium conditions. The merging of the reference systems produces a landscape evolution model which has a tendency toward relief reduction occurring as long as rock persists above base level combined with a tendency for erosional slopes to be adjusted to lithology as landform evolution continues. Within the model, hillslope gradients and form as well as local relief change as the budget between rock mass entering and leaving the system (crossing base level) changes; forms similar to those of waxing development characterize a positive budget and forms similar to those of waning development characterize a negative budget with, at all times, modification in form resulting from lithologic adjustments.

This paper does not concern itself with the details and mechanics of slope evolution or of drainage basin evolution. It concedes that many of the details in Davis and Penck are not longer considered valid, but maintains that the invalidity of a detail does not render either the model or the reference system invalid.

REFERENCES CITED

Bretz, J H., 1962, Dynamic equilibrium and the Ozark land forms. Am. Jour. Sci., v. 260, p. 427-438.

Carson, M. A., and M. J. Kirkby, 1972, Hillslope Form and Process. Cambridge University Press, Cambridge, 475 p.

Carter, C. A., and R. J. Chorley, 1961, Early slope development in an expanding stream system. Geol. Mag. v. 98, p. 117-30.

Chorley, R. J., 1962, Geomorphology and general systems theory. U. S. Geol. Sur. Prof. Paper 500-B, 10 p.

Davis, W. M., 1909, The systematic description of landforms. Geog. Jour. v. 34, p. 300-314.

———————, 1930, Rock floors in arid and in humid climates, Jour. Geol. v. 38, p. 1-27, 136-158.

———————, 1932, Piedmont benchland and Primarrumpfe. Bull. Geol. Soc. Am. v. 43, p. 399-440.

———————, 1954, Geographical Essays. (Ed. by D. W. Johnson). Dover, New York, 777 p.

Fairbridge, R. W., 1961, Eustatic changes in sea level, *in* Ahrens, L. W. and others (eds). Physics and Chemistry of the Earth, v. 4, p. 99-185.

Flemal, R. C., 1971, The attack on the Davisian system of geomorphology: a synopsis. Jour. of Geol. Ed., v. 19, p. 3-13.

Garner, H. F., 1974, The Origin of Landscapes: A Synthesis of Geomorphology. Oxford Univ. Press. New York. 734 p.

Gilbert, G. K., 1877, Report on the geology of the Henry Mountains. U. S. Geog. and Geol. Survey. Washington, 160 p.

Gilluly, J., 1949, Distribution of mountain-building in geologic time. Bull. Geol. Soc. Am. v. 60, p. 561-90.

Hack, J. T., 1957, Studies of longitudinal stream profiles in Virginia and Maryland. U. S. Geol. Sur. Prof. Paper 294-B, p. 49-97.

———————, 1960, Interpretation of erosional topography in humid temperate regions. Am. Jour. Sci. v. 258A, p. 80-97.

———————, 1965, Geomorphology of the Shenandoah Valley, Virginia and West Virginia and origin of the residual ore deposits. U. S. Geol. Sur. Prof. Paper 484, 84 p.

———————, and J. C. Goodlett, 1960, Geomorphology and forest ecology of a mountain region in the Central Appalachians. U. S. Geol. Sur. Prof. Paper 347, 66 p.

Harris, S. A., and C. R. Twidale, 1968, Geomorphic Cycles, *in* Fairbridge, R. W. (ed), 1968, Encyclopedia of Geomorphology: the Encyclopedia of Earth Science, v. 3. Reinhold, New York, 1295 p.

Horton, R. E., 1945, Erosional development of streams and their drainage basins: hydrophysical approach to quantitative morphology. Bull. Geol. Soc. Am. v. 56, p. 275-370.

Howard, A. D., 1965, Geomorphical systems — equilibrium and dynamics. Am. Jour. Sci., v. 263, p. 302-312.

King, P. B., 1965, Tectonics of Quaternary Time in Middle North America, *in* Wright, H. E., and D. G. Frey (eds.) The Quaternary of the United States,

Princeton Univ. Press. Princeton. 922 p.
King, L. C., 1962, The Morphology of the Earth: A Study and Synthesis of World Landscapes. Hafner. New York. 699 p.
Leopold, L. B., M. B. Wolman, and J. P. Miller, 1964, Fluvial Processes in Geomorphology. Freeman, San Francisco. 522 p.
Lobeck, A. K., 1939, Geomorphology: An Introduction to the Study of Landscapes. McGraw-Hill. New York. 731 p.
Melton, M. A., 1957, An analysis of the relations among elements of climate, surface properties, and geomorphology. Technical Rept 11, Proj. NR 389-042, ONR, Columbia Univ., 102 p.
Penck, W., 1953, Morphological Analysis of Landforms, (trans. by H. Czech and K. C. Boswell), MacMillan, London. 429 p.
Ruhe, R. V., 1950, Graphic Analysis of drift topographies. Am. Jour. Sci. v. 248, p. 435-443.
———————, 1959, Quaternary Landscapes in Iowa. Iowa State Univ. Press. Ames. 255 p.
———————, 1975, Geomorphology: Geomorphic Processes and Surficial Geology. Houghton Mifflin. Boston. 246 p.
Schumm, S. A., 1956, Evolution of drainage systems and slopes on badlands at Perth Amboy, New Jersey. Bull. Geol. Soc. Am. v. 67, p. 597-646.
———————, 1963, The disparity between present rates of denudation and orogeny. U. S. Geol. Sur. Prof. Paper 454-H, 13 p.
———————, R. W. Lichty, 1965, Time, space and causality in geomorphology. Am. Jour. Sci., v. 263, p. 110-19.
Simons, M., 1962, The morphological analysis of landforms, a new review of the work of Walter Penck (1888-1923). Trans. Institute of Brit. Geog., v. 31, p. 1-14.
Strahler, A. N., 1950, Equilibrium theory of erosional slopes approached by frequency distribution analysis. Am. Jour. Sci., v. 248, p. 673-96, 800-14.
———————, 1952, Hypsometric analysis of erosional topography, Bull. Geol. Soc. Am., v. 63, p. 1117-42.
Tanner, W. F., 1968, Equilibrium in Geomorphology, *in* Fairbridge, R. W. (ed), 1968, The Encyclopedia of Geomorphology: Encyclopedia of Earth Science, v. 3. Reinhold, New York, 1295 p.
Thornbury, W. D., 1969, Principles of Geomorphology, Wiley, New York. 594 p.
Trowbridge, A. C., 1921, The erosional history of the Driftless Area. Iowa Univ. Studies, Studies in Natural History, v. 9, no. 3, 127 p.
Woldenberg, M. J., 1968, Open systems — allometric growth, *in* Fairbridge, R. W. (ed), 1968, The Encyclopedia of Geomorphology: Encyclopedia of Earth Science, v. 3. Reinhold, New York. 1295 p.
Young, A., 1972, Slopes. Oliver and Boyd. Edinburgh. 288 p.

CHAPTER 10

CONCEPT OF THE GRADED STREAM

James C. Knox

ABSTRACT

The concept of grade has implicit within it the notion that the functional relationship between discharge and sediment load can remain relatively constant over a fixed period of time. Hydraulic geometry studies have shown that interactions among hydraulic variables are similar in both graded and ungraded streams and that stream channels adjust quickly to environmental constraints. W. M. Davis' concept of a "youthful stage characterized by disequilibrium" is wrong. Most streams are graded in the traditional sense of the definition; ungraded streams should probably be defined as those undergoing relatively rapid morphologic changes involving longitudinal profiles and/or channel cross section characteristics. Using "rapid morphologic change" as the basis for defining ungraded streams results in a reassignment of the relative importance of factors traditionally described as responsible for ungraded streams. For example, tectonic and eustatic events, which are relatively low intensity and long duration, compared to lag response times of channels, are probably associated with an ungraded condition only after the cumulative effects of slowly changing environmental conditions result in the breaking of a threshold of channel stability. Climate change and man's activities, on the other hand, frequently have strong direct impacts on the functional relationship between discharge and sediment yield and are major factors causing ungraded channels. While the concept of grade in the Davisian cycle of erosion is not supported by present knowledge of stream processes, it is true that long periods of relative stability have occurred in natural channels. Discontinuities between stable periods are to a large degree related to relatively abrupt changes of climate on Holocene and longer time scales.

INTRODUCTION

The concept of grade, as applied to stream systems, indicates that stream channels evolve morphologic properties which represent a state of balance between erosion and deposition. Since the concept was widely introduced into geomorphology by W. M. Davis near the end of the nineteenth century, much debate has ensued regarding its validity. This paper represents an attempt to trace key developments in the evolution of the concept and to show that the concept differs from other stream equilibrium concepts. It is suggested that the concept of grade extends beyond other equilibrium concepts, (e.g., the concept of quasi-equilibrium, because grade implies morphologic stability over time, a condition which is not essential to other equilibrium concepts that express the self adjustment tendencies of natural channels). A principal fault of the concept, as originally introduced by W. M. Davis, relates to its implication that graded streams are initially achieved through a long-term (geologic time) evolutionary process. It is now understood that morphologic evolution takes place relatively rapidly. Therefore, it is suggested that a graded stream be defined as one in which the relationship between process and form is stationary and the morphology of the system remains relatively constant over time. The revised definition changes the relative importance of factors traditionally viewed as responsible for the occurrence of ungraded streams. Climate change and anthropogenic factors rival the importance of base level adjustments as causes for ungraded streams. Because the influence of climate change on fluvial episodes has been characterized by persistent controversy, special attention is given to its role as a cause of ungraded channel conditions.

HISTORICAL DEVELOPMENT OF THE CONCEPT

It has long been recognized that streams tend to develop morphologic characteristics which approximate a type of stability based upon prevailing hydrologic conditions. The concept of the graded stream was widely introduced into geomorphology by W. M. Davis in 1899 and 1902, but understanding of the principle that streams tend to develop a "slope of equilibrium" was recognized by engineers at least as early as the late seventeenth century. Evolution of the equilibrium principle, as related to the concept of grade, can be divided into three stages which include: 1) a pre-Davisian period where developments were primarily oriented to the study of contemporary stream processes, 2) a period of Davisian dominance where applications of the concept were primarily in the context of natural landscape evolution emphasizing morphologic characteristics of stream systems, and 3) a post-Davisian re-emphasis of fluvial processes. A key turning point from cyclical Davisian applications to the current re-emphasis on fluvial processes is Leopold and Maddock's (1953, p. 51) introduction of hydraulic geometry to describe the tendency of all stream morphologies, graded or ungraded, to undergo adjustments among the hydraulic variables which reflect the inputs of discharge and sediment. The following three subsections present a generalized breakdown of the evolution of the concept of grade. No attempt is made to provide a complete review of all earlier studies involving the concept of grade because such reviews are available in the papers by Baulig (1926), Kesseli (1941), Mackin (1948), Woodford (1951) and Dury (1966). However, key historical studies have been selected to show that there has long been an understanding of the basic principle that stream channels tend to develop a stable morphology that is based on prevailing discharge and sediment load characteristics. It will also be shown that a principal cause of debates about the usefulness of the concept of grade is related to meaning of "balance between erosion and deposition".

The Concept of Equilibrium Before Davis.

The earliest mention of an equilibrium concept as applied to streams appears to be the work of Guglielmini, an Italian engineer active in the late 1600's. Guglielmini, in 1697 (quoted in Rouse and Ince, 1957, p. 71) made the important observation that:

> "A stream with sufficient velocity scours its bed, and with the increase in depth the slope is lessened; and late in its motion, if it runs turbid, the stream will deposit sediment on the bed. Hence I can conceive of no other reason to seek what slope would be necessary for a stream than to be certain that it would not cover its bed with deposits, or, if the slope were greater than necessary, that it would not scour excessively.
>
> It is certain that a stream widens and deepens in proportion to the violence of the motion which erodes and carries away the earth that forms its sides and bottom; it is therefore necessary that the scouring force be greater than the resistance of the earth or other material that forms the bed, because otherwise, if the one were equal to the other, there would be no excavation. . . . It is always necessary to say that in the scouring process of a stream either the force of the water gradually decreases or the resistance of the soil increases . . . until some sort of equilibrium is reached."

The equilibrium principle was again expressed in Surell's classic *Etude sur les torrents des Hautes-Alps* published in 1841. Surell made the important observation that all streams tend to develop a "limiting gradient" as determined by the type of transported materials. Surell [as quoted by Kesseli (1941, p. 567-568)] suggested a three stage process by which a river established a stable regime, which included:

> "1) A period of corrasion and upbuilding which prepares the bottom of the valley and everywhere distributes gradients which bring an equilibrium between the resistance of the ground and the friction of the water; 2) A period of divagation, whereby the water seeks the form of section and the bends which correspond to the greatest stability; for the straight course is not always the

> most stable, firstly because it has a steeper gradient, and secondly because it does not necessarily lead the current toward the points where the bank is most solid. The action of the water is restricted to the shifting of a badly defined bed within one and the same plane, without noticeable wear or accretion of the bottom [of the valley]; it is the liquid mass rather than the ground which changes location. The result of this second period consists in fixing the alignment of the course, or, in other words, in determining its plan; and 3) Finally, a period of regimen, during which the water overflows and returns into an invariable bed."

Following Surell, Dausse, a French engineer, published important papers in 1857 and 1872 stressing equilibrium concepts which are related to stream channel morphology. Dausse, for example, in 1872 (quoted in Kesseli, 1941, p. 568) stated:

> ". . . the gradient of the stream and the form of each of its sections are the result of an equilibrium between the force of the current and the resistance of the materials of the river bed. . . ."

Another major milestone in the recognition of equilibrium tendencies in streams which later served as a basis for the concept of grade was the research of Gilbert, much of which is included in his classic on the *Geology of the Henry Mountains* published in 1877. A few of the relevant statements of Gilbert include:

> (on page 111) "Where a stream has all the load of a given degree of comminution which it is capable of carrying, the entire energy of the descending water and load is consumed in the translation of the water and load and there is none applied to corrasion. If it has an excess of load its velocity is thereby diminished so as to lessen its competence and a portion is dropped. If it has less than a full load it is in condition to receive more and it corrades its bottom. A fully loaded stream is one the verge between corrasion and deposition . . . it may wear the walls of its channel, but its wear of one wall will be accompanied by an addition to the opposite wall. The work of transportation may thus monopolize a stream to the exclusion of corrasion, or the two works may be carried forward at the same time."

and,

> (on page 126) "It has been shown in the discussion of the relations of transportation and corrasion that downward wear ceases when the load equals the capacity for transportation. Whenever the load reduces the downward corrasion to little or nothing, lateral corrasion becomes relatively and actually of importance."

The quotations from Guglielmini, Surell, Dausse, and Gilbert reveal attempts to characterize the multivariate nature of fluvial processes that produce a stable channel system. While special significance is given to recognition of the slope of the equilibrium channel, the authors stress that channel cross sections are also involved in the development of a condition of equilibrium. In contrast, when Davis (1899, 1902) later introduced the concept of grade, principal emphasis was given to the development of the longitudinal profile.

Davis' Impact on the Concept of Equilibrium.

Davis, in his important papers of 1899 on *The Geographical Cycle* and of 1902 on *Base-Level, Grade, and the Peneplain*, provided a wide introduction of the concept of grade in North America and Europe. Davis (1902, pp. 89-90) attributed his adoption of the term grade to a suggestion from G. K. Gilbert. Davis (1902, pp. 89-90) stated that Gilbert had introduced the term grade as early as 1876, that W. J. McGee had employed the term in 1891, and that he adopted the term in 1894. However, Davis' definition of grade was different from previous usage. Davis (1899; 1902) envisioned a graded stream as one in which a balance between erosion and deposition was attained in the "mature" stage of the "geographical cycle of landscape evolution" and persisted through the "old age" stage of an uninterrupted cycle. Whereas the earlier workers cited the recognized equilibrium between channel form and contemporary processes, Davis suggested that a period of geologic time was required for the first attainment of the equilibrium profile. The shift of the emphasis to the scale of geologic time is evident in the following quotations from Davis' (1902, pp. 86-87) publication:

> "... we find that the balance between erosion and deposition, attained by mature rivers, introduces one of the most important problems that is encountered in the discussion of the geographical cycle. The development of this balanced condition is brought about by changes in the capacity of a river to do work, and in the quantity of work that the river has to do. The changes continue until the two quantities at first unequal, reach equality; and then the river may be said to be graded, or to have reached the condition of grade.
>
> The condition of grade must not be confused with the limiting undersurface of erosion, with respect to which the graded condition is developed; the name for this surface is 'base-level.' Nor must it be confused with the stage in the history of river development in which the graded condition is reached; 'maturity' is the name for that stage but it may be noted in passing that the graded condition persists all through the old age as well as maturity of an uninterrupted cycle. 'Grade', meaning a condition of balance, must not be confused with the same word used with

> another meaning, namely, the slope or declivity of the river when the graded condition is reached; for 'grade', meaning slope, varies in place and in time; while 'grade', meaning balance, always implies an equality of the two quantities."

Davis' publications on the concept of grade received favorable acceptance by many geomorphologists of the day because his application of the concept to a geologic time scale conformed with concept of natural landscape evolution. The application of the equilibrium principle to a geologic time scale resulted in confusion regarding the meaning of "balance between erosion and deposition". One of the first major criticisms of the concept of grade was that of Kesseli (1941, p. 587) who argued:

> "Balance between the power of a stream and its load is maintained neither during floods nor during stages of low water, neither along the entire length of a stream nor along limited stretches, neither in youth, maturity, nor old age. River action consists of a combination of erosion, transportation, and deposition, all three of which are likely to take place simultaneously in any river section."

Although Davis (1899, pp. 488-490; 1902, p. 87) defined a graded stream as one which evolved in the "mature stage" and persisted, in an uninterrupted cycle, through the "old age stage", his criteria for grade were not as rigid as many of his critics imply. For example, Davis' (1902, pp. 95-98) incorporation of statements of others with his own revealed that he did not interpret grade to mean a perfect balance between erosion and deposition, for he indicated (pp. 97-98) that the graded stream:

> ". . . continually alters its inclination. There is a slow departure from equilibrium, and there is a closely following readjustment toward recovery of it."

and,

> ". . . if a stream be deprived of the greater part of its load by some abnormal changes, it would at once attack its former slope of equilibrium, and rapidly, though at a progressively slower rate, lower it. On the other hand, its load largely increased, the stream would rapidly build up the slope. In either case, however, it would come to a stand at a new grade of equilibrium.
>
> The conception of grade must therefore include the conception of different and changing slopes in large and small streams, in mature and old streams, in streams dissecting weak and strong rocks, in streams of arid and humid regions."

The above quotations from Davis suggest that the narrow view of "balance between erosion and deposition" by Kesseli was an incorrect interpretation of Davis' definition of grade. A few years after Kesseli's attack, Mackin (1948,

p. 471) attempted to clarify the meaning of "balance between erosion and deposition" as related to the concept of grade whereby he defined a graded stream as:

> ". . . one in which, over a period of years, slope is delicately adjusted to provide, with available discharge and with prevailing channel characteristics, just the velocity required for the transportation of the load supplied from the drainage basin. The graded stream is a system in equilibrium; its diagnostic characteristic is that any change in any of the controlling factors will cause a displacement of the equilibrium in a direction that will tend to absorb the effect of the change."

Mackin (1948, p. 477) introduced the term "shifting equilibrium" to describe the maintenance of approximate adjustment of stream channels to the occurrence of either relatively slow aggradation or degradation. Unfortunately, the concept of "shifting equilibrium" offered little more interpretive value than Davis' (1902, p. 98) concept of "different and changing slopes".

Most debate on the concept of grade from the time of its introduction by W. M. Davis at about the turn of the century through the time of J. H. Mackin's paper of 1948 focused upon problems related to the definition of limits for the "balance between erosion and deposition". It is ironic that at approximately the same time Davis defined the concept of the graded stream, R. G. Kennedy in 1895 (cited in Lane, 1937) published the first set of hydraulic diagrams to aid in the design of non-silting channels. Kennedy's work represented a breakthrough in the quantification of natural processes which determine equilibrium conditions in channels. However, Kennedy's hydraulic diagrams had little impact on geomorphic thought of the day which was primarily oriented to longer time scales of landscape evolution. Continued dominance by "Davisian thinking" during the first half of the twentieth century resulted in other pioneering engineering papers by Griffith (1927), Lacey (1930), and Lane (1937) suffering similar fates as the work of Kennedy. The classic paper by Gilbert (1914) on *The Transportation of Debris by Running Water* and an equally important paper by Rubey (1933) on *Equilibrium Conditions in Debris-Laden Streams* also had relatively little impact on geomorphic thought of the day. These papers later became important reference bases for many of the equations of hydraulic geometry. It was the introduction of the concept of hydraulic geometry by Leopold and Maddock in 1953 that has been most responsible for the current understanding that morphologic response of a stream channel to external controls is a relatively rapid process.

The Concept of Equilibrium and Hydraulic Geometry.

Hydraulic geometry was introduced by Leopold and Maddock (1953) to describe the response of selected parameters of channel cross sections and longitudinal profiles to changes in stream discharge. At first it was anticipated that hydraulic geometry analyses would reveal the basic differences between graded and ungraded streams. A theoretical experiment by Leopold and Maddock (1953, pp. 49-51) led them to suggest that:

> "The equilibrium profile differs from other possible profiles in the relative rates of increase of velocity and depth in the downstream direction."

and,

> "... a particular rate of increase of both velocity and depth downstream is necessary for maintenance of approximate equilibrium in a channel, inasmuch as the drainage area produces sediment and water in a characteristic manner."

However, studies of natural streams resulted in the conclusion by Leopold and Maddock (1953, p. 47) that:

> "Graded reaches of a river are shown to have width, depth, velocity, and discharge relations similar to those of reaches not known to be graded."

Wolman (1955) also indicated that little difference existed between the interactions of the hydraulic variables in graded and ungraded streams. He stated (p. 48):

> "... the mutual adjustments of the variables take place whether the stream is at grade, aggrading, or degrading. Hence the processes of aggradation and degradation are not recognizable at any given time by maladjustments in the variables. Save over a long period of time during which finite changes may be detected in the shape or profile of a given reach, the differences between the river at grade, degrading, and aggrading may be indistinguishable."

Studies by Hack (1960), of Appalachian streams, resulted in his suggestion (p. 84) that it is:

> "... doubtful that streams reach a balanced condition through any evolutionary sequence involving a gradual reduction in slope."

He also concluded (p. 84) that:

> "... equilibrium can be achieved under many conditions and is arrived at very quickly, almost immediately, in the development of a valley. The uniform, or regular concave-upward longitudinal profile that is characteristic of many streams and has been called 'the profile of equilibrium' results not from the attainment of a certain stage in the evolution of a valley, but merely from the regular change downstream in some of the many variables involved in channel equilibrium. Most important of these is probably discharge that increases downstream as a consequence of a regular enlargement of the drainage area."

Leopold and Maddock (1953, p. 51) introduced the term "quasi-equilibrium" to describe the strong tendency of all streams, graded or ungraded, to undergo an adjustment among the hydraulic variables which reflects the inputs of discharge and sediment. Later, Langbein and Leopold (1964, pp. 793-794) specifically defined quasi-equilibrium:

> "The conception of the graded condition as a condition of quasi-equilibrium or the most probable state becomes clearer when it is realized that quasi-equilibrium is a mean or some intermediate position between two opposing tendencies concerning energy utilization."

The development of the concept of quasi-equilibrium led in the 1960's and early 1970's to many papers on probability and statistical characteristics of stream systems. While these studies have proven that certain morphologies of channel cross sections and longitudinal profiles are more probable than others (see for example, Leopold and Langbein, 1962; Chorley, 1962; Langbein and Leopold, 1964; Langbein, 1964; Scheidegger and Langbein, 1966; and Smith, 1974), they fail to identify differences between graded and ungraded streams.

In summary, the initial enthusiasm that hydraulic geometry studies would result in a clear definition of the difference between graded and ungraded streams was not fulfilled. Hydraulic geometry and related studies have shown, however, that most streams are adjusted to the discharge and sediment load regimes which prevail in their respective watersheds. It would thus appear that the many debates concerning Davis' meaning of "balance between erosion and deposition" were somewhat in vain because Davis' postulate that a stream becomes graded only in the "mature" stage of landscape evolution is incorrect.

THE CONCEPT OF GRADE IN PERSPECTIVE

The concept of grade has been shown to be based upon a long recognized principle that stream channels tend to develop morphologic characteristics which are adjusted to prevailing hydrologic conditions. When Davis (1902, p. 89) indicated that a stream graded its course:

> ". . . by a process of cutting and filling, until an equable slope is developed along which the transportation of its load is most effectively accomplished, . . ."

he merely restated a general well-known hydraulic principle.

The reader may recall from quotations presented earlier in this paper that Guglielmini's "sort of equilibrium", Surell's "period of regimen", Dausse's "equilibrium between . . . force of current . . . and . . . resistance of materials", Gilbert's "capacity for transportation", Mackin's "shifting equilibrium", and Leopold and Maddock's "quasi-equilibrium" are similar to Davis' concept of "different and changing slopes" in that all refer to the tendency for natural channels to develop cross section and longitudinal profile characteristics which are adjusted to the prevailing discharge and sediment load characteristics. Rubey's (1952, p. 130) "approximate balance" and Blench's (1957, pp. 2-3) "in regime" were not previously quoted, but their concepts also express the same principle. However, Davis differed from others who had applied the principle of self-adjustment because he attempted to designate the amount of time required for a stream system to reach a condition of stability (Davis, 1902, pp. 91-95). It is now known that the "process of cutting and filling", which would lead to the most efficient slope for transportation of sediment load, does not evolve slowly over geologic time as envisioned by Davis (1902, p. 91), but occurs relatively rapidly, perhaps on a time scale measured in years. Davis' gross overestimation of the length of time required for a stream system to undergo the process of self adjustment has led to some individuals calling for abandonment of the concept of grade (Dury, 1966, p. 231). It is unlikely, however, that the term "grade" will disappear from usage because the concept is deeply infused in the literature of geomorphology. Furthermore, the concept of a graded stream also permeates the engineering literature. Engineering interpretation of the concept is somewhat similar to the definition of Mackin (1948) as is indicated by a recent statement from the Committee on Channel Stabilization for the Corps of Engineers (Lindner, 1969, p. I-5):

> "At any given time there will be filling or bank and bed cutting at various locations showing that at that time at those specific locations there was not a balance between energy and transport; but when this equalizes over a period of time so that on the average the energy of the stream is just equal to the task of transporting the material supplied to it, the stream is graded."

Those who currently call for dismissal of the concept of grade generally base their arguments on observations emanating from hydraulic geometry studies. The hydraulic geometry studies have revealed that nearly all natural streams tend to display interactions among their hydraulic variables which indicate a more or less stable self-adjusting system, characterized by the condition that Leopold and Maddock (1953, p. 51) have described as "quasi-equilibrium". It is apparent that the concept of grade differs from the concept of quasi-equilibrium because historically grade has generally been applied with a morphologic connotation, whereas quasi-equilibrium has represented a process orientation. However, the concept of grade is similar to the concept of quasi-equilibrium from the standpoint that both imply that a stream system will tend to develop morphologic properties which are adjusted to prevailing hydrologic conditions in the watershed. But, the concept of grade extends beyond the concept of quasi-equilibrium in that grade implies morphologic stability over time, a condition which may or may not occur with a state of quasi-equilibrium. The different connotations of the two concepts can be illustrated by considering a situation involving stream channel entrenchment. For example, the self-adjusting tendencies of hydraulic variables imply that a condition of quasi-equilibrium would probably be maintained during a period of channel entrenchment from the surface of an alluvial terrace to the general level of a modern stable floodplain. On the other hand, in the context of the traditional interpretation of the concept of grade, the terrace level would be interpreted as representing the approximate slope of a former graded stream, and the stream during the period of relatively rapid entrenchment would have been characterized as ungraded.

A graded system, therefore, should not reveal trends of change in its morphologic properties when averaged over a "period of years" in the sense of Mackin (1948, p. 471). The suggested limitations on morphologic change for definition of a graded stream is similar to Blench's (1957, pp. 2-3) requirements for a stream "in regime". Blench suggested that a stream may be considered as being "in regime" if the average behavior of stream flow, sediment load, and channel morphologic properties do not change greatly over a period of twenty to forty years.

In summary, a graded stream is here defined as one in which the relationship between process and form is stationary and the morphology of the system remains relatively constant over time. A graded stream, in the Davisian framework of natural landscape evolution, was viewed as developing over geologic time, and great emphasis was given to change of base level as a cause for streams becoming ungraded. Other causes of ungraded streams, such as climate change and activities of man, were acknowledged, but were considered to be of lesser importance than base level. Current knowledge of stream processes suggests that climate change and anthropogenic factors certainly equal if not exceed the relative importance of base level adjustments as causes of ungraded conditions. Changes of base level are primarily induced by tectonic and eustatic events, both of which are relatively low intensity and long duration processes compared to the response and recovery times of natural channels. Base level changes probably have their greatest impact on morphologic adjustments in channel systems only after thresholds of geomorphic stability have been exceeded. Climate change and activities of man, on the other hand, frequently have strong direct impacts on both discharge and sediment yield characteristics and are major contributive causes of streams becoming ungraded. The effect of climate change, as a control on stream system stability, has been the source of much confusion and debate. In comparison to the effects of tectonic and anthropogenic factors, the role of climate is poorly understood, even though it rivals the other factors in importance as a cause for streams to become ungraded. The last portion of this paper is therefore devoted to an examination of the role of climate as a cause of ungraded streams.

CLIMATE CHANGE AS A CAUSE OF UNGRADED STREAMS

The poor understanding of the role of climate is most unfortunate because climatic change can have great impact on fluvial morphology through control of the relative relationship between inputs of discharge and sediment loads to stream systems. Although Davis (1902, p. 97) gave principal attention to base level effects on channel stability, he also acknowledged that climate was an important cause of disequilibrium between erosional and depositional forces, for he stated:

> ". . . if at any time in the cycle a change of climate should occur, new slopes would have to be developed by the streams in order to bring about a new balance between erosion and transportation under the new relation of load and volume. If, for example, the changes were from humid to arid conditions, all the valley floors would have to be steepened by aggradation. If from arid to humid, the graded valley floors would be sharply trenched, and in time reduced to lower slopes."

Another early advocate of the relative importance of climate variation to affect slopes of graded rivers was W. D. Johnson (1901, p. 628) who stated in reference to rivers of the High Plains:

> ". . . it is not necessary, in order to account for change in behavior of the traversing streams, to appeal to deformation. A sufficient cause may be looked for in change of climate. There is record of erosion, with reversal to deposition and rebuilding, and reversal again finally to erosion, and there is reason for believing that this series of interruptions of the gradation cycle was an effect of climatic oscillation rather than of earth movement."

Critics of climatic hypotheses often point to the fact that erosion and deposition do not begin everywhere at the same time and that fluvial activity in tributaries and trunk streams are frequently out-of-phase with each other. However, it is important to recognize that many of the events that appear as out-of-phase on historical time scales would appear as synchronous on geologic time scales. It has also been noted that instances of exceeding thresholds of geomorphic stability frequently appear to be time transgressive on a historical scale but they, too, are likely to be clustered on the scale of geologic time.

Other critics of climatic hypotheses indicate that regional comparisons of radiocarbon dated alluvial chronologies frequently do not suggest an absolute synchroneity of fluvial events. The lack of synchroneity between regions should not be surprising because fluvial events in any given area are either directly or indirectly controlled by meteorological conditions which reflect preferred patterns of large scale upper atmospheric circulation regimes. The prevalence of given circulation patterns over extended periods of time determine whether different regions within the same general latitudinal limits behave in an environmentally synchronous or asynchronous fashion. Furthermore, different environmental characteristics of latitudinal belts reflect the effects of large scale circulation regimes to displace boundaries of regional climatic zones which, in turn, may vary from region to region.

In the following sections of this paper it will be demonstrated how large scale climate changes can cause streams to become ungraded. The focus will be directed to episodes of valley alluviation and degradation within the general time frame of the last 10,000 years. The last ten millennia, hereafter termed the Holocene, not only provide the best available record of radiocarbon dated alluvial chronologies, but they also provide a basis for testing the impact of relatively modest (in the sense of Pleistocene scale) climatic changes.

One of the more well-known environmental subdivisions of the Holocene Epoch is the Blytt-Sernander sequence of climatic episodes based upon events in northern Europe (Zeuner, 1952). Recently, Wendland and Bryson (1974) found that the Blytt-Sernander subdivisions of the Holocene could be applied on a global scale. On the basis of a statistical study of radiocarbon dates representing major geologic and botanic discontinuities, Wendland and Bryson proposed specific dates for bracketing the Blytt-Sernander environmental episodes (Table 1). If it is to be demonstrated that climatic change can cause a graded stream to become ungraded and involve an episode of alluviation or degradation or both, it must therefore be demonstrated that alluvial episodes bear some resemblance to the Blytt-Sernander sequence of environmental episodes.

TABLE 1

Holocene Climatic Episodes*

Episode Name	Transition Date ^{14}C Years B.P.
	850
	1680
Sub-Atlantic	
	2760
Sub-Boreal	
	5060
Atlantic	
	8490
Boreal	
	9300
Pre-Boreal	
	10030
Late-Glacial	

*(After Wendland and Bryson, 1974)

How Climate Changes Influence Graded Streams.

Climate is the driving force in any hydrologic system, and through its influence on vegetation, is especially important in controlling rates of erosion, transportation, and deposition of sediments. The importance of climate as a cause for streams to become ungraded, once graded, depends largely on the recurrence characteristics of floods, especially large floods. The importance of floods as determinants of morphologic characteristics of channel systems was identified long ago by Gilbert (1877, p. 113). More recently, Rubey (1952, p. 130) and Wolman and Miller (1960, p. 66) conducted pioneering work which established that the size of a channel cross section is highly correlated with the volume of water transported during relatively frequently recurring flood flows. The morphologic properties of most channel systems tend to show stronger correlations with the discharge characteristics of moderate magnitude floods than with relatively large magnitude floods. The good association with moderate floods reflects the ability of streams to quickly adjust their morphologies to relatively frequent events which have sufficient energy to transport and erode and deposit sediments. In the absence of large floods of high energy, a stream maintains morphologic stability over time and is graded. The occurrence of a large flood may cause a stream to become temporarily ungraded if thresholds of geomorphic stability are exceeded, but if no additional large floods occur over the time span of the next several years, the channel system will soon recover a stable morphology adjusted to the characteristics of moderate magnitude and relatively frequent floods.

The ability of channel systems to recover rapidly from the damaging effects of a large flood (having a recurrence interval of at least 200 years) has been illustrated for Piedmont channels of eastern United States by Costa (1974). The flood, produced by Hurricane Agnes in June 1972, was found by Costa (1974, pp. 109-111) to have produced major widening and scouring of Piedmont channels. Costa concluded that post-Agnes reshaping and modification of the flood-scarred channels is taking place so rapidly that in a few years, provided another catastrophic flood does not occur, the only evidence of the Agnes flood will be gravel lenses in alluvium at tributary junctures and low berms or steps in the floodplain where slumped banks have gathered sediment and revegetated. Gupta and Fox (1974) have also studied post-flood recovery of channel forms for Piedmont streams and reached a conclusion similar to that of Costa.

Relatively frequent recurrence of large floods can be particularly devastating to the preservation of morphologic stability in channel systems. An excellent example of channel and floodplain responses to an increase in the recurrence frequency of large floods is the study by Schumm and Lichty (1963) of the Cimarron River in southwestern Kansas. Schumm and Lichty reported that a large flood in 1914 converted a previously stable narrow, deep, and sinuous channel into a wide, shallow, and braided channel system. Unlike the Piedmont channels described above, the Cimarron River did not recover its pre-flood form because the three decades following 1914 were characterized by relatively frequent recurrences of large floods. The frequent large floods of the period 1914 through 1942 nearly completely destroyed the Cimarron floodplain (pp. 74-76). Schumm and Lichty indicated that after 1942 and until about 1954, the Cimarron system was characterized by floods of low to moderate magnitude which resulted in the termination of floodplain destruction and the occurrence of channel narrowing and floodplain construction. The findings of Schumm and Lichty were supported by the study of Stevens *et al.* (1975), who concluded that river systems which have a wide range of peak-flood discharges are susceptible to frequent changes of form.

Although almost any flood has the potential to cause a stream to become temporarily ungraded, it is important to remember that a stream will soon regrade itself if another significant flood does not soon follow. The recurrence intervals of large floods are highly influenced by the patterns of large scale upper atmospheric circulation regimes (Knox *et al.*, 1975). The upper air circulation patterns not only produce the synoptic conditions which favor the occurrence of excessive precipitation, but they also influence flood characteristics through effects on antecedent conditions related to surface cover and moisture storage.

In general, large floods tend to be more frequent during regimes of large scale upper atmospheric circulation which are characterized by a strong meridional (north-south/south-north) component. The influence of meridional circulation to influence both the incidence of large floods and general variability of stream flow is especially marked in the middle latitudes where large contrasts between adjacent air masses often occur. Meridional regimes also tend to increase the variability of climatic and hydrologic events in the high and low latitudes, but the effects are modest compared to the middle latitudes.

The principal characteristics of the large scale upper atmospheric circulation are represented in the pattern of waves in the upper air westerly circulation (Fig. 1). The large waves of the circumpolar vortex depicted in Figure 1 normally vary between three and seven in number. Low wave numbers represent a strong westerly component to the large scale circulation, whereas high wave numbers denote strong north-south/south-north components to the large scale circulation. The strong meridional element associated with high wave number regimes favors deep penetration of polar air masses into low latitudes and equally deep penetration of tropical air masses into high latitudes. The extreme latitudinal ranges of the air masses during regimes of meridional circulation frequently result in the development of intense cyclones and storms in the middle latitudes followed by large floods. In contrast, the strong westerly component of zonal circulation regimes tends to diminish the likelihood of intense cyclones and storms in the middle latitudes. As a consequence, many regions which are relatively far removed from moisture source regions tend to become drier and experience fewer large floods during periods of zonal dominance.

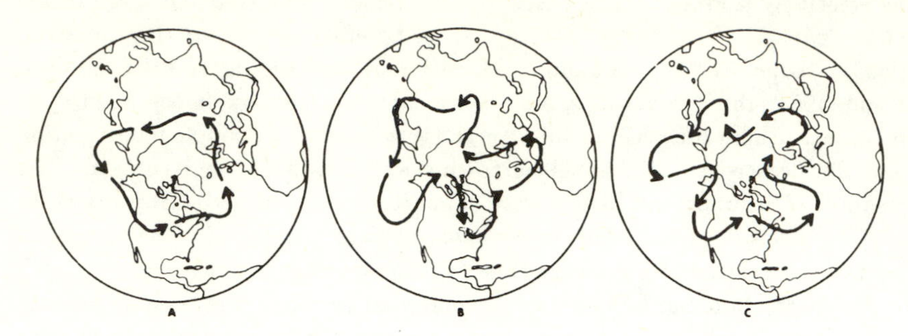

Figure 1. Isopleth of 500 millibar surface illustrating regimes of large scale upper atmospheric westerly circulation. The illustrations include: a low amplitude and long wave zonal pattern (A) and two five-wave high amplitude and short wave length meridional patterns (B & C). The two meridional patterns are distinguished from each other according to different associated positions of cold troughs and warm ridges of pressure. (Illustration based on Wangenheim/Girs classification, presented by Lamb, 1972, p. 267).

The relative importance of changes in the characteristics of the large scale upper atmospheric circulation to influence the recurrence intervals of large floods is illustrated by the annual flood series for the Mississippi River at Keokuk, Iowa (Fig. 2). The flood series reveals that large floods were relatively common in the early part of the record and again in recent years, but that most of the first half of the twentieth century was a time of moderate to relatively low magnitude floods. Knox *et al.* (1975, pp. 29-30) found that a large number of years of the first half of the twentieth century were characterized by upper atmospheric circulation regimes which had a strong zonal component whereas many years of the last two and one-half decades have been dominated by circulation regimes having a relatively strong meridional component. Although data for characteristics of large scale upper atmospheric circulation regimes are not available for the period of large floods in the late 1800s, the principal features of the large scale circulation of that period can be inferred from maps showing regional distributions of temperature and precipitation. Maps of temperature and precipitation characteristics for selected decades of the nineteenth century have been produced by Wahl (1968) and Wahl and Lawson (1970) and show climatic patterns which suggest that large scale circulation regimes were most likely characterized by relatively strong meridional components as might be anticipated from the flood series of Figure 2.

Although the increased magnitude of floods on the Mississippi River in recent years is partly due to the effects of man (Belt, 1975), the increased volumes of discharges for many of the recent years suggest that climate is an overriding factor. In fact, the increased importance of the meridional

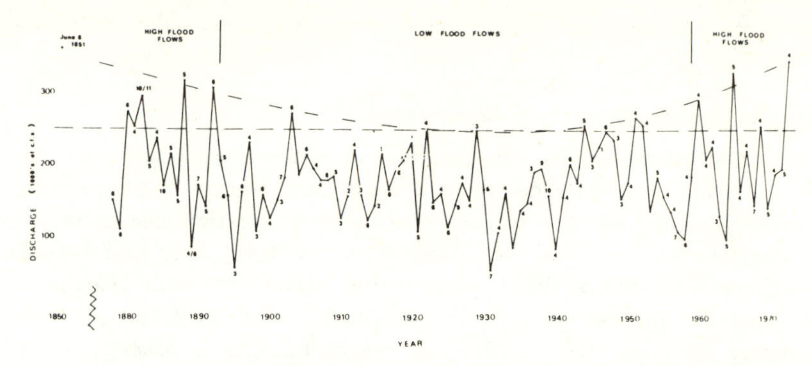

Figure 2. Recurrence characteristics of floods on the upper Mississippi River at Keokuk, Iowa. The short-term relative variability of the magnitude of the annual floods was large in the late 1800's and since about 1950, but was relatively small during much of the first half of the twentieth century when the largest flood was only about 250,000 cubic feet per second for the daily peak discharge. The non-random pattern of the flood series is strongly related to climatic variations caused by changes in the patterns of the large scale upper atmospheric circulation regimes (see text for discussion).

component to the large scale circulation of recent years has been responsible for a significant increase in the number of large floods in many middle latitude drainage basins. For example, Lamb (1974, pp. 19-20) noted:

> "... there has been an intensification of the meridional (north-south) currents of the mean wind flow (except over Asia). These aspects account for the fact that some areas in middle latitudes, notably California, the U.S. Middle West and parts of Europe, have also experienced in these same years great floods, high levels of the lakes and rivers, and avalanches. Long-continued rains associated with slow-moving cyclones at times of weak or very distorted flow of the upper westerlies have been the characteristic feature of these episodes."

In summary, maintenance of a graded condition in fluvial systems depends to a large degree upon the recurrence intervals of large floods. The recurrence intervals of large floods are in turn strongly dependent upon the recurrence of unique patterns of large scale upper atmospheric circulation regimes. Lamb (1966, pp. 58-112) has shown that climatic epochs of many time scales, including the Holocene, tend to be associated with certain preferred large scale circulation patterns. Since streams quickly adjust to new climatic and hydrologic regimes, occurrences of ungraded streams, as represented by major morphologic adjustments and/or episodes of alluviation and degradation should closely follow occurrences of climatic discontinuities. On the scale of the Holocene, this implies that ungraded streams, caused by climatically induced morphologic instability, should closely follow dates representing breaks between episodes of the Blytt-Sernander chronology (Table 1).

Holocene Climate Change and Fluvial Episodes.

Alluvial stratigraphic units representing the Holocene are usually incomplete and poorly bracketed by radiocarbon dates in most regions. Although there have been many attempts to correlate Holocene stratigraphic units with climatic episodes (e.g., see Vita-Finzi, 1973, pp. 79-93), only a few studies have focused on alluvial chronologies which extend over wide regions and span the entire Holocene record (e.g., see Starkel, 1966, 1972; Haynes, 1968). In nearly all cases the number of radiocarbon dates associated with stratigraphic sequences has been less than desirable. The insufficient quantity of detailed dating of Holocene alluvial chronologies prompted this writer to undertake a special study of published radiocarbon dates representing alluvial environments in an attempt to assess the role of climate change as a factor affecting the behavior of stream systems.

The study, which focused upon the statistical properties of approximately 800 radiocarbon dates published in volumes 8-14 of *Radiocarbon*, was based upon the premise that large scale upper atmospheric circulation regimes exert a strong influence on the recurrence frequencies of large floods and that climatic episodes tend to be characterized by certain preferred patterns of large scale circulation. Bryson and Wendland's (1967) tentative reconstructions of Holocene air mass regimes for eastern North America suggest that substantial differences occurred between dominant large scale circulation characteristics of Holocene climatic episodes. Wendland and Bryson (1974) have presented quantitative data which indicated that a change from the dominant circulation regimes of one Holocene climate episode to the dominant circulation regimes of a following episode usually occurred abruptly during a relatively short transition period. Therefore, given the relationship between flood characteristics and large scale circulation regimes, it is reasonable to expect flood magnitudes and frequencies to reflect climatic adjustments. Also, the lag response of vegetation to abrupt shifts of climate could further enhance the effects of runoff to produce morphologic instability (Knox, 1972, pp. 408-409).

The statistical study of the 800+ radiocarbon dates was based on the hypothesis that a histogram of radiocarbon dates representing fluvial environments should be multimodal in character with peaks and troughs reflecting responses to climatic stresses. It was thought that episodes of valley alluviation should enhance the likelihood of burial and preservation of datable materials and be associated with frequency maxima on the histogram, whereas episodes of stability or of valley degradation should diminish the likelihood of datable materials being buried or preserved and be associated with frequency minima on the histogram. If climatic change were not a factor, then a histogram of radiocarbon dates representing fluvial environments should appear either as a rectangular distribution or as a multimodal distribution with peaks and troughs randomly distributed in time. The actual histogram of 802 dates proved to be multimodal as expected (Fig. 3).

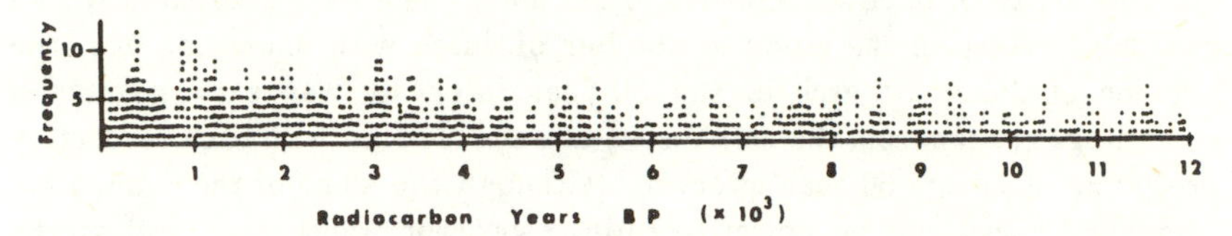

Figure 3. The effects of climate changes on fluvial episodes as indicated by statistical properties of a histogram of radiocarbon dates representing alluvial environments.

The radiocarbon dates which compose the histogram of Figure 3 are mostly representative of attempts to date stratigraphic horizons in floodplains, terraces, and alluvial fans and include relatively few archeaological dates. Care was taken to not include archaelogical dates which were associated with disturbed sites (e.g., refuse pits, fire pits, etc.). The radiocarbon dates are primarily representative of the Northern Hemisphere middle latitudes. Sixty-eight percent of the dates were associated with latitudes between 30°-60° N and 83 percent of the dates were associated with sites northward of 30° N. Very few dates represented sites poleward of 80° and only 11 percent of the dates represented sites equatorward of 30° north and south latitudes. The strong bias toward middle latitude sites indicates that the statistical properties of the dates should be particularly sensitive to the effects of changes in the dominant patterns of large scale upper atmospheric circulation regimes (Fig. 1).

The radiocarbon dates were plotted at 50-year intervals to better define millennium scale discontinuities in the histogram. No attempt was made to incorporate the plus and minus limits of the dates because the exercise would add little information to the mean statistical properties of the histogram. Obviously, not all maxima and minima shown on Figure 3 represent climate related events. Some may reflect the absence of dating in particular time intervals, and others may be related to other factors of environmental change, including the impact of man. However, within the last eight to nine millennia, the distribution of dates are sufficiently dense to warrant selective comparisons of histogram characteristics with presumed climatic episodes.

Two principal modifications of the radiocarbon data were undertaken to render the histogram more interpretable. First, the histogram frequencies in each interval were smoothed with a bionomial filter of the form:

$$N_i = 0.25N_{i-1} + 0.50N_i + 0.25N_{i+1},$$

where N_i is the number of dates that fell in the ith interval. The binomial filter was used in place of the running mean for smoothing because it resulted in minimal displacement of histogram peaks and troughs from their true locations. The second transformation of the data involved presenting the radiocarbon dates of the filtered histogram as a cumulative frequency curve of a form analogous to the *mass diagram* of hydrologic investigations (Wisler and Brater, 1959, p. 284). In order to minimize the effects associated with a general decrease in the absolute number of dates with increasing age, the number of dates for each of the 50-year intervals was expressed as a percentage change from the number of dates associated with each respective preceding older age 50-year interval. Although the slope of the cumulative frequency curve can be upward or downward on short time scales, the general orientation must be upward to the left from the starting base at *ca.* 12000 radiocarbon years B.P. because the summation of percentage changes were expressed relative to the number of dates in a given interval and its adjacent older interval (Fig. 4).

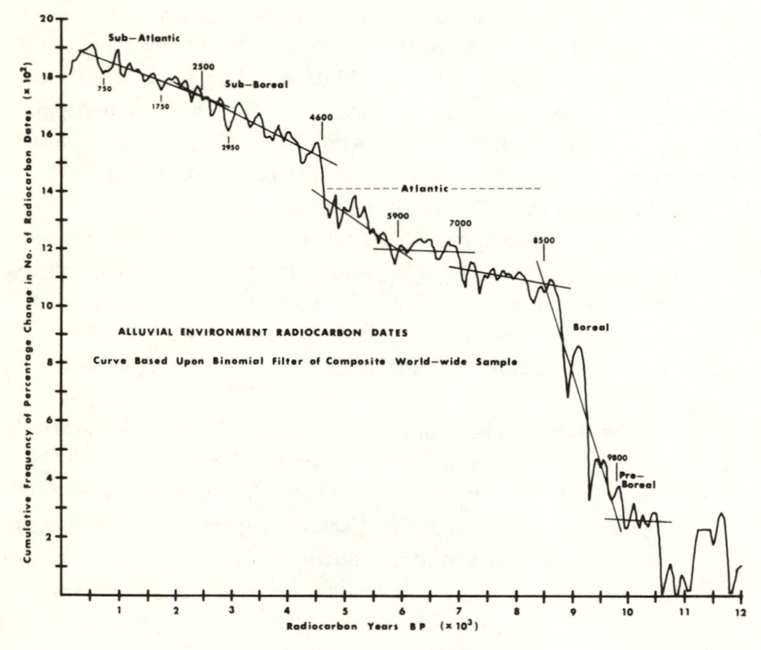

Figure 4. Fluvial episodes suggested by discontinuities in a cumulative frequency distribution (mass curve) of radiocarbon dates representing alluvial environments (Figure 3).

Interpretation of the cumulative frequency curve of radiocarbon dates is similar to the interpretation of a *mass diagram*. That is, the slope of the curve represents the general rate of change in the relative number of radiocarbon frequencies with time. Discontinuities on the cumulative frequency curve were determined by drawing straight lines through segments of the curve which had a relatively constant rate of slope.

Inspection of Figure 4 reveals that the method of line segments to identify discontinuities results in major breaks in the cumulative frequency curve at the approximate dates of 9800, 8500, 4600, and 2500 radiocarbon years B.P. Although the date at *ca.* 2500 B.P. appears as very modest on the cumulative frequency curve (Fig. 4), it was justified in relation to the relatively large absolute number of dates associated with the histogram intervals for the last 400 radiocarbon years. Conversely, the breaks in slope occurring at approximately 7000 and 5900 radiocarbon years B.P. were not given major status because of the relatively small absolute number of dates associated with the period between 7500 and 5500 radiocarbon years B.P.

A comparison of the major discontinuities on the cumulative frequency curve of Figure 4 with discontinuity dates separating the Blytt-Sernander climatic episodes suggested by Wendland and Bryson (Table 1) indicates relatively close agreement (Fig. 5). Principal exceptions are in the Sub-Atlantic episode, for which Wendland and Bryson have recognized two additional major discontinuities. Although the discontinuities between Blytt-Sernander episodes of Wendland and Bryson (1974) were also determined from an analysis of radiocarbon dates biased toward representing middle latitudes of the Northern Hemisphere, those dates were primarily associated with paleobotanical changes and involve negligible overlap with the radiocarbon dates represented in the present determination of alluvial episodes. The good agreement for dates of discontinuities within the two independent data sets supports the concept of global synchroneity of climatic shifts. Furthermore, it implies that alluvial episodes and the stability of stream channels are strongly influenced by climatic change.

Climate changes, while tending to be globally synchronous in time of occurrence, normally produce different hydrologic impacts among widely separated geographic regions. The relative degree of between region synchroneity or asynchroneity depends upon the large scale wave characteristics of the upper atmospheric circulation (Fig. 1). On short time scales (e.g: years, decades, and perhaps even centuries) a comparison of hydrologic events between regions of the same geographic latitude shows that asynchroneity is a common occurrence (Kalinin, 1968, pp. 91-121; Smirnov, 1973, p. 120). On the other hand, the integrated climatic effects of long episodes characterized by circulation regimes having a strong meridional component often result in many region to region hydrologic similarities, especially with regard to the occurrence of large floods as was noted earlier. The tentative climatic patterns, which Bryson and Wendland (1967) associated with different Blytt-Sernander episodes, suggest that the relative strength of dominance by meridional and zonal components of the large scale circulation experienced considerable variation between Holocene climate episodes. While it is unlikely that any given Blytt-Sernander episode was entirely

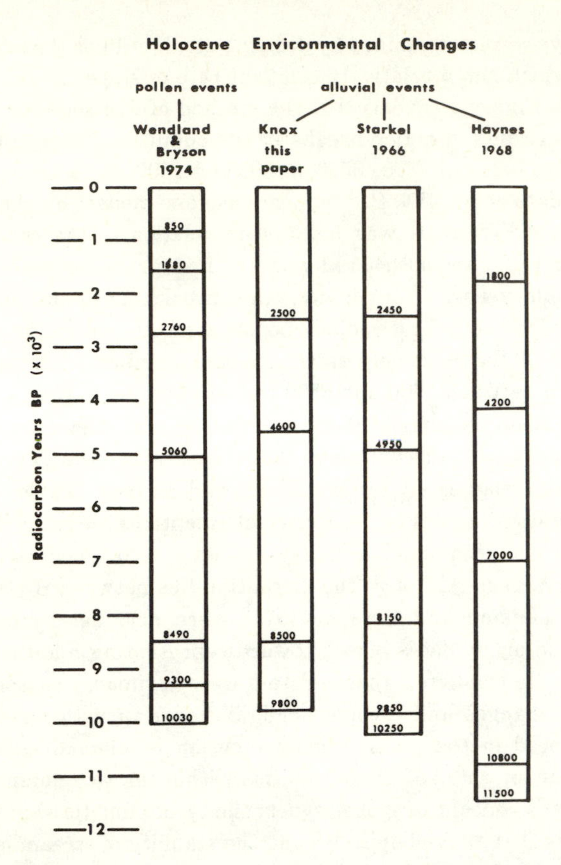

Figure 5. Discontinuities in paleobotanical and alluvial records. Discontinuities shown in the first two columns are biased toward representing northern hemisphere events. The discontinuities attributed to Starkel represent paleobotanical and fluvial episodes in Europe. The discontinuities attributed to Haynes represent the tops of depositional units found in the western Great Plains and Southwest United States.

dominated by any particular circulation regime, it is generally agreed that certain preferred circulation patterns often dominated for extended periods (Lamb, 1966, pp. 59-68). The close agreement between the discontinuities of the pollen based data of Wendland and Bryson (1974) and the alluvial data of this paper (Fig. 5) suggests that the cumulative frequency curve of Figure 4 is a reasonable index of fluvial response to these climate episodes.

The slope of the line segment and the oscillations of the cumulative frequency curve (Fig. 4) probably have geomorphic and hydrologic significance, but it is too hazardous to offer general interpretations at this time. Many of the short period characteristics of the cumulative frequency curve undoubtedly reflect meaningless summations of dates resulting from combining geographic regions which are out-of-phase with each other. Some tentative results of analyses of histograms of alluvial environment radiocarbon dates categorized according to climatic regions indicated that short period out-of-phase behavior is definitely the case (Knox, 1974.) Other short period oscillations on the frequency curve may be a product of the non-progressive character of channel response to external events. For example, Schumm (1973) along with his students (Schumm and Kahn, 1972; Schumm and Parker, 1973; Patton and Schumm, 1975), has shown in both laboratory and field studies that morphologic change in channel systems, including episodes of alluviation and degradation, often occur abruptly after a threshold of stability has been exceeded. Schumm (1973, p. 307) has also shown that one external event can trigger a complex reaction resulting in multiple geomorphic surfaces and stratigraphic units. It is important to remember, however, that thresholds are especially vulnerable during the occurrences of large floods. In turn, large floods have been shown to be strongly dependent upon the recurrence of particular large scale circulation regimes (Knox *et al.*, 1975). Therefore, on a Holocene time scale the breaking of thresholds may be a clustered phenomenon strongly related to climatic episodes.

In spite of the problems related to the aggregation of dates, many of the larger oscillations on the cumulative frequency curves (Fig. 4) seem to be related to relatively widespread environmental changes. For example, the pollen based discontinuities at 2760, 1680, and 850 radiocarbon years B.P. are reasonably close to unusually sharp depressions from the general trend on the cumulative frequency curve of alluvial radiocarbon dates (see dates 2950, 1750, and 750 on Fig. 4). It is also significant to note that reconstructions of late-Holocene positions of treeline in Keewatin and Mackenzie, N.W.T., Canada indicate that treeline was located south of its present position at *ca.* 2900, 1800, and 800 radiocarbon years B.P. (Sorenson and Knox, 1973). Sorenson and Knox (1973, pp. 192-194) associated southward migrations of treeline with increases in the strength of the meridional component of the large scale upper atmospheric circulation. The reasonably close agreement among the three sets of data may be a product of chance, considering the small number of events, but it is interesting that strengthening of the meridional component, as suggested by the southward migrations of treeline, is compatible with high climatic variability and morphologic instability of channels and floodplains in the middle latitudes.

The ultimate proof of the importance of climate change to affect the morphologic stability and the alluvial history of river systems depends on well documented field evidence. The association between fluvial episodes and climate episodes has been reasonably well documented for Europe (especially the valleys of the Polish Carpathians) by Starkel (1966, 1972) and by Klimek and Starkel (1974). Starkel (1972, p. 128) indicated that periods at the close of the Glacial, during the Atlantic, and at the beginning of the Sub-Atlantic were characterized by accumulations of sediments on alluvial fans and deposition of sands and gravels in wide channels (1-2 km). On the other hand, Starkel indicated the periods of the Boreal and Sub-Boreal, as well as late-Roman to early medieval time, were generally characterized by low magnitude supplies of discharge and sediment. He associated the reduced supplies with the development of deep and meandering channels. The occurrence of large floods was found to be particularly significant in understanding the alluvial chronologies of Carpathian valley fills. For example, Klimek and Starkel (1974), p. 185) stated:

> "... the observations of present-day fluvial processes as well as inserted series of Holocene alluvia indicate that in the Post-glacial time the periods of major floods have led to a simultaneous intense erosion (mainly lateral) and the accumulation of channel and floodplain—mud—deposits. They are divided by periods of the reduced activity of fluvial processes when only narrow channels, highly sinuous with low width-depth ratio are being formed."

The dates which Starkel (1966, pp. 24-27) suggested for bracketing Holocene climatic phases of importance to the formation of European relief are presented in the third column of Figure 5. A comparison of Starkel's discontinuities with those suggested by the pollen data of Wendland and Bryson and the alluvial data of this paper reveals relatively close agreement (Fig. 5). Unfortunately, Starkel's (1966, p. 24) dates of discontinuities were based upon European paleobotanical conditions and therefore do not strictly represent alluvial discontinuities. In spite of the paleobotanical basis for designation of the discontinuities, Starkel (1966, pp. 24-27) has shown that the designated episodes were characterized by considerable differences in fluvial activity.

The Holocene alluvial chronology of the Southwest United States has been documented by Haynes (1968) who has shown that climate changes have been responsible for the occurrences of widespread synchroneity in sequences of deposition, soil formation, and erosion throughout the region. I have generalized the dates which separate five principal depositional units recognized by Haynes and I have plotted them in Column four of Figure 5. Inspection of Figure 5 indicates that the number of discontinuities and their general location in time are weakly reflective of the Blytt-Sernander episodes. The general tendency for the major discontinuities to be younger than dates of comparable discontinuities in the other series represented on Figure 5 may reflect time-transgressive effects of the Southwest data being biased toward the lower middle latitudes whereas the other data were biased toward the middle and high middle latitudes. The younger dates for the discontinuities are also undoubtedly influenced by the tendency for the other series (represented in Fig. 5) to record climatic events more quickly than would events associated with dates representing tops of depositional units. For example, termination of deposition associated with the discontinuity at *ca.* 4200 radiocarbon years B.P., relates to increasing soil moisture and soil formation initiated as early as *ca.* 4500 radiocarbon years B.P. (Haynes, 1968, pp. 611-612). Inspection of Figure 5 shows that the date 4500 B.P. more closely approximates the discontinuities of the other series than the general date of 4200 B.P. which is associated with the top of a depositional series. Similar comparisons for other dates of discontinuities are less convincing, but a comparison of many specific alluvial events cited by Haynes with major oscillations on the cumulative frequency curve of Figure 4 shows a general tendency for temporal agreement. It is therefore tentatively concluded that the climatically induced alluvial episodes of the Southwest reflect global scale adjustments in the upper atmospheric circulation regimes.

Detailed field investigations have also been conducted to determine the impact of Holocene climate changes on the alluvial chronology of the Driftless Area of southwestern Wisconsin. Major alluvial activity involving morphologic adjustments of channels appears to have taken place between *ca.* 10,500 and 9500 B.P., *ca.* 6000 B.P., and in the period between 5000 and 4500 B.P. (Knox and Johnson, 1974). Continued studies have led Johnson (personal communication) to recognize three additional minor episodes of morphologic instability of channels since *ca.* 4500 radiocarbon years B.P. In radiocarbon years, the first occurs between 2700 and 2500 B.P., the second is between 2000 and 1600 B.P., and the third is at *ca.* 1000 B.P. Both the major and minor alluvial episodes, which are based upon about 55 radiocarbon dates, appear to be strongly related to the Holocene paleoclimatic events described for the upper Midwest by Webb and Bryson (1972). A comparison of the dates representing active morphologic adjustments in channels and floodplains of the Driftless Area with the index of global scale fluvial activity (Fig. 4) also shows that fluvial activity in the Driftless Area often reflects global scale natural events.

In summary, tentative results of a statistical analysis of approximately 800 radiocarbon dates representing alluvial environments suggest that the Holocene Epoch can be subdivided into a sequence of fluvial episodes. The discontinuities between the episodes occur at the approximate dates of 9800, 8500, 4600, and 2500 radiocarbon years B.P. and are in relatively close temporal agreement with dates proposed by Bryson and Wendland (1974) to separate global scale climate episodes. A comparison of results of the statistical study with stratigraphic studies of Holocene alluvium in the Polish Carpathians, Southwest United States, and the Driftless Area of southwestern Wisconsin provided further support for recognition of climatically induced fluvial episodes. In general, each of the fluvial episodes tends to represent a form of biogeomorphic adjustment to the dominant climate characteristics of that period. The relatively short intervals of time between episodes tend to be characterized by morphologic instability as a fluvial system attempts to adjust to dominance by a new climate episode. Therefore, with respect to climate control, it is likely that channels of a river system would maintain a graded condition for relatively long periods representing the duration of dominance by a given climate episode. On the other hand, ungraded conditions might be expected during the transition period between climate episodes because rapid morphologic changes would be anticipated for biogeomorphic response to dominance by a new climatic regime. Climatically controlled episodic behavior of fluvial systems is suggested by the three stratigraphic studies cited above. The concept of episodic behavior of fluvial systems appears to be applicable to many time scales other than the Holocene as illustrated above.

CONCLUSIONS

The concept of the graded stream as originally defined by Davis (1899, 1902) has two major components. First, it recognized the tendency for natural channels to develop and maintain a relatively stable morphology which represented a general balance between erosion and deposition. Second, it indicated that the length of time required for initial development of graded streams on a newly uplifted landscape would consume the geologic period of "youth" of Davis' (1899) "geographical cycle" of landscape evolution. The first part of the concept has stood the test of time, but the second part is now known to be false. The development of a relatively stable morphology does not take place through any long-term evolutionary process, but tends to occur rather quickly. Some have suggested that the concept of grade should be replaced by the concept of "quasi-equilibrium" which Leopold and Maddock (1953, p. 51) introduced to describe the strong tendency of all streams, graded or ungraded, to undergo an adjustment among the hydraulic variables which reflects inputs of sediment and discharge. It is true that both the concept of grade and the concept of quasi-equilibrium imply that a stream will tend to develop morphologic properties which are adjusted to prevailing hydrologic conditions. However, this writer is of the opinion that the concept of grade extends beyond the concept of quasi-equilibrium because the traditional definition of grade implied morphologic stability over time, a condition which may or may not exist with a state of quasi-equilibrium. Therefore, the concept of the graded stream should be retained, and modified to correct the evolutionary connotation given to it by Davis (1899, 1902).

It is suggested that a graded stream be defined as one in which the relationship between process and form is stationary and the morphology of the system remains relatively constant over "a period of years" in the sense of Mackin's (1948) definition of grade. Conversely, ungraded streams should probably be defined as those undergoing relatively rapid morphologic changes involving properties of longitudinal profiles and/or cross sections.

The use of "rapid morphologic change" as a basis for defining ungraded streams, results in a reassignment of the relative importance of factors traditionally described as responsible for ungraded streams. Tectonic and eustatic events, which were assigned great importance in the Davisian context, are associated with relatively low intensity and long duration processes compared to lag response times of channels. Therefore, tectonic and eustatic events are probably associated with an ungraded condition only after the cumulative effects of slowly changing environmental conditions result in the breaking of a threshold of channel stability. Climate change and man's activities, on the other hand, frequently have strong direct impacts on the functional relationship between discharge and sediment yield and are major factors causing ungraded channels. The contribution of climate change to produce fluvial episodes and related morphologic adjustments in drainage systems has long been a source of debate, and lack of agreement concerning its importance is persistent (Vita-Finzi, 1973, pp. 85-88). However, the results of this study indicate that climate change, as a cause of ungraded streams, rivals the importance of the other traditional factors. On some time scales, such as illustrated in this paper for the Holocene, it is probably the most important factor. Climatically induced fluvial episodes involving rapid morphologic adjustments in channels appear to reflect the meteorological effects of the dominant large scale upper atmospheric circulation patterns to control both the general characteristics of runoff and sediment yield and the recurrence frequency of large floods. On longer time scales, it appears that the breaking of thresholds of channel stability is a clustered phenomenon strongly related to climatic episodes.

Non-progressive, or episodic, adjustments are normal in fluvial systems because major morphologic changes, including episodes of alluviation and degradation often occur abruptly after a threshold of stability has been exceeded. The quasi-equilibrium principle indicates that fluvial adjustments must immediately follow abrupt changes of external watershed variables, including the effects of changes in base level, land use, and climate. During a period of adjustment to a new stable channel morphology, the stream system or parts of the stream system would experience relatively rapid morphologic changes and be ungraded. After the adjustment phase, the stream system would be graded to the new environmental regime. The stream system would have maintained quasi-equilibrium throughout the entire process.

REFERENCES CITED

Baulig, H., 1926, La Notion de profil d'ēquilibre: histoire et critique: C. R. Congress Internat. de Geography (1927); v. 3, pp. 51-63.

Belt, C. B., 1975, The 1973 flood and man's constriction of the Mississippi River: Science, v. 189, pp. 681-684.

Blench, T., 1957, *Regime Behavior of Canals and Rivers:* Butterworth's Scientific Publications, London.

Bryson, R. A. and Wendland, W. M., 1967, Tentative climatic patterns for some late-glacial and post-glacial episodes in central North America: In: *Life, Land, and Water* (W. J. Mayer-Oakes, Ed.), Univ. of Manitoba Press, Winnipeg, pp.271-298.

Chorley, R. J., 1962, Geomorphology and general systems theory: U. S. Geological Survey Prof. Paper 500 B, 10 pp.

Costa, J. E., 1974, Response and recovery of a Piedmont watershed from tropical storm Agnes, June 1972: Water Resources Research, v. 10, pp. 106-112.

Davis, W. M., 1899, The geographical cycle: The Geographical Journal, v. 14, pp. 481-504.

———————, 1902, Base-level, grade, and peneplain: Journal of Geology, v. 10, pp. 77-111.

Dury, G. H., 1966, The concept of grade: In: *Essays in Geomorphology* (G. H. Dury, Ed.), Heinemann, London, pp. 211-233.

Gilbert, G. K., 1877, Report on the geology of the Henry Mountains: U. S. Geographical and Geological Survey of the Rocky Mountain Region, Washington, 160 pp.

———————, 1914, The transportation of debris by running water: U. S. Geological Survey Prof. Paper 86, 259 pp.

Griffith, W. M., 1927, A theory of silt and scour: Institute Civil Engineers Proceedings, v. 223, pp. 243-314.

Gupta, A., and Fox, H., 1974, Effects of high-magnitude floods on channel form: a case study in Maryland Piedmont: Water Resources Research, v. 10, pp. 499-509.

Hack, J. T., 1960, Interpretation of erosional topography in humid temperate regions: American Journal Science, v. 258-A (Bradley Volume), pp. 89-97.

Haynes, C. V., 1968, Geochronology of late-Quaternary alluvium: In: *Means of Correlation of Quaternary Successions* (R. B. Morrison and H. E. Wright, Eds.), Proceedings VII INQUA Congress, Univ. of Utah Press, Salt Lake City, v. 8, pp. 591-631.

Johnson, W. C., 1975, Personal communication, Department of Geography, Univ. of Wisconsin, Madison.

Johnson, W. D., 1901, The High Plains and their utilization: U. S. Geological Survey, 11th Annual Report, pt. IV, pp. 601-741.

Kalinin, G. P., 1971, *Global Hydrology:* Israel Program for Scientific Translations, Jerusalem, 311 pp.

Kesseli, J. E., 1941, The concept of the graded river: Journal Geology, v. 49, pp. 561-588.

Klimek, K., and Starkel, L., 1974, History and actual tendency of flood-plain development at the border of the Polish Carpathians: In: *Report of the Commission on Present-day Processes* (H. Poser, Ed.), International Geographical Union, Vandenhoeck and Ruprecht, Gottingen, pp. 185-196.

Knox, J. C., 1972, Valley alluviation in southwestern Wisconsin: Annals of the Association of American Geographers, v. 62, pp. 401-410.

———————, 1974, Holocene alluvial response to climatic change: American Quaternary Association, Abstracts Third Biennial Meeting, Madison, p. 65.

Knox, J. C. and Johnson, W. C., 1974, Late Quaternary valley alluviation in the Driftless Area of southwestern Wisconsin: In: *Late Quaternary Environments of Wisconsin* (J. C. Knox and D. M. Mickelson, Eds.), American Quaternary Association Guide Book for Third Biennial Meeting, Madison, pp. 134-162.

Knox, J. C., Bartlein, P. J., Hirschboeck, K. K., and Muckenhirn, R. J., 1975, The response of floods and sediment yields to climate variation and land use in the Upper Mississippi Valley: University of Wisconsin, Madison, Institute for Environmental Studies Report No. 52, 75 pp.

Lacey, G., 1930, Stable channels in alluvium: Institute Civil Engineers Proceedings, v. 229, pp. 259-384.

Lamb, H. H., 1966, On the nature of certain climatic epochs which differed from the modern (1900-1939) normal: In: *The Changing Climate* (H. H. Lamb, Ed.), Methuen and Co., Ltd., London, pp. 58-112.

———————, 1972, *Climate: Present, Past, and Future:* Volume 1, *Fundamentals and Climate Now:* Methuen and Co., London, 613 pp.

———————, 1974, The current trend of world climate — a report on the early 1970's and a perspective: University of East Anglia, Norwich, Climatic Research Unit Research Publication No. 3, 28 pp.

Lane, E. W., 1937, Stable channels in erodible materials: American Society Civil Engineers, Transactions, No. 102, pp. 123-194.

Langbein, W. B., 1964, Geometry of river channels: Proceedings American Society Civil Engineers, Journal Hyd. Division, HY2, v. 90, pp. 301-312.

Langbein, W. B., and Leopold, L. B., 1964, Quasi-equilibrium states in channel morphology: American Journal Science, v. 262, pp. 782-794.

Leopold, L. B., and Langbein, W. B., 1962, The concept of entropy in landscape evolution: U. S. Geological Survey Prof. Paper 500-A, 20 pp.

Leopold, L. B., and Maddock, T., 1953, The hydraulic geometry of stream channels and some physiographic implications: U. S. Geological Survey Prof. Paper 252, 57 pp.

Lindner, C. P., 1969, Geomorphology: In: *State of Knowledge of Channel Stabilization in Major Alluvial Rivers* (G. B. Fenwick, Ed.), Committee on Channel Stabilization, Corps of Engineers, U. S. Army, Vicksburg, pp. I 1-I 25.

Mackin, J. H., 1948, Concept of the graded river: Bulletin Geological Survey of America, v. 59, pp. 463-512.

Patton, P. C. and Schumm, S. A., 1975, Gully erosion, northwestern Colorado: a threshold problem: Geology, v. 3, pp. 88-90.

Rouse, Hunter and Ince, Simon, 1957, *History of Hydraulics:* Iowa Institute of Hydraulic Research, State University of Iowa, Iowa City, 269 pp.

Rubey, W. W., 1933, Equilibrium conditions in debris-laden streams: American Geophysical Union Transactions, v. 14, pp. 497-505.

———————, 1952, Geology and mineral resources of the Hardin and Brussels Quadrangles (in Illinois): U. S. Geological Survey Prof. Paper 218, 175 pp.

Scheidegger, A. E., and Langbein, W. B., 1966, Probability concepts in geomorphology: U. S. Geological Survey Prof. Paper 500-C, 14 pp.

Schumm, S. A., 1973, Geomorphic thresholds and complex response of drainage systems: In: *Fluvial Geomorphology* (M. Morisawa, Ed.), Proceedings Volume, 4th Annual Geomorphology Symposium, Binghamton, N. Y., pp. 299-310.

Schumm, S. A., and Lichty, R. W., 1963, Channel widening and flood-plain construction along Cimarron River in southwestern Kansas: U. S. Geological Survey Prof. Paper 352-D, pp. 71-88.

Schumm, S. A., and Khan, H. R., 1972, Experimental study of channel patterns: Bulletin Geological Society of America, v. 85, pp. 1755-1770.

Schumm, S. A., and Parker, R. S., 1973, Implications of complex response of drainage systems for Quaternary alluvial stratigraphy: Nature (Physical Science), v. 243, pp. 99-100.

Smirnov, N. P., 1973, Spatial patterns of long-period streamflow fluctuations in the European USSR: Soviet Hydrology (Water Resources) Issue No. 2, pp. 21-32.

Smith, T. R., 1974, A derivation of the hydraulic geometry of steady state channels from conservation principles and sediment transport laws: Journal of Geology, v. 82, pp. 98-104.

Sorenson, C. J., and Knox, J. C., 1973, Paleosols and paleoclimates related to late Holocene forest/tundra border migrations: Mackenzie and Keewatin, N. W. T., Canada: In: Proceedings of *International Conference on the Prehistory and Paleoecology of the Western North America Arctic and Sub-Arctic*, (S. Raymond and P. Schlederman, Eds.), University of Calgary, Calgary, pp. 187-203.

Starkel, L., 1966, Post-glacial climate and the moulding of European relief: In:

World Climate from 8000 to 0 B. C., Royal Meteorologic Society, London, pp. 15-33.

———————, 1972, Trends of development of valley floors of mountain areas and submontane depressions in the Holocene: Studia Geomorphologica Carpatho-Balcanica, v. 6, pp. 121-133.

Stevens, M. A., Simons, D. B., and Richardson, E. V., 1975, Nonequilibrium river form: Proceedings American Society Civil Engineers, Journal Hyd. Division, HY5, v. 101, pp. 551-566.

Vita-Finzi, C., 1973, *Recent Earth History:* John Wiley & Sons, New York, 138 pp.

Wahl, E. W., 1968, A comparison of the climate of the eastern United States during the 1830's with the current normals: Monthly Weather Review, v. 96, pp. 73-82.

Wahl, E. W., and Lawson, T. L., 1970, The climate of the midnineteenth century United States compared to the current normals: Monthly Weather Review, v. 98, pp. 259-265.

Webb, T. and Bryson, R. A., 1972, Late- and post-glacial climatic change in the northern Midwest, USA: Quantitative estimates derived from fossil pollen spectra by multivariate statistical analysis: Quaternary Research, v. 2, pp. 70-115.

Wendland, W. M. and Bryson, R. A., 1974, Dating climatic episodes of the Holocene: Quaternary Research, v. 4, pp. 9-24.

Wisler, C. V., and Brater, E. F., 1959, *Hydrology:* J. Wiley & Sons, New York, 408 pp.

Wolman, M. G., 1955, The natural channel of Brandywine Creek, Pennsylvania: U. S. Geological Survey Prof. Paper 271, 56 pp.

Wolman, M. G., and Miller, J. P., 1960, Magnitude and frequency of forces in geomorphic processes: Journal of Geology, v. 68, pp. 54-74.

Woodford, A. O., 1951, Stream gradients and the Monterey Sea Valley: Bulletin Geological Society of America, v. 62, pp. 799-852.

Zeuner, F. E., 1952, *Dating the past: An Introduction to Geochronology:* Methuen and Co., London, 3rd Ed., 495 pp.

CHAPTER 11

TECTONICS AND GEOMORPHIC MODELS

Marie Morisawa

ABSTRACT

Both Davis and Penck realized that landforms evolved by the interaction of endogenetic and exogenetic forces working on the earth's surface. The Davis model was simplified to an initial uplift of a landmass which was then modified by denudational processes during a stillstand. Uplift and denudation were successive, although uplift may intermittently recur.

Penck, on the other hand, saw landforms as an expression of the continuous interaction of the tectonic and degradational processes. His landscape models depended on the comparative rates of uplift as against the rates of denudation.

Erosion, initiated by uplift of the land leads to a positive feedback situation as progressive erosion reduces elevation and promotes isostatic readjustment to cause further rising. If there is no delay in the feedback, the Penckian model results. If there is a delay in the feedback, where isostasy requires a threshold value, the Davisian model with intermittent upward movement is supported.

Current global tectonic theory considers the world's surface to be divided into plates which are rafted about. Vertical movements vary with spreading rates. Subduction, continental collision and transform faulting have affected oceanic and continental configuration. Models which combine global tectonics, erosion and sedimentation are needed to work out the geomorphic history of regions such as the Appalachians.

INTRODUCTION

American geomorphologists, with a few exceptions, have not been concerned with the influence of active tectonism on evolution of landforms. Rather, they have tended to look at the effects of inactive structures, such as folds, faults, domes and basins, on geomorphic forms. It is the purpose of this paper to examine some ideas on the relationship of active tectonics to landform development.

Both Davis and Penck realized that landforms evolved by the interaction of endogenetic and exogenetic forces working on the Earth's surface. The Davisian model (Fig. 1) was simplified for pedagogic purposes to an initial uplift of some dimension which was then modified by denudational processes during a tectonic stillstand. Uplift and denudation are successive events, although uplift may occur intermittently, rejuvenating a landscape over and over. It should be emphasized that Davis did realize that changes did occur during uplift. However, for convenience his model was built upon an hypothesized rapid uplift and stillstand. In this way, he eliminated many complications. As illustrated in Figure 1 and as elucidated by Davis (1909) and his many followers, an individual landscape is essentially dependent upon time, i.e., stage, since changes in landforms occur progressively and systematically.

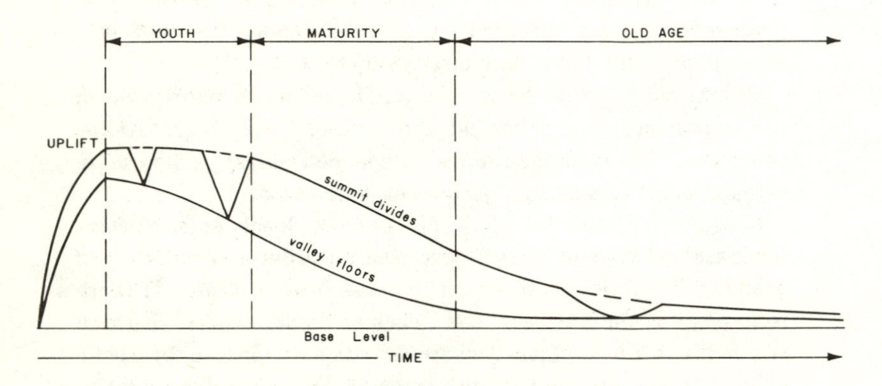

Figure 1. The Davisian model. After Davis (1909).

Penck, on the other hand, saw landforms as an expression of the continuous interaction of the tectonic and gradational forces, with both events occurring at the same time. His landscape models (Fig. 2) depend upon comparative rates of uplift and degradation. Penck (1953) stated that elevation occurs only if uplift is more rapid than denudation, that fault scarps are evident only if faulting takes place more quickly than erosion. The individuality of landforms arises from the ratio of the intensity of these two forces, one tectonic, the other gradational. In particular, slopes reflect the ratio of the endogenetic/exogenetic processes. Waxing slopes indicate that endogenetic forces are greater than the exogenetic ones. Waning slopes result when the endogenetic forces are less effective than the exogenetic. Equal retreat of slope occurs when the two opposing forces are equal in intensity or when tectonism has ceased. Interaction of the tectonic and gradational processes results in a continuously modified landscape. The Penckian model is one of unequal activity.

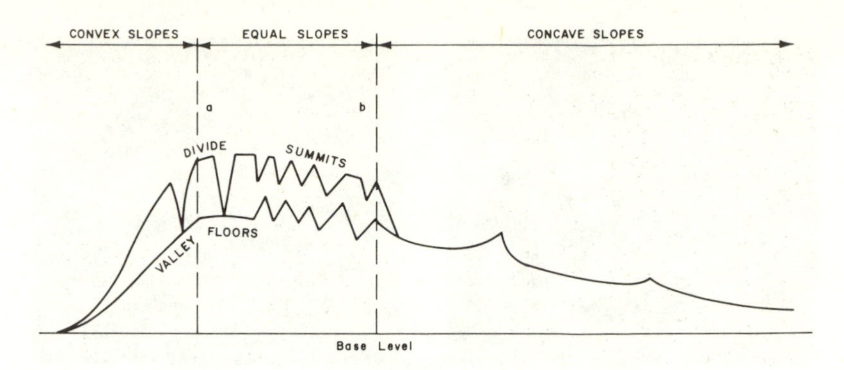

Figure 2. The Penckian model. After von Engeln.

IMMEDIACY OF MODIFICATION

If we look at the world around us, it is difficult to find a land surface which has been uplifted or recently formed which is unmodified by denudational processes. Still active volcanoes such as Yakedake, Japan show erosion scars (Fig. 3). Richards (1965) noted that slopes of a newly formed volcano, Barcena, Mexico, was well gullied only a month and a half after its creation.

Figure 3. Gullies on the slopes of Yakedake, Japan, an active volcano.

Fault scarps, such as those formed in 1959 by the Hebgen earthquake in northwestern Wyoming, were quickly changed (Fig. 4). They were so quickly changed that the Forest Service was worried lest the newly created Earthquake Park would lose its best examples of tectonic activity. Shorelines and former lake bottom, upraised by the quake, were quickly dissected (Fig. 5). A knickpoint, formed where the fault scarp cut across Cabin Creek,

Figure 4. Changes in the fault scarp, Cabin Creek campground, Hebgen Earthquake area, Montana.
A. One month after the quake.
B. Two years after the quake.
C. Four years after the quake.

was eroded almost completely away by the next spring snowmelt as it cascaded down the stream channel (Fig. 6).

From these examples it seems evident that when land is newly formed or newly uplifted, it is usually modified at once by exogenetic processes. The rate and extent of modification depends upon factors to be discussed.

INFLUENCE OF ELEVATION AND RELIEF

One would expect the exogenetic processes to act faster when uplift is great since elevation provides potential energy for erosion (Fig. 7). Glock's (1931) term "available relief" refers to the elevation difference in a drainage

Figure 5. Dissection of the shoreline and former lake bottom which was exposed by the Hebgen earthquake. Photograph taken in 1960 (July).

basin. In the Hebgen area it was discovered that streams with the greatest available relief developed on the upraised lake bottom more rapidly and enlarged their valleys more quickly than those with little relief (Morisawa, 1964).

In a study of a number of mid-latitude river basins, Ahnert (1970) determined that the mean denudation rate was highly correlated with basin relief. Statistically, 96% of the variation in denudation rates of the watersheds studied was attributed to mean relief. Ruxton and McDougall (1967) in evaluating erosion of volcanoes on Papua found that at 760 m relief, denudation occurred at a rate of 75 cm/1000 years. At 60 m relief, erosion took place at a much slower rate (8 cm/1000 years).

According to Yoshikawa (1974) denudation rates in Japan are closely related to rates of uplift in such a way that rate of degradation increases with maximum amounts of Quaternary uplift. In most of the catchment basins he studied, rates of uplift are higher than the rates of denudation. However, he found that denudation rates are higher than the rates of tectonic uplift in those areas which are located in the highest mountain regions of Japan. It is interesting that his data show the average rate of present-day denudation to be 0.840 M/1000 years, approximately equal to the modern average rate of uplift (0.863 M/1000 years). Isacks, *et al* (1973) estimated average rates of vertical movement at the present time in the Himalayas to be about 0.3 mm/1000 years. Using an average of Holeman's (1968) data for a number of rivers in southern Asia, the mean denudation rate of these streams which drain the Himalayas is 0.3 mm/1000 years.

A.

B.

Figure 6. Knickpoint on Cabin Creek. A. One month after the quake (Sept., 1959). B. Summer after the quake (July, 1960).

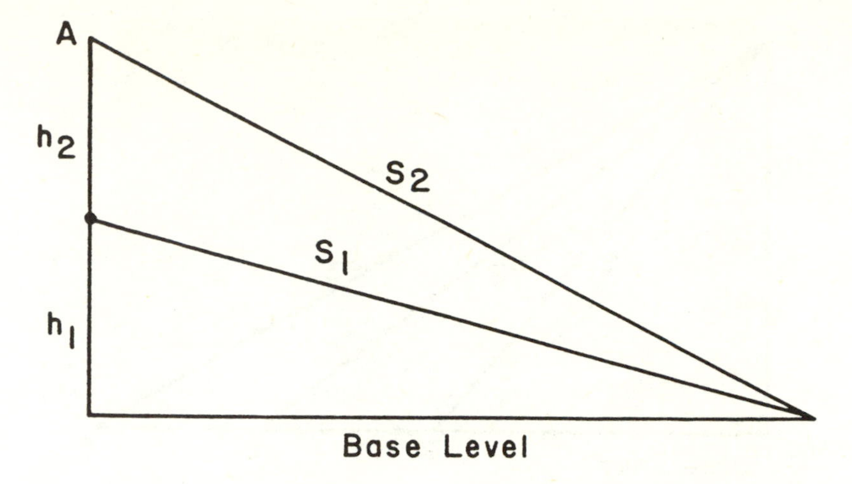

Figure 7. Difference in potential energy of two streams having different elevations. If both have the same base level, the available energy differs.

UNEQUAL FORCE AND RESISTANCE

Yoshikawa (1971) noticed the scatter of points on his diagram and attributed it to factors such as rainfall, lithology and vegetation. Hence, it is recognized that factors other than elevation or relief play a part in determining rates of surface changes. Landforms and their evolution may be distinctive because of inequalities in force or in resistance.

Three different stream profiles are represented in Figure 8. Imagine that all have the same discharge and the same elevation, but different slopes. The change in potential energy is the same, but the final kinetic energy available for work depends upon the length traversed during energy conversion. The greater the length (i.e., the gentler the slope) the greater is the loss of kinetic energy available for erosion and transportation. In such a case, the forces are unequal even though the elevation, or even relief, are the same. If, on the other hand, the elevation and slopes were the same, but the mass (discharge) differed, the kinetic energy ($\frac{1}{2}MV^2$) would also differ. Again the forces would be unequal although the elevation and relief might be the same.

Three knickpoints, formed during the 1959 Hebgen earthquake, were monitored for changes over a period of years. The knicks were on Cabin Creek, Red Canyon and an un-named tributary above Corey's Spring. Both the ten-foot scarp at Cabin Creek and the 20-foot scarp on Red Canyon were created on bouldery material. The third knickpoint, approximately 2 feet in height, was composed of small boulders, sand and silt. The streams are not gaged but a comparison of discharges can be made by examining the watershed areas, since a close relationship has been established between

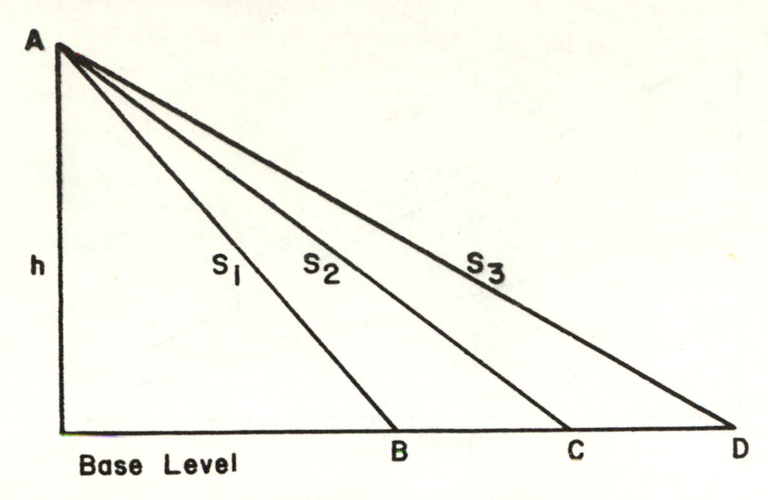

Figure 8. Difference in kinetic energy in drainage basins where streams have the same elevation and relief, but different gradients.

Table 1

Area, relief and elevation
Watersheds with knickpoints, Hebgen Lake

	Correy's Spring	Red Canyon	Cabin Creek
Area	0.41 mi^2	3.30 mi^2	33.2 mi^2
Relief	800 ft	1000 ft	1000 ft
Elevation (maximum)	7597 ft	9785 ft	10,664 ft

discharge and drainage basin area. Because of large snowmelt and dry summers, all the rivers have a widely fluctuating discharge. From the data (Table 1) Cabin Creek has approximately ten times the discharge of Red Canyon. Red Canyon Creek in turn, has about ten times the discharge of the third tributary, which is dry most of the time. As might be expected, the knickpoint on Cabin Creek was almost completely washed away the spring following the quake (Fig. 6). On the other hand, Red Canyon Creek still had a recognizable cascade at the scarp locality (although reaching further downstream) ten years after the quake (Fig. 9). The small tributary had changed little over the years (Fig. 10). In these cases, unequal force and unequal resistance resulted in unequal rates of change.

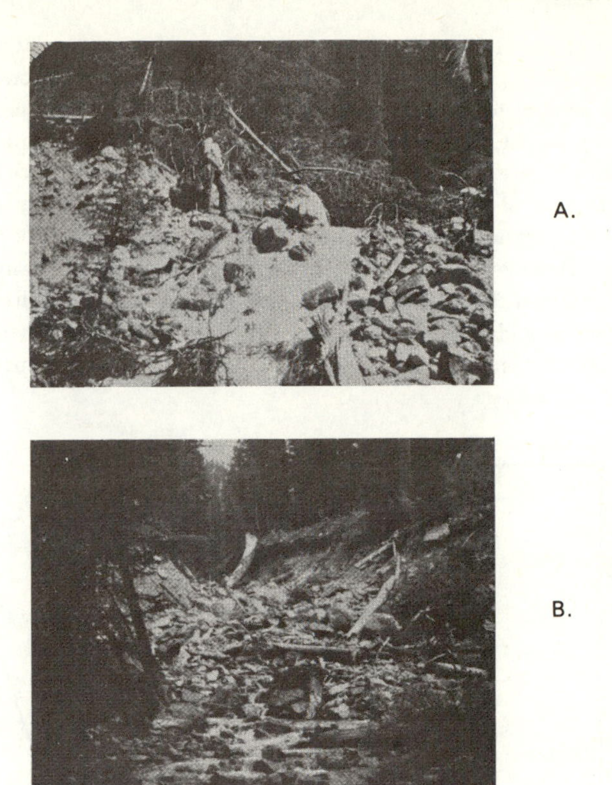

Figure 9. Knickpoint on Red Canyon Creek. A. Summer of 1960. B. Summer of 1969.

Figure 10. Knickpoint on tributary above Corey's Spring in the summer of 1972. Tributary comes in from the left and cuts across the scarp which extends to the right. Knickpoint is lowered and changed very little.

A study of the Hebgen scarps as they changed with time was also made (Morisawa, 1962). Statistical analysis by multiple regressions of various factors on amount of retreat (Table 2) indicated that ground slope above the scarp, scarp height, and percent of fines in the material account for most of the variance in rate of retreat. Figure 11 illustrates the distinct difference in rates of retreat of the scarps in alluvium and those in deeply weathered bedrock. At very low ground slopes above the scarps, rates of retreat were the same. However, weathered bedrock scarps with steep upper ground slopes maintained themselves at the same rate as alluvium with low ground slopes. One would expect this since the weathered bedrock material was angular and thus more stable (resistant) than the rounded material in the alluvium.

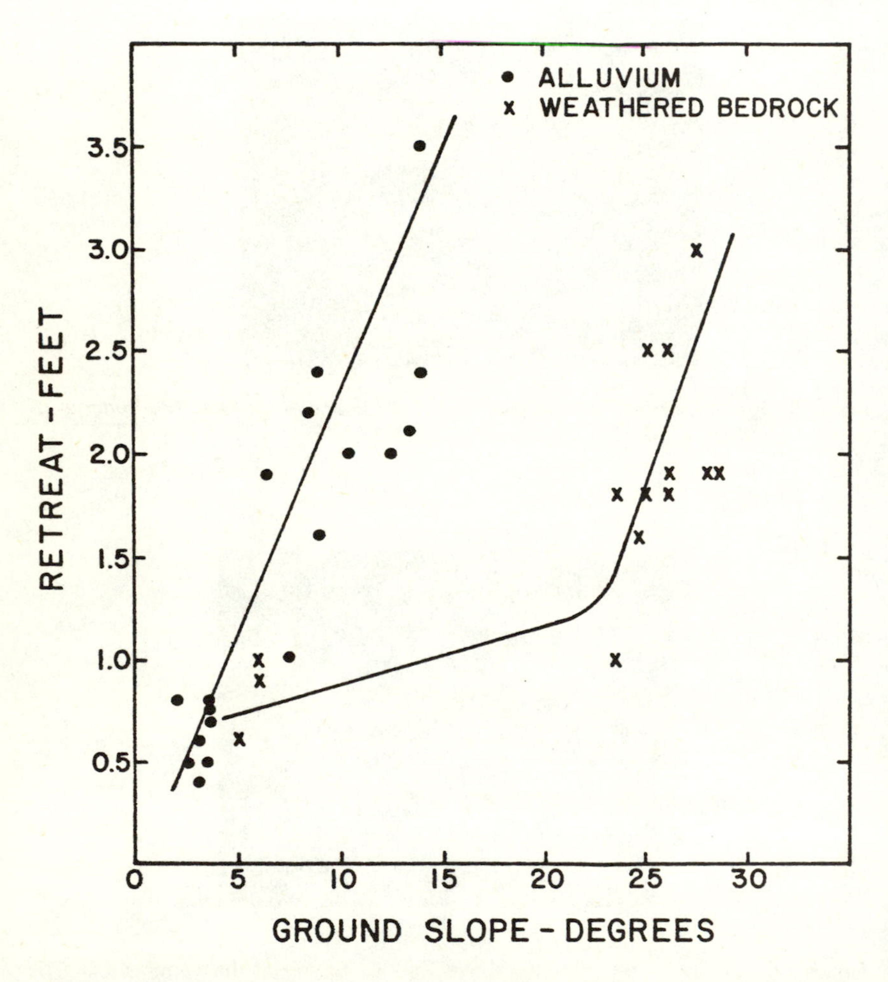

Figure 11. Retreat of earthquake scarps at Hebgen Lake plotted against upper ground slope. From Morisawa, 1962.

Table 2

Correlation coefficients for regressions of factors involved in retreat of scarps*

Factor	Correlation coefficients	
	Alluvium	Weathered bedrock
Ground slope above scarp (X_2)	.8871	.6935
Scarp height (X_3)	.4842	.3778
Percent fines (X_4)	.4436	.2163
X_2X_3	.8875	.6943
$X_2X_3X_4$	.9374	.8245

*From Morisawa, 1962

Some of the scarps formed by the Hebgen earthquake were rapidly eroded and affected by mass movement so that only a few years later they were healed and unrecognizable except to the expert eye (Morisawa, 1962). However, near the Hebgen area, on the west side of the Madison Range, there were older scarps which appear fresh and unchanged by weathering and erosion (Pardee, 1950). Slemmons, *et al* (1959) found that scarps formed in the Nevada quake of 1954 were older-looking, more weathered and obliterated than those created during earlier quakes. Along the western edge of the Wasatch Mountains, there are some extremely clean and unweathered scarps (Fig. 12). These are certainly older than either those studied at Hebgen or by Slemmons since they were referred to by Russell in 1885 and by Gilbert in 1890 (Fig. 13). They are still remarkably fresh-looking.

These examples illustrate that unequal forces, or unequal resistance to the same force will result in different rates of denudation. Unequal forces at work, or unequal resistance to the same force results in individuality and variety of landforms.

TENDENCY TOWARD EQUILIBRIUM

When a tectonic force is less than or greater than the denudational processes, there is a disequilibrium of action. The disequilibrium is temporary since there will be a tendency for equilibration of these two opposing forces. If uplift is rapid denudation lags and relief increases quickly. Energy from the growing relief causes denudational forces to increase until they are in balance with tectonic forces.

Figure 12. Scarp along the western edge of the Wasatch Mountains north of Nephi, Utah.

Figure 13. Scarp in Pleistocene moraine, at the mouth of Bell's Canyon, Utah. Photo by G. K. Gilbert, USGS.

On the other hand, if the denudational forces are stronger than the tectonic forces, degradation will gradually slow down because of diminishing relief and energy and finally reach an equilibrium of action with tectonism. If the figures quoted previously are correct, denudational and tectonic forces in Japan and in the Himalayas have reached an equilibrium of action at the present.

This equilibrium state may occur only while tectonism is waning and degradation is increasing, or vice versa. It is, thus, a transitory state since the earth is so dynamic. Nor is the equilibration static, because of the continuous influence of thermal, chemical, magnetic and other types of forces acting on and within the earth. One illustration is the effects of isostasy.

Erosion, initiated by uplift, reduces the land surface and provides material which is deposited in lower areas. This results in a positive feedback from isostatic readjustment where denuded continental areas continue to rise and basins of deposition continue to subside. Thus both erosion and deposition are reinforced, as are the vertical crustal movements. Isostatic readjustments may be immediate or delayed. A lag in feedback occurs when isostatic readjustment requires a threshold value. Uplift by delayed isostatic feedback causing renewed erosion, supports the Davisian model where intermittent upward movement results in erosion levels at different elevations. Continuous isostatic feedback supports the Penckian model of continuously changing ratios of uplift to denudation. According to this present analysis, both models may be expected in the geomorphic history of a region.

When the erosional forces act upon earth materials of differing resistance, there may be a temporary disequilibrium of action and form. However, there is a tendency toward establishing an equilibrium form for force and resistance. For example, a river adjusts its slope to provide just the energy required to move the load eroded. Although stream gradients may differ on different types of material, an equilibrium of action between resistance and energy is reached. Likewise, on a beach an equilibrium of wave energy and resistance of beach material is reached, and demonstrated by the slope of the beach. Steeper slopes are found on beaches where materials are more resistant to movement.

Thus where materials are resistant, they cause a temporary increase in energy. This results in a rise in force to equal the high resistance. Where materials are low in resistance, high energy is quickly reduced to equal the low resistance.

It seems, then, that there is a tendency toward equilibrium in nature. The equilibrium is between forces, or between force and resistance and is expressed by landforms.

GLOBAL TECTONICS AND LANDFORM CHANGES

Current theories of global tectonics consider the world's surface to be divided into plates which are moved about (Dietz and Holden, 1970).

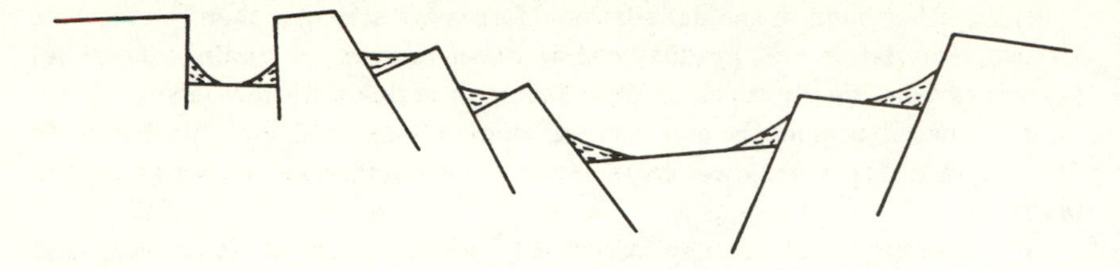

Figure 14. Block faulting and graben formed by rifting.

Comparative movement of plates at junctions may be divergent (away from each other), convergent (towards each other) or obliquely past each other as transform faults. Divergent plate junctions create rifting within the ocean or continent, or at an ocean-continent boundary. An ocean rift is the site of generation of new sea floor, as is presently occurring in the mid-Atlantic ridge. Continental rifting forms downfaulted structures such as the Rhine graben, Lake Baikal or the East African rift valley, and may eventually result in the break-up of a continent.

Rifting causes major subsidence with block faulting and the formation of troughs and grabens (Fig. 14). Tectonic movement is accompanied by basaltic flows. Geomorphic effects will include entrenchment of streams across the uptilted edges of the fault blocks and erosion of deep gorges. Drainage reversals, stream piracy and water gaps may occur as erosion proceeds. The overall stream pattern may become oriented along fault block edges in a parallel alignment. Internal drainage basins may be created in the downthrown areas between blocks. Aggradation will occur in these subsiding basins.

Converging plates give rise to different types of terrain, depending on whether the boundary lies between an island arc and oceanic plate, an oceanic and continental plate, or two continental plates. The convergence of a denser oceanic plate with a lighter island arc or continental plate results in subduction and underthrusting of the denser plate beneath the lighter. This causes an orogenic welt with doming, folding and volcanism on the lighter plate (Dewey and Bird, 1970). A Cordilleran type mountain range (Andes Mountains) rises where a continental edge is underthrust by the leading edge of an oceanic plate. Such tectonism gives high elevations which are maintained as subduction continues.

Geomorphic effects of such a convergence are noticeable primarily along the coasts. In such regions landforms such as raised coastal terraces, and uplifted sea stacks and sea arches are common. These are evident on island arcs such as Japan, New Zealand and New Guinea as well as along the shorelines of both western North and South America. Increase in the erosive ability of rivers will cause channel incisement, high level terraces and deep bedrock canyons. Because of continued movement, many of the terraces will be deformed, and numerous knickpoints will mar the smoothness of longitudinal river profiles.

The convergence of two continental plates results in the welding of the two continents with a compressive orogenesis (Dewey and Bird, 1970). This is occurring today as India collides with the Tibetan plateau (Molnar and Tapponnier, 1975) and the Himalaya Mountains rise. A continental collision in the past produced the Appalachian Mountain range (Dewey and Bird, 1970). The high altitudes and uplift have extensive geomorphological effects. Rapid dissection and incisement of rivers takes place with the production of high level terraces and huge deltaic deposits; continued tectonism will cause stepped features such as a series of terraces, benches or knickpoints. Major rivers already well established may be able to maintain their courses across the rising land surface resulting in antecedent drainage. According to King (1967) the Indus River is such an example of an antecedent river which downcut its channel as the Himalayas rose across its path. By analogy, some of the major Appalachian rivers may also be antecedent.

Transform fault boundaries, such as the San Andreas rift zone (Anderson, 1971) are characterized by fault scarps and long lineaments. Other noteworthy geomorphic features are disrupted drainage with sag ponds, misaligned stream courses and water gaps.

Some gross continental morphologic features may also be ascribed to plate tectonics. The low east coasts of North and South America are stretched-out, trailing edges of continental plates moving westward. The topographic asymmetry of these two continents result from this and the additional fact that both have west coasts which are leading edges of continents being underthrust by oceanic plates. The drainage patterns of both American continents are also guided by global tectonics. The major drainage of South America is towards the trailing edge of the continent, whereas streams flowing west to the Pacific are short. In North America the major drainage is southward to the Gulf reflecting the fact that the continent rifted apart from Gondwana first by opening up the present Gulf of Mexico (Le Pichon and Fox, 1971).

India has this same geomorphic pattern. Topographic asymmetry is present, with the northern leading edge being very high and the southern trailing edge strung out and low. Major drainage is southward through incised canyons and over low plains.

The Appalachian region exemplifies the relation between plate tectonics and geomorphic history and landform development. During the upper Paleozoic Gondwanaland was formed by the collision and welding together of North and South America and Europe (Dewey and Bird, 1970). The Appalachians rose as part of a broad domal uplift formed at the suture (Fig. 15). Drainage would have been off and away from this high mountain area. On the North American continent this would have been toward what is now the mid-continent. Rivers would entrench themselves into the rising

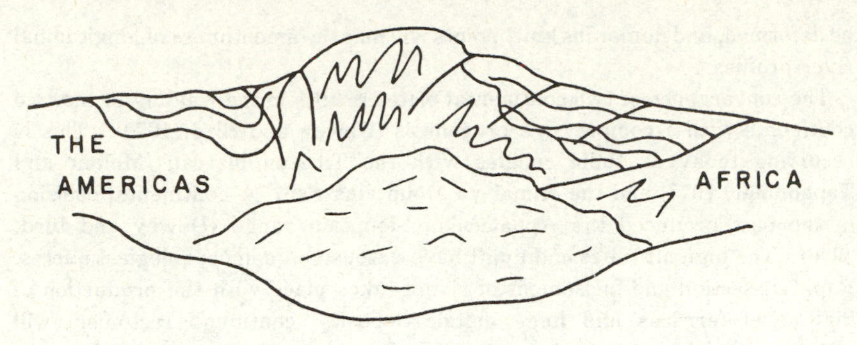

Figure 15. Uplift and mountains formed by the welding together of two continental plates.

mountains with attendant terraces, knickpoints, canyons and thick deltaic and alluvial flood plain deposits.

During the Mesozoic rifting began and the mountains were downfaulted in a system of blocks. This tectonic event caused rejuvenation of many rivers as they flowed across the tilted blocks (Fig. 14). Drainage divides and stream courses shifted and may even have been reversed as stream piracy occurred. Thick sequences deposited in the downfaulted block basins attest to the vigor of erosion during this time while Gondwanaland was drifting apart. Cretaceous onlap onto the subsiding Atlantic Coast reflects the widening of the gap between the continents. Although lowered in intensity and pace, these tectonic and geomorphic events continued during the Tertiary.

Thus, the skeleton of the geomorphic history can be sketched but the details are lacking. For example, can the knickpoints in the longitudinal profiles of rivers or the "peneplains" of the Appalachians be integrated with the sediment wedges of the coastal plain and with plate movements in a tectono-geomorphic model (Morisawa, 1974)? These are the kinds of details that need to be explored so that the geomorphic history of the Appalachians, or any other region, can be clearly understood.

Thus the new global tectonics can explain some of the general landscape features of the earth. But, although general statements can be made about the relation of plate tectonics to geomorphic development (Garner, 1974), more detailed regional models are needed. The application of the concepts of plate tectonics should solve some of our old geomorphic problems.

CONCLUSIONS

1. Landforms are the result of inequality of force and/or the inequality of resistance. These principles provide for much of the variety in the landscape. The natural consequence of the inequality of the rates of the endogenetic and exogenetic processes at work on different earth materials is individuality of landforms.

2. Nature is striving to reach an equilibrium of force and resistance but cannot always do so because the earth is so dynamic. Because conditions are always changing there is only a *tendency* towards equilibrium. A vital part of the dynamic earth system is the isostatic feedback which operates to affect rates of erosion/uplift and deposition/subsidence. The landscape which we see about us is the result of the differences in the ratio of endogenetic/exogenetic activity as it varies from time to time and place to place. Thus landforms are often composite and complex making it difficult to understand their origins and evolution.

3. Plate tectonics can help to explain some of the landscape features of the earth. However, more detailed regional studies are needed in order to integrate tectonic and geomorphic events into a reasonable evolution of the present landscape.

REFERENCES CITED

Ahnert, F. (1970) Functional relationships between denudation relief and uplift in large, mid-latitude drainage basins: Amer. Jour. Sci. 268, pp. 243-263.

Anderson, D. L. (1971) The San Andreas fault: Scientific Amer., 225, (5), pp. 66-75.

Davis, W. M. (1909) Geographical Essays. Ginn, Boston (republished 1954, Dover, N. Y.), 777 p.

Dewey, J. F. and Bird, J. M. (1970) Mountain belts and the new global tectonics: Jour. Geophys. Research 75 (4), pp. 2625-2647.

Dietz, R. S. and Holden, J. C. (1970) Reconstruction of Pangaea: break-up and dispersion of continents, Permian to present: Jour. Geophys. Research 75, pp. 4939-4956.

Garner, H. F. (1974) The Origin of Landscapes. Oxford Univ. Press, N. Y., 734 p.

Gilbert, G. K. (1890) Lake Bonneville. U. S. Geol. Survey, Monograph I, 438 p.

Glock, W. S. (1931) The development of drainage systems: a synoptic view: Geog. Rev. 21, pp. 74-83.

Holeman, J. N. (1968) The sediment yield of major rivers of the world: Water Resources Research 4, (4), pp. 737-747.

Isacks, B., Mueller, I. I., Walcott, R. I. and Talwani, M. (1973) Vertical crustal motions and their causes: slow vertical movements in continental interiors: EOS 54, (12), pp. 1257-1259.

King, L. C. (1967) The Morphology of the Earth. Hafner, New York, 699 p.

Le Pichon, X. and Fox, P. J. (1971) Marginal offsets, fracture zones and the early opening of the North Atlantic: Jour. Geophys. Research 76, (26), pp. 6294-6308.

Molnar, P. and Tapponnier, P. (1975) Cenozoic tectonics of Asia: effects of a continental collision: Science 189, (4201), pp. 419-426.

Morisawa, M. (1962) A study of geomorphic changes in earthquake features, Hebgen Lake, Montana: Final Report, ONR Contract Nonr-3254 (00), Project NR 389-130. Montana State Univ., Missoula, Mont.

——————— (1964) Development of drainage systems on an upraised lake floor: Amer. Jour. Sci. 262, pp. 340-354.

——————— (1974) Plate tectonics and geomorphology: Recent Researches in Geology, Dept. of Geology, Univ. of Delhi, India, pp. 269-282.

Pardee, J. T. (1950) Late Cenozoic block faulting in western Montana: Geol. Soc. Amer. Bull. 61, pp. 359-406.

Penck, W. (1953) Morphological Analysis of Land Forms. Translated by H. Czech and K. C. Boswell. Macmillan, London, 429 p.

Richards, A. F. (1965) Geology of the Islas Revillogigedo, effects of erosion on Isla San Benedicto, 1952-61 following the birth of the volcano Barcena: Bull. Volc. 28, pp. 381-403.

Russell, I. C. (1885) Geological history of Lake Lahontan. U. S. Geol. Survey, Monograph XI, 288 p.

Ruxton, B. P. and McDougall, I. (1967) Denudation rates in northeast Papua from potassium-argon dating of lavas: Amer. Jour. Sci. 265, pp. 545-561.

Slemmons, D., Steinbrugge, K. V., Tocher, D., Oakeshott, G. B., and Gianella, V. P. (1959) Wonder, Nevada earthquake of 1903: Bull. Seismol. Soc. Amer. 49, pp. 251-265.

von Engeln, O. D. (1942) Geomorphology. Macmillan, New York, 655 p.

Yoshikawa, T. (1974) Denudation and tectonic movement in contemporary Japan: Bull., Dept. of Geog., Univ. of Tokyo, No. 6, pp. 1-14.

CHAPTER 12

THEORIES OF THE DEVELOPMENT OF KARST TOPOGRAPHY

Richard L. Powell

ABSTRACT

Early researchers of karst or carbonate bedrock terrains included hydrologists, mining engineers, and geographically and geologically oriented geomorphologists. Many descriptive papers, mostly of a combined physiographic and geomorphic content, were assimilated periodically in a rather purely geomorphic context. Simultaneous evolution of dual theories about karst landform development and groundwater movement in carbonate rocks resulted in an application of the geographical cycle concept to karst landscapes. There is an abundance of recent literature in the United States on the topics of cavern development and geohydrology of carbonate bedrock, whereas research on karst features *per se* has declined.

A dual goal of karst geomorphic research should be to first explain the origin of different karst forms by the various processes which form them under different climatic regimes and, secondly, to correlate the evolution of the karst landscape with geomorphic surfaces and temporal events in adjacent, non-karst landscapes.

The concept of control factors such as structure, process, and stage may still be used in karst landform analysis if a more liberal meaning of these terms is accepted. Structure should include all properties of carbonate and non-carbonate earth materials on which karst landscape develops. The intensity of all aggradational and degradational processes that cause development of a particular karst landscape should be studied rather than studying solution processes alone. Stages or episodes, albeit non-cyclic, that may be recognized in any karst landscape should be correlated with postulated episodes in adjacent non-karst areas and the climatic conditions prevalent during each episode or stage should be indicated.

INTRODUCTION

The study of karst landforms or topography is considered a branch of geomorphology concerned with landforms developed on soluble rocks and characterized to some degree by internal or subterranean drainage.

This simple definition is inadequate, however, in that it fails to specifically encompass other branches of geology such as structural geology, stratigraphy, carbonate petrology, geohydrology and geochemistry, as well as speleology, that should be a part of modern studies related to karst or carbonate terrains. Jennings (1971, p. 10-22) for example, recognized the large number of rock types capable of supporting karst features and utilized the classification of carbonate rocks by Folk (1959). He emphasized that karst terrain has "distinctive characteristics of relief and drainage arising primarily from a higher degree of rock solubility in natural waters than is found elsewhere," (Jennings, 1971, p. 1). Sweeting (1973, p. 1) initially claimed, "all karst regions are areas of massive limestones", but her misuse of the term limestone to include all carbonate rock types was clarified somewhat later when she referred to the classifications of different carbonate rock types by Folk (1959). The use of the term massive is incorrect, in that several classical karst areas contain other than massive strata.

Thornbury (1969, p. 303-304) stated that "karst is a comprehensive term applied to limestone or dolomite areas that possess a topography peculiar to and dependent upon underground solution and diversion of surface waters to underground routes". He indicated that solution forms could develop in gypsum or salt, but that areas underlain with such rocks are of limited areal extent and therefore are minor in comparison to more significant, notable karst areas developed on limestones or dolomites. He also stated that caverns alone do not constitute karst topography.

Not all areas of readily soluble bedrock are characterized by normal, well developed, or significant karst features, yet they are all characterized by some ground water condition dependent on the development of internal drainage denoted in part by the terms carbonate or limestone porosity or permeability. Many such soluble rock areas are characterized by minor or poorly developed karst forms. Some areas of soluble strata characterized by caverns are overlain with insoluble rocks, yet collapse features breach the surface, or, the insoluble rocks are an integral part of the ground water system contributing to cavern development. Thus, there is sufficient reason to consider the studies of karst geomorphology, carbonate petrology and stratigraphy, carbonate geochemistry, karst or carbonate groundwater hydrology, and speleogenesis, as closely allied branches of geology.

PURPOSE

A primary purpose of this paper is to briefly summarize research in various fields related to karst studies in the United States. A particular emphasis is placed on early American works to indicate the extent of knowledge related to karst or cavernous terrains in order to point out some deficiencies in subsequent works. The intention is to show shortcomings and successes of American karst researches to indicate necessary areas of future study rather than to make a comparison with foreign works. Some recent papers are mentioned as examples to indicate the significance of current research in fields such as carbonate petrology, hydrogeology, geochemistry, and speleology, related to karst geomorphology.

The purpose of citation here is to offer examples. Bibliographies and references are available for researchers wishing to pursue some aspect of study in detail (for example, LaMoreaux, Raymond and Joiner, 1970; Warren and Moore, 1975; Herak and Stringfield, 1972; Sweeting, 1973, p. 336-352; Jennings, 1971, p. 299-241; Stringfield and LeGrand, 1969, p. 410-417; and LeGrand and Stringfield, 1973, p. 117-120).

Research restricted to karst geomorphology is generally lacking in the United States in recent decades. Contrarily, considerable American researches on carbonate terranes have evolved from earlier studies in hydrology and geochemistry and in speleology, all of which, are directly related to karst geomorphology in a broad sense.

A study of karst geomorphology must include a familiarity with background literature concerning the distribution and description concerning the karst features *per se*, but should also include allied reports on groundwater conditions, geologic structure, stratigraphy or lithology, hydrology, regional geomorphic history and speleology. Certainly much of the reason for argument related to development of universal models of karst and cavern development is that each major carbonate bedrock area has a somewhat unique structure, lithology, climate, hydrologic and geomorphic history that imparts special conditions to that region. The published descriptions are adequate proof that other than the simplest statement will not fit all conditions. A summary of some landmark papers follows, somewhat in chronological order, to show the simultaneous development of karst geomorphology, carbonate groundwater hydrology and speleology.

REVIEW OF KARST-RELATED LITERATURE

One of the earliest reports concerning the origin of karst features must be attributed to Brown (1854, p. 309) regarding the origin of sinkholes:

> "The principle on which these excavations are made is a very simple one. At first, a mere fissure exists in the stone beneath, through which the rainwater from the surface, fully saturated with carbonic acid, finds it way downward. But water charged with free carbonic acid, is a rapid solvent of limestone. By this process the exposed surfaces of the fissure are dissolved and carried away, thus at once saturating all the spring water with

> carbonate of lime, and widening the fissure, until the aperture becomes so large that the incumbent earth begins to fall into it, which in turn is borne away by the increased volume of water collected by the funnel-shaped depression on the surface thus produced, until the face of the country becomes dotted over with these "sink holes," and subterranean caverns abound."

Many of the earliest works are mostly descriptive, and are commonly based on observations rather than measurements. Many are more concerned with caverns than karst features. Hovey (1882, p. 15-17), for instance, stated that rainwater, which is capable of dissolving limestone absorbs carbonic acid from vegetation as it sinks into the soil. He indicated that sinkholes were circular depressions leading to crevices that channel surface waters into an underlying cavern which is the gathering bed for the accumulated waters. The acidulated water entering sinkholes carries pebbles which cut through the successive floors of the cavern, and thus enlarge its dimensions. The cave streams emerge from hillsides as springs.

Some of the early lucid statements concerned with karst and cavern development in general are those of Posepny (1894, p. 212-216). He noted that a portion of the atmospheric precipitation sinks through open fissures or directly through permeable rock and fills them to a certain level, called the water level or groundwater level. This level is actually an inclined plane sloping towards the lowest point of the surface topography or to where impermeable rock crops out. This water is not stagnant, but moves towards a surface outlet at a rate depending upon the height it stands above the outlet and the size of the intervening openings.

> "For the part of the subterranean circulation, bounded by the water-level, and called the vadose or shallow underground circulation, the law of the descending movement holds good in all cases, even in those complicated ones which show ascending currents in parts. The total difference in altitude between the water-level and the surface-outlet is always the controlling factor (Posepny, 1894, p. 213).
>
> Peculiar conditions are created by the occurrence of relatively soluble rocks, such as rock-salt, gypsum, limestone and dolomite, in which, by the penetration of meteoric waters and the circulation of the ground-water, connected cavities are formed, constituting complete channels for the vadose circulation. . . . The water flowing at the bottom of a cave in limestone is unquestionably ground-water; and it follows that the whole complex group of cavities has been eaten out by it. If in another limestone cave we see no flowing water, the current must have found some lower outlet; and the cave represents for us an ancient ground-water channel," (Posepny, 1894, p. 214).

Caverns are formed in limestone, as well as other soluble rocks, when a route of maximum circulation exists between the entrance of water into the limestone and above the exit. The circulation follows open fissures, enlarging them by solution, not necessarily along a parabolic course as with homogenous materials, but influenced by stratification, unequal solubility, or intermixture of impermeable rocks, causing a change in direction, even an upward inclination in places. A change in the position of the outlet, such as progressive erosion of valleys, can cause a change in level and direction of the new channel (Posepny, 1894, p. 216).

The groundwater movement concepts of Posepny, and others, were classified into three zones by Finch (1904, p. 201-209), in which Zone I was the zone of accumulation or water gathering in fissures or joints, the "belt of weathering" of Van Hise (1901); Zone II includes the part of the belt of saturation, in any region, which has a means of horizontal flow and discharge that results in a continuous condition of gravitative flow along prominent fissures in the rock; and Zone III consisting of essentially static water. The position of the boundary between Zones I and II, the top of the saturated zone or the water table, varies with the topography, rainfall and the nature of the rock. Thus, we assume, karst features would therefore form in Zone I, the belt of weathering, while cavern development would most rapidly progress in Zone II.

The concept of weathering as the process of karst and cavern development was indicated by Cumings (1906), although he stated that solution was the predominant weathering process. Cumings demonstrated that the limestone bedrock of Indiana was furrowed with solution-widened joints and sinkholes and drained in part by caverns owing to three conditions in soluble rocks: the presence of an elaborate system of joints, numerous relatively impermeable layers, and adequate drainage provided by entrenched streams. Cumings (1906, p. 89) apparently originated the somewhat erroneous concept that cavern development is more extensive in fine-grained, conspicuously jointed limestone (Mitchell Limestones) in contrast to more porous limestone with wider joint spacing (Salem Limestone). Green (1909), first among many to reiterate the concepts of Cumings, indicated the development of sinkholes by subterranean water infiltrating down to a level where caverns are formed. Greene also discussed the collapse of cavern passages by lateral erosion to form large rooms and collapse sinkholes, the up-slope progression of a cave entrance by collapse, straightening of angular passages by erosion, differential solubility, and the solutional development of pits or domes by trickling water.

Bowman (1907, p. 10-11), described the significance and independent nature of individual sinkholes as collecting basins for surface water, apparently using the term "karst" for the first time in American literature. Purdue (1907), outlined the development of sinkholes formed by solution at the bedrock surface as tributaries to caverns by way of tubular passages, in contrast to collapse sinkholes. Sellards (1907 and 1908), however, argued that data from Florida demonstrated that many sinkholes there resulted from sudden collapse of strata. The development of solution-widened joints, called cutters, under a mantle of phosphatic residuum and soil was described by Hook (1914, p. 64-65). Logan (1918) indicated that sinkholes in gently dipping strata commonly were unsymmetrical, steep and rock-walled in the downdip direction, more gently sloping on the other sides, and commonly elongated parallel to the dip direction. Swinnerton (1937) however, found that sinkholes in moderately steeply dipping strata were commonly steepest on the downdip side and were elongated parallel to the strike of the strata.

Although William Morris Davis is credited with the concept of the geographic cycle, the application of this concept in American literature to both karst landscapes and the development of cavern systems must be attributed to Beede (1911), apparently independently from the work of Grund (1903). Beede interpreted stages of karst and cavern development and attempted to relate each episode with entrenchment of surface streams and the establishment of a lower base level below a peneplain. Beede postulated a youthful stage characterized by progressive subterranean capture of surface streams and scattered solution sinkholes; a mature stage in which the drainage is mostly underground and the surface valleys of youth are entirely segmented by solution sinks and some collapse sinkholes progressing towards the collapse of subterranean channels; and, a stage of old age in which a plain drained by normal streams is formed as the last caverns are obliterated by collapse and erosion (Fig. 1).

Sanders (1921) presented a translation of a scheme of cyclic erosion in a karst region by Cvijic (1918). The paper presented many Yugoslavian terms for karst features and presented block diagrams to depict the structural and hydrologic conditions of the classical karst area, including a set of four diagrams to illustrate Cvijic's four stages of evolution of karst topography. Karst features such as scattered dolines begin to appear accompanied by a progressive loss of surface drainage during the youthful stage of development. Maximum cavern development and many sinkholes characterize the mature stage, with collapse of caverns and the formation of valleys by cavern collapse marking the late mature stage. The old age stage is characterized by erosional removal of the limestone with only scattered residual limestone hills remaining. Wray (1922) presented a descriptive paper on the karst of Yugoslavia in which he interpreted the landscape in terms of geologic events rather than by cyclical evolution.

Geologic structure, that is, the inclination or dip of the strata in contrast to lithology, was considered to be an important factor influencing the direction of ground water movement. Beede (1911, p. 101) indicated that the dip of the limestone strata favored the flow of subterranean water from a higher surface drainage basin to a lower entrenched surface stream. This example of subterranean stream piracy was elaborated upon by Malott (1922, p. 197-203). Malott correlated the geographic stages of karst landform and

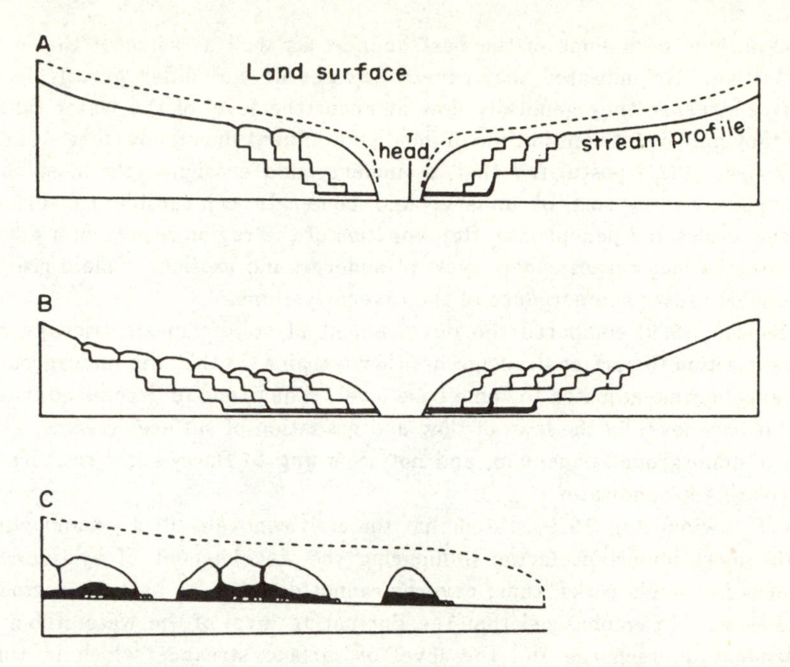

Figure 1. Stages of karst and cavern development according to Beede (1911, p. 83 and 85) showing profile of youthful stage — A, mature stage — B, and old age stage — C (enlarged).

cavern development with major episodes of rejuvenation, that is, development progressed during stream entrenchment of a peneplain and eventual establishment of graded stream conditions at a lower level.

J. M. Weller (1927) roughly correlated multiple cavern levels and karsted hanging valleys in Kentucky with stages of stream entrenchment below a peneplain, adding that each level was accompanied or followed by a lowering of the water table. He also recognized that sedimentation of the cavern passages was caused by erosion and transportation of surface materials through sinkholes, and, that dripstone deposits are formed in air-filled passages by water seeping into the cave passage.

The general principle of ground water movement and types of ground water were outlined in two papers by O. E. Meinzer (1923a and b), essentially following the classification of Daly (1917, p. 495-499). He defined vadose water as ground water above the ground water table and phreatic water as water below the water table in the zone of saturation, but yet within the zone of rock fracture. The term "phreatic" was invented by Daubree (1887, p. 19) to denote water that would supply an ordinary well, generally synonymous with groundwater immediately below the water table (Daly, 1917, p. 494-495). Meinzer (1923a, p. 34-35) also recognized the cycles of fluctuation of the water table, which he called phreatic fluctuation, at various, somewhat simultaneous, short to long intervals. Meinzer (123b, p. 131) recognized that carbonate rocks include both some of the best aquifers as well as some of the least productive. He indicated that cavern streams do not differ greatly from surface streams; they generally flow at about the level of the water table, and they may be influent in time of flood and effluent during low flow stages.

Meinzer (1927) postulated that an underground erosion cycle in soluble rocks produces a sort of underground peneplain comparable to surface erosion cycles and peneplains. Rejuvenation of the region results in a lower base level which causes a new cycle of underground erosion, while a rise in base level causes submergence of the cavern systems.

Meinzer (1929) compared the development of subterranean drainage by solvent action to surface drainage in other terrains, in that the underground streams become adjusted to some base level. They tend to become adjusted to this base level by the laws of flow and gradation of surface streams, as a sort of underground peneplain, and not according to Darcy's law relating to percolating groundwater.

A. C. Swinnerton (1929) stated that the active movement of groundwater is the most important factor influencing the development of systems of caverns in soluble rocks, thus, caverns cannot develop far beneath regional base level. He emphasized that the fluctuating level of the water table is dependent on recharge and the level of surface streams, which in turn depends on climate and the regional base level.

Many of the factors important to the development of karst features and caverns were discussed by Addington (1929, p. 252-253). He indicated that the topography of a region is not solely due to the process operating on it, but is conditioned to some extent by climate, by the attitude, exposure and lithology of the bedrock, the altitude and degree of stream entrenchment, and by the amount of time the process of denudation has been active. Addington (1929, p. 254-260) mentioned the development of sinkholes, sinkhole ponds, rejuvenated sinkholes, subterranean stream piracy, cavern development, and limestone springs in a region which has undergone drainage adjustments from a former peneplain surface to alluvial terrace levels and filled valley conditions.

Although the concept that some karst areas were sculptured by fluvial processes prior to the development of the karst features was implied by some authors and stated by a few, this view was not everywhere considered applicable. Ward (1930) proposed six reasons for why he believed the Somerville peneplain of eastern Pennsylvania to be a result of solutional lowering and beveling rather than the generally accepted schemes of base level beveling by streams. Karst features, cavern development, and geohydrology are not discussed in detail, but sinkhole development was considered in general as a predominant process.

The development of depressions or sinkholes in predominantly non-karst and non-carbonate landscapes was demonstrated in several papers in the late 1920's and 1930's. Russell (1929) had argued that depressions in western

Kansas were caused by deformation, faulting and subsequent collapse, but Elias (1930 and 1931) indicated solution at depth was responsible for the extensive, progressive collapse of overlying strata. A sinkhole formed by collapse of about 800 feet of non-carbonate strata above a postulated limestone cavern was described by Stockdale (1936). Such instances of sinkhole development indicates the extent to which cavern development may cause "karst" features in otherwise non-karst terrain.

Perhaps no single karst feature in the United States has been as controversial as the Natural Bridge in Virginia. Reviews of the literature were presented by Malott and Shrock (1930) as well as Wright (1936), both of whom presented subterranean stream piracy theories with different geomorphic histories. These theories were adopted, more or less, by Woodward (1936) to also explain Natural Tunnel in Virginia.

William Morris Davis (1930), in the first compendium of American cavern literature, challenged existing theories of cavern development. Davis (1930, p. 481) had translated part of a paper by Grund (1910) as follows:

> "Water-filled caverns give us an indication of the origin of dry caverns. They also were once at the level of karst water and were formed by it. The karst water level sank in consequence of valley erosion or regional elevation and the water-filled caves were thereby laid dry. The partial filling of such caves with dripstones shows that they were not dissolved out by vadose water."

Davis postulated what has become known as the two-cycle concept of cavern development. He indicated that two cavern cycles occurred during only two epochs or stages of the four stages of the karst cycle that he proposed. The first cycle of cavern development took place by random deep phreatic flow below the water table during an old age or peneplain stage of a limestone bedrock area (Fig. 2). The second cycle of cavern development, that of dripstone deposition, took place in a young stage of karst development in the zone of aeration caused by rejuvenation of the limestone mass owing to regional uplift and lowering of the water table. Davis' primary thesis was that dripstone deposits were emplaced in cavern passages only after their excavation had been completed. He stated that two distinct epochs of cavern development are thus recorded, an early epoch of solutional or corrasional excavation followed by a later epoch of deposition, and that adequate reason should be found for the change. He also indicated that American authors have not considered the meaning of the network arrangement other than branching arrangement of large cavern galleries (Davis, 1930, p. 481).

Davis made several assumptions basic to his thesis that were obviously erroneous so far as contemporary general knowledge was concerned: i.e., he 1) assumed a thick mass of homogenous rock, 2) postulated that large caverns formed as networks of channels and were later modified to form a branching network, 3) ignored anisotropic movement of ground water along

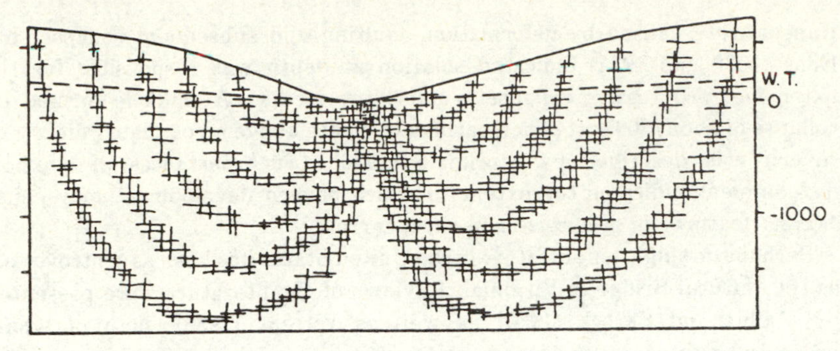

Figure 2. Ideal paths of groundwater movement in limestone according to Davis (1930, p. 549).

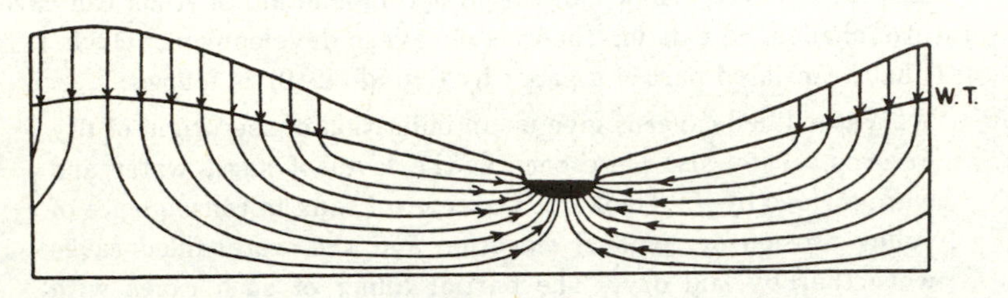

Figure 3. Theoretical paths of groundwater flow in homogenous drift (after King, 1899, p. 99).

joints, and 4) misapplied a model by King (1899) designed for shallow flow in glacial drift (Fig. 3).

A. C. Swinnerton (1932) lauded the attempt by Davis to explain the genetic relationship of cavern development to the physiographic development of the region in which they occur, and added that caverns may be useful as physiographic indexes. Swinnerton's major purpose, however, was to point out the numerous inadequacies of the Davis hypothesis; he then outlined the nature of ground water conditions in massive, jointed rocks. Swinnerton (1932, p. 668) preferred the dynamic concept of classification of groundwater movement by Finch (1904) to the static classification of Meinzer (1923). He explained the normal fluctuation of the water table and lateral ground water movement along joints; high rates of solution caused by large volumes of infiltrating vadose water as contrasted to small volumes of phreatic water; and how the solution channels and cavern ceilings are dissolved by solution during flood periods, only to be emptied as the passages drain.

Swinnerton suggested that if a limestone landscape of low relief lies at a low level and is drained by poorly integrated surface streams, then groundwater will lie near the surface and deep-seated circulation will be slow owing to low hydrostatic head. If the low-level surface becomes uplifted there is a slow development of surface drainage and valley incision accompanied by an increased ground water circulation (Fig. 4). The water table exhibits considerable relief as the uplifted limestone surface becomes dissected by gorge-like valleys and the lateral ground water flow increases towards the lowest point of discharge, enlarging integrated rock openings, particularly during times of abundant precipitation. The dissection of the uplifted plain increases with maturity, runoff increases with the ramification of the surface drainage system, and the extension of valley slopes and the water table is reduced in relief as the underground drainage becomes well integrated. The water table would form a "peneplain" long before the land surface.

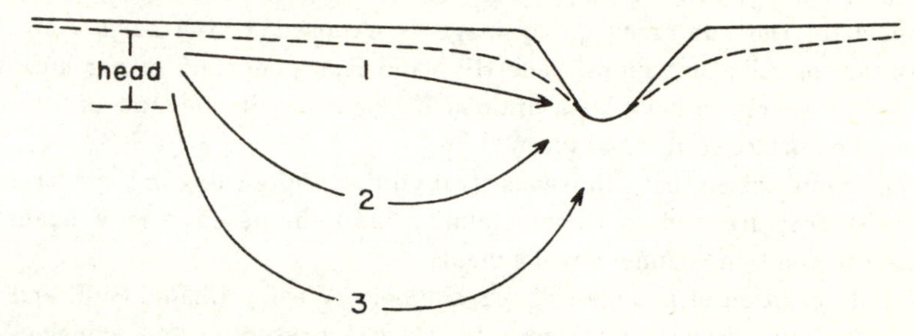

Figure 4. Water flow potentials in homogenous rock according to Swinnerton (1932, p. 675) would have least resistance along path 1, most along path 3.

Swinnerton noted that there is a topographic evolution involving the deepening and intergrowth of sinkholes along with subsurface development. He also indicated that uplift at any time would interrupt this normal development and cause new water table outlets at lower elevations.

Piper (1932, p. 78-82) indicated that Beede's stages of the erosion cycle defined in terms of development of minor landforms by solution were somewhat unsatisfactory because they imply that certain minor forms invariably accompany each major stage of development. Piper considered it more rational to analyze the cyclic history of the surface and subsurface drainage systems independently, though by analogous stages which may not be contemporaneous except at the beginning and end. Piper presented a description of stages of surface drainage development and upland reduction within which he described the stages of cavern development. During the first or youthful stage solution channels are first developed along the deeper youthful valleys and extend themselves headward beneath the uplands. Most of the upland drainage is carried by surface streams. During the second or mature stage the solution channels which gather the largest amounts of surface water and discharge at the lowest points develop as caverns and lower the divides betweeen underground drainage systems as new channels are formed at lower altitudes. Since groundwater circulation is controlled by the point of discharge into surface streams, the profile of solution channeling is adjusted to the local equilibrium profile of erosion.

Where the rocks are equally jointed and equally soluble the groundwater conduits tends to adjust to the level of surface streams, but where an impermeable bed or one that is less soluble is present, the equilibrium profile of solution may be established in a more soluble, overlying bed, temporarily or permanently, depending upon the character of the restraining bed. Collapse sinkholes reach the surface during this stage. In the final or old age stage of cavern development the equilibrium profile of solution channeling is developed to the limits imposed by stratigraphy and structure, producing a network of large solution channels defined as a peneplain of solution (after Meinzer, 1927, p. 215). Collapse sinks are most numerous during this stage as much of the subterranean drainage is exposed at the surface by the formation of collapse valleys. The Highland Rim peneplain of north-central Tennessee seems to have been drained by solution channels about 100 feet below the surface in its final form.

Piper emphasized that calcareous strata differ appreciably in their strength to resist fracture and in their solubility, and discussed briefly examples related to solution channel development.

The description of a large collapse sinkhole, Wesley Chapel Gulf, and the associated subterranean passage network was presented and explained by Malott (1932) as a part of the Lost River subterranean drainage system of Indiana. Malott (1932, p. 314) indicated five factors led to the development of the subterranean network system and the gulf: 1) two intersecting joint systems in the limestone, 2) groundwater percolation and movement below the water table dissolved out three-dimensional embryonic passages in particularly soluble beds sufficiently large to permit free flow of waters through them, 3) this network is maintained above the water table by flood waters which fill all of the passages in the network, enlarging it as a whole, 4) the network is occupied only by high water periods as the water table is lowered into lower primitive routes, and 5) ponding of the underground system behind barriers, such as collapsed rock in the gulf, caused waters to follow other routes. Malott presented no reason for the postulated change in position of the water table between factors two and three.

Karst windows were described by Malott (1932, p. 290) as collapse sinkholes that expose the cave stream flowing across the bottom of the sink hole. He indicated that they are smaller than the large collapse sinkholes called gulfs which have rising and sinking streams in their floors.

The definitive physiography of the karst terrain and carbonate bedrock areas of the Pennyroyal Plateau of Kentucky by Sauer (1927) was followed by the popularized work of Lobeck (1928) concerning the geology and physiography of the Mammoth Cave area. A report by Dicken (1935) was the first to describe various types of sinkholes, called karst types, in part of the area, but the types were revised and presented again by Dicken and Brown (1938, p. 8-15): 1) cistern sinks were described as well-like pits into which a sandstone cap rock has collapsed; 2) cave-in depressions are broad, elongate, valley-like sink holes formed by collapse of cavern passage beneath valleys; 3) doline karst consists of solution-formed, mostly symmetrical, either round or oval shaped sinkholes in which the underground channels remove the insoluble sediment, resulting in a shape like the bell of a trumpet; and, 4) basin karst, which consists of dolines nearly filled with soil. They related their sinkhole types to specific geologic formations, but no significant mention of related groundwater conditions nor caverns was included.

Gardner (1935) indicated that large integrated cavern systems develop along permeable zones in gently dipping strata by a form of subterranean stream piracy. Downcutting by a surface stream into the strata down dip from a higher surface stream allows release of static water from the zone and initiates cavern development. Underground streams in general flow with the dipping strata, rarely along the strike, but in no case for any appreciable distance in a direction against the dip.

Theis (1936, p. 43-44) described stages of ground water circulation in limestone in which the initial stage of groundwater flow through the rock is compared to the sheet flow of surface drainage, although more water will flow through more fractured or porous parts of the rock above or near the water table. The favored larger openings grow more rapidly in proportion to smaller ones as a sort of underground piracy. Thus, the underground drainage develops trunk streams with both horizontal and vertical tributaries, some of which give rise to sinkholes. The solution passages are partly enlarged by mechanical erosion as the subterranean stream velocities increase with increasing ease of water movement. Collapse sinkholes develop as the passage enlarges to allow subsidence during the late stage, with progressive cavern foundering leading to re-establishment of surface streams and the development of a lower peneplain.

Bretz (1942) modified the two cycle theory of Davis by interpolating a phase of cavern filling with sediment prior to uplift and drainage of the ground water. His major contribution was basically the identification and classification of cavern features formed when the passage is water filled (phreatic features) and those formed by underfit cave streams (vadose features). Bretz insisted that phreatic features were proof that the cavern was formed below a permanent water table during an early cycle, and that vadose features were formed during a late cycle following rejuvenation of the surface.

Meinzer (1939, p. 675) presented a reclassification of subterranean water types. He divided the zone of saturation (Zone 2) into two types, "(2a) water that occurs in a permeable bed with no confining bed between it and the water-table and that has a pressure-head the same or nearly the same as the pressure-head of the water at the water-table", and (2b) water confined by a bed such that there is a pressure-head sufficient to raise the water level above the confining bed. These two conditions are applicable to strata in various carbonate terrains. Meinzer's zone 2a is equivalent to the terms "free water", "unconfined water", "nonartesian water", and phreatic as defined by Daubree (1887, p. 19), and as yet used by the French.

Stubbs (1940, p. 161) described karst geohydrology in Florida and explained that the active sinking streams, sinkhole development and dynamic flow systems feeding deep springs originated during low stands of sea level during Pleistocene time, but continued to enlarge during the high stands, as at the present time, under artesian conditions.

Hubbert (1940, p. 930) presented a diagram to show theoretical flow paths for groundwater in uniformly permeable material that has been cited by some carbonate hydrologists and speleologists. The model has no intended application to non-uniform, lithologically variable sequences of strata.

Rhodes and Sinacori (1941) reviewed earlier groundwater models and presented a new theory based on their experience with subterranean water in carbonate rocks of the Tennessee Valley area. They argued that this theory was somewhat a reconciliation of the deep flow theory of Davis (1930) and shallow flow theory of Swinnerton (1932). They postulated that the groundwater flow pattern would adjust to headward growth of enlarged cavern passageways such that the upstream end of the master conduit would serve as the groundwater outlet (Fig. 5). They concluded, however, that: "Progressive concentration of solution in the upper part of the zone of saturation produces master-conduits and causes the eventual diminution of flow and solution at deeper levels" (Rhodes and Sinacori, 1941, p. 794).

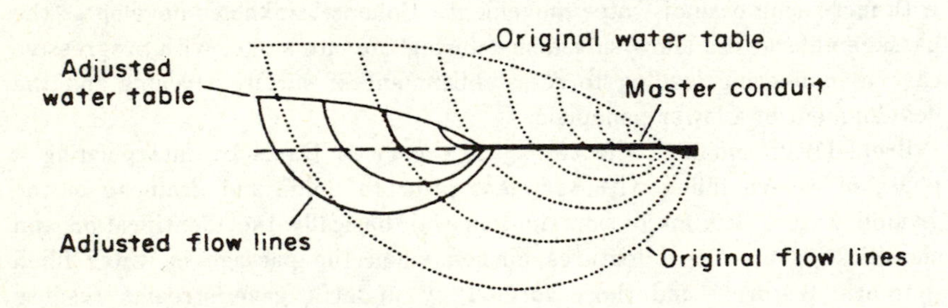

Figure 5. Flow paths in a homogenous permeable material (dotted lines) and the flow paths (solid lines) adjusted to the development of a master conduit according to Rhoades and Sinacori (1941, p. 789 and 793).

Swinnerton (1942) marshalled many of the current working concepts of geohydrology of limestone terrain, with some examples, into one brief paper, but karst features and caverns are only barely mentioned.

The development of small scale karst features, such as solution widened joints or grikes, lapies, solution furrows, rillensteine, solution facets, and tinajitas or solution pans were described from an area in Texas (Smith and Albritton, 1941). They indicated that the difference in origins of some forms was a function of slopes onto which the meteoric water falls.

Strahler (1944) reinterpreted so-called mature stream valleys of Davis and others on the Kaibab Plateau as abandoned valleys drained by sinkholes and modified by slopewash deposits. He related the conditions to existing climatic zones on the plateau.

A description of joint induced groundwater permeability to produce karst features and caverns in perched water bodies in carbonate strata separated by impermeable units was presented by Hamilton (1948). Hamilton indicated that the surface topography and karst features were developed by solution along joint-oriented flow paths, aligned either accordantly or discordantly with the joint orientation and that the water table in the perched water bodies was somewhat accordant to the land surface (Hamilton, 1948, p. 29-40).

The structural control of "solutioning", including both the development of solution channels or caverns and sinkholes along joint and fault trends, was described for anticlinal and synclinal structures in Texas by Kiersch and Hughes (1952). The jointing pattern was indicated to be related to structural trends. They generally related structural controlled fractures and stratigraphic lithologic changes to various groundwater conditions and applied the concepts to dam and reservoir engineering.

The origins of various types of depressions on the Great Plains of Kansas, some of which are solution induced sinkholes and collapse sinkholes, were catalogued by Frye (1950), and a comprehensive paper by Merriam and Mann (1957) specifically described sinkholes in Kansas and contains an extensive bibliography.

One of the largest sinkholes in the United States, Grassy Cove, in Tennessee, was described by Lane (1952 and 1953) as an uvala formed by extensive collapse of non-carbonate strata into cavern systems. Lane (1957) also postulated the headward growth of Sequatchie Valley, a breached anticline, by solution and sapping of non-soluble strata into cavernous systems by coalescing of sinkholes and uvalas, although the cycle was initiated by fluvial erosion.

An elaborate description of the karst features related to subterranean drainage of the Lost River drainage basin was posthumously published (Malott, 1952). Observed characteristics of the flow conditions were presented and interpreted to indicate the geomorphic history and dynamics of the system. Many of Malott's conclusions regarding underground stream courses have been proven by recent fluorescein tests (Murdock and Powell, 1968; Bassett and Ruhe, 1973).

A descriptive and interpretative presentation on the bedrock surface beneath residuum in Missouri indicated that the covered karst forms of solution enlarged joints should be termed cutters, while denuded bedrock forms should be termed lapies (Fellows, 1965). Numerous examples of cutters exposed by road construction and soil erosion are shown.

Karst features, predominantly sinkholes and solution widened fissures have been reported in thick- to thin-bedded limestones of the Madison

Limestone of the Gros Ventre Mountains of Wyoming (Keefer, 1963). Both solution-formed and collapse sinkholes are present. The karst features are the result of solution by snow-meltwater. Karst features, as well as cavern development in alpine terrain have recently been described in the Scapegoat Wilderness area of Montana (Campbell, 1973 and Munthe, 1974). Most of the karst features are karren and sinkholes that have developed along joints in a glaciated limestone plateau.

The preceding review of literature should demonstrate the wide range of disciplines involved in a thorough study of karst features and karst landscape development. Although many references must be classified as geohydrologic or speleologic, the relation of these topics to the origin of karst features is obvious if the tenet is valid that karst features are the surface expression of diversion of surface runoff to subsurface conduits and caverns. Perhaps, also, the review will be useful as a guide to background reading. The review also shows that historically the triad of structure, process and stage has served as an outline for geomorphic study in karst as in some other geomorphic endeavors.

STRUCTURE

Structure was intended to include all aspects of the Earth materials upon which processes form a landscape. Thus, the physical and chemical characteristics and attitude of lithologic materials were considered as a controlling factor of the rate at which a process could create a landform of a predictable shape in an interval of time.

Studies of the effect of detailed lithologic variations on karst and cavern development are meager. Although caverns offer excellent sites to investigate stratigraphic variation, few such studies have been conducted (Currens, 1975; Palmer, 1975, p. 41-46). Lateral variation and vertical stratigraphic successions are common in most carbonate bedrock terranes. The effect of these changes should be investigated in association with processes of solution, erosion, weathering, mass wasting and rock mechanics which combined are partly responsible for karst landscape and cavern development.

The type of karst research that is generally lacking is that of Pluhar and Ford (1970) concerning the development of karren on Niagaran dolomite to the extent that they could differentiate forms developed along joints or bedding planes, and demonstrate small differences in rock solubility by petrographic and geochemical methods. Massive data involving petrographic and geochemical identifications of karsted rocks and their relative and absolute solubilities as compared to relatively non-soluble carbonate rocks need to be determined. Rauch and White (1970) demonstrated the techniques that may be applied to carbonate strata in Pennsylvania, although the relationship of the data to cavern cross sections is poorly presented by comparison.

Various schemes of classification of porosity and permeability in carbonate rocks need to be re-examined (Imbt, 1947). Some classifications are more genetic than empirical, yet the genetic schemes do not always follow a rational order. For example: *if* joints are a secondary form of permeability it logically follows that conduits formed by solution along joints must be a tertiary form of permeability.

Structural geology *per se* has been often discussed with regard to hydrology and cavern development as a controlling factor of ground water movement, but the distinction between a structurally controlled erosion surface and a stripped plain or structural bench has not been clarified. This problem is paramount to the establishment of erosion levels in gently dipping strata. Howard (1968, p. 112-113) and Quinlan (1968, p. 237) indicated that the sinkhole plain of Kentucky south of Mammoth Cave consists of structurally controlled escarpments and plains that are a stripped surface developed on chert horizons and that evidence of past episodes of base leveling is lacking. Howard admitted that caverns, owing to their protected position, may preserve evidence of base level positions inasmuch as base level often closely controls the position of the water table. Miotke and Palmer (1972, p. 49-53) proved that a correlation of cavern level and major stream base levels does exist in the area. The karst area of Indiana, an area of somewhat similar lithology, structure and geomorphic history, contains base leveled surfaces which led Powell (1964) and Palmer and Palmer (1975) to conclude that although the dip of the strata strongly influenced the direction of drainage, present and ancient stream gradients did not coincide with the rate of dip of the strata.

PROCESS

Perhaps the simple definition of a karst terrane or cavern in carbonate strata that indicates that the unique forms are formed by solution of soluble rocks by water can be little improved without making argument-provoking statements. This simple introductory phrase is not sufficient, however, for any serious study of origin and development of either karst landforms or caverns in a specific area. The combined effects of corrosion in conjunction with chemical and physical rock weathering, rock mechanics, soil formation, mass wasting, corrasion, and sediment transport and deposition need to be given far more importance in karst and cavern studies instead of mere reliance on solution processes to create the complex forms present. For example, most sinkholes are formed in rocks and soils (engineering sense) by combined effects of solution, mass wasting and other processes. Also, fluviatile forms and processes within karstic terranes are little mentioned in karst literature in the United States. Characteristics of underground flood-plains were discussed by Jones (1971) who demonstrated cavern forms analogous to some surface forms.

The failure of karst geomorphologists to adequately study subterranean drainage and speleologists to take into account tributary surface drainage areas is damaging to their conclusions. For example: karst researchers assume that the total carbonate removed from a carbonate bedrock area amounts to a solutional lowering or denudation of the land surface by a rate in units per year (Sweeting, 1966, p. 205-206 and Pitty, 1968, p. 174), while speleologists may calculate that a cave with a given cross section and a certain length is formed in a given time from the same type of data (Swinnerton, 1932, p. 679, and others).

Much of the geochemical data to determine constituents of carbonate strata is very well executed work. There is considerable doubt, however, that rocks macerated to 200 mesh retain any characteristics of carbonate rocks, (such as grain types, size and amounts, cement, porosity, and permeability) that are germane to determining solution rates of various lithologies. Mere comparison of chemical analysis, petrographic descriptions and observed solution forms cannot be considered as adequate without including some experimental work on solution rates of bulk rock samples.

The delimitation of karst or cavern drainage basins and time-travel studies greatly improves an understanding of the hydrology of groundwater in karst areas or carbonate terrains. Numerous recent papers have reported the results of subterranean water tracing from karst sinkholes or swallowholes to spring resurgences; i.e., Murdock and Powell (1968) and Miotke and Papenberg (1972). Techniques of stream tracing have become sophisticated from the early use of fluorescein reported by Dole (1906). Various dyes, such as fluorescein and rhodomine, salts, radioactive tracers, lycopodum, bacteria, etc., have been described (Hass, 1959; Buchtela, *et. al.*, 1968; Brown, 1968, p. 33-62; and Wilson, 1968) which generally are more useful than sawdust or computer chips.

The number of published papers which adequately relate cavern development to their overlying karst landscapes or related geomorphic surfaces as well as the detailed geologic setting and bedrock stratigraphy are limited, but several excellent examples appear to do so: i.e., Bates (1932), Deike (1960), Hack and Durloo (1962), Malott (1932), Palmer (1972), Thrailkill (1960), and Varnedoe (1964).

STAGE

The ultimate goal of karst and cavern research, so far as geomorphology is concerned, is to postulate, or perhaps even prove, a scheme of landscape evolution. This commitment demands only that a sequence of events or a series of time transects be made. Hopefully, however, a chronology can be devised to equate obvious, somewhat static base levels, graded surfaces, depositional surfaces or materials, and associated groundwater conditions with surficial deposits and geomorphic surfaces of known age in the karst area or adjacent areas. Two basic assumptions are considered important: 1) the water table conditions which are associated with cavern development may be equated with base level, and 2) evidence for base levels or equilibrium profiles in caverns may be better preserved and thus make a more easily interpretable record than erosional and depositional levels on the surface.

The cyclical concept of genetic age (youth, maturity and old age) must be abandoned in favor of cause and effect related to degree to which a process has affected various rock types over a period of time. Some strata are less competent than others, for example, therefore collapse sinkholes would more likely occur than in a more competent limestone. Generally, sinkholes are more common or deeper along major cavern passages than above small tributary subterranean conduits, yet, they are probably nearly the same age, the difference is possibly caused by the relative amount of water transmissivity involved.

In consideration of the early concepts that indicated cavern development and contemporaneous karst development was directly related to base level and associated water table conditions, the field evidence to prove or disprove cavern level development relative to valley bottom level, berm, strath or terrace levels, as indicators of former base levels of erosion, was obviously postponed by the peneplain concepts of Davis (1931) and Bretz (1942 and 1953). Correlations of base level or graded surfaces with caverns have been made by Davies (1957) and Wolfe (1964) in the Appalachian area, Miotke and Palmer (1972) in the Mammoth Cave area of the United States, and by Sweeting (1950) and Waltham (1970) in the Ingleborough District of Great Britain. Certainly the relationship of the highly fluctuating water table in various carbonate terrains to master streams or base level and relationships to subterranean drainage systems and surface drainage basins need additional study. Such studies would require long-term data from water level observation wells and stream flow records, which are commonly lacking or inadequate in most karst areas in the United States.

Perhaps the fact that Davis (1931) established separate karst stages or epochs and cavern cycles that were not precisely correlative is partially responsible for the limitation of modern interdisciplinary work on karst and caverns by karst geomorphologists and speleologists. Many speleologists lack an interest in geomorphology in general, or, at least, indicate no interest in geomorphic history beyond that of sequential stages of cavern development. Some of the separation of interests, however, has been owing to the different type of people attracted to either field — (1) field oriented, but academically trained geomorphologists (either geographers or geologists) who study karst as a result of formal training and subsequent interest, in contrast to (2) the more daring speleologist (an educated spelunker) who maintains an academic interest in caverns as a result of an initial interest in cave exploration as a sport.

CLIMATE

The possible effects of variable climates throughout the Pleistocene Epoch and the Tertiary Period have been virtually ignored in regard to differential rates and amounts of erosion to produce karst landforms, caverns and associated depositional features in the United States. Certainly the variations of annual precipitation, rain or snow, as well as climatic changes in annual rainfall amounts must be accounted for in regard to fluctuations of the water table, average height of the water table, stream entrenchment or establishment of graded streams in conjunction with rejuvenation or raising base level, misfit streams on the surface and in cavern passages, and changes in regimen that cause deposition and later reworking of sedimentary deposits. An assumption of even distribution of precipitation through space and time is no more acceptable than regional structural stability or uniformity of amount or rate of erosion over a span of time. More effort, both theoretical and physical, needs to be expended towards an understanding of climatic changes and their geomorphic effects in karst terrain.

TERMINOLOGY

A portion of any research effort must be devoted to semantics, terminology and grammar. We are fraught by new terms and phrases which are not always of obvious meaning or as concise as older terminology (i.e. use of speleothem to collectively include all deposits of dripstone, flowstone, rimstone, etc.), but some of which alleviate confusion (i.e., use of speleothem as a replacement for the phrase "cave formation" which had been interpreted several ways). Some terms, such as groundwater, water table, vadose water and phreatic water have obviously had different meanings to various writers, and certainly their application to karst hydrology and the origin of caverns and karst features have been debated. Much of the argument is due to a lack of rational understanding of the relationships of water to all types of carbonate reservoirs, while some of the argument stems from either a too precise or too indefinite original definition, and some argument is due to misuse of an incorrect term when a proper term is available.

Considerable argument has ensued regarding cavern development in relation to the water table. Some researchers argue development in the vadose, shallow phreatic, or phreatic zones, or, at or near, above or below the water table. There can be little doubt that in many cases authors are using different terms for the same subterranean water situation. For example: (1) Is "shallow phreatic" the same as "near the water table" if the latter includes the possibility of a phreatic rise? (2) Can "phreatic rise" be synonomous with "vadose rise" if it is assumed that the temporary rise of the water table is caused by descending meteoric or infiltrating vadose water? (3) Does phreatic water consist only of that water in the groundwater zone of permanent saturation? (4) Can a permanent zone of saturation be defined and mapped as the non-fluctuating entity so often shown on idealized cross section diagrams?

Contrary to the classical concept that the term karst refers to particular landform types, somewhat as expressed by Thornbury (1969, p. 303-305), the term has been used as a synonym for carbonate terrane, particularly in the titles of papers that are mostly concerned with carbonate geohydrology or speleology and contain very little mention of karst features.

Much of the geologic, hydrologic and speleologic terminology related to karst and cavern features and development was recently compiled in a useful glossary (W. H. Monroe, 1970), which was reviewed by many karst geomorphologists, including 18 who made suggestions and added definitions (Monroe, 1975, personal communication). Yet, this glossary contains incorrect definitions for two important terms, cave system and conduit, and does not include the terms channel, solution channel, ramifying channels, meteoric water, subterranean water, underground water, artesian, gravity flow, phreatic rise, hydrostatic, piracy, subterranean piracy, or solution peneplain, which are common to many textbooks and older works.

CONCLUSIONS

Geomorphologists, concerned with karst development, are not publishing very many reports. Geohydrologists, geochemists and speleologists are contributing valuable additions to knowledge, but their reasoning and conclusions commonly indicate insufficient background to fully relate to geomorphic principles. There is a great need for integration of knowledge and efforts of all research specialists, as well as greater effort for combined related karst and cavern research. Some karst researchers need to do their homework in carbonate petrography, stratigraphy, geomorphology, and hydrology to gain more insight into multiple working hypotheses, while others need to do more field work before proposing theoretical solutions or overstating their conclusions. Certainly clearer, readable, more precise reporting is a worthwhile goal.

REFERENCES CITED

Addington, A. R., 1929, Special topographic features and the physiographic background of the Bloomington, Indiana, Quadrangle: Indiana Acad. Sci., v. 38, p. 247-261, 7 figs.

Bassett, J. L., and Ruhe, R. V., 1973, Fluvial geomorphology in karst terrain: *in* Fluvial Geomorphology, M. Morisawa, ed., p. 74-89, 8 figs., 3 tbls.

Bates, R. E., 1932, Underground features of Sinking Creek, Washington County, Indiana: Indiana Acad. Sci., v. 41, p. 263-268, 2 figs.

Beede, J. W., 1911, The cycle of subterranean drainage as illustrated in the Bloomington, Indiana, Quadrangle: Indiana Acad. Sci., v. 26, p. 81-111, 32 figs.

Bowman, I., 1907, Water resources of the East St. Louis District: Ill. Geol. Surv., Bull. 5, 128 p., 11 figs., 4 pls.

Bretz, J. H., 1942, Vadose and phreatic features of limestone caverns: Jour. Geol., v. 50, p. 675-811, 55 figs.

———————, 1953, Genetic relations of caves to peneplains and big springs in the Ozarks: American Jour. Sci., v. 251, p. 1-24, 3 figs.

Brown, M. L., 1968, Karst hydrology of the Lower Maligne Basin, Jasper, Alberta: Cave Studies, No. 13, 84 p., 47 figs., 8 pls.

Brown, R. T., 1854, Geological Survey of the State of Indiana: Indiana State Board Agricul., Ann. Rept. 3, p. 299-332, 1 fig.

Buchtela, K., Mairhofer, J., Maurin, V., Papadimitropoulos, T., and Zotl, J., 1968, Comparative investigations into recent methods of tracing subterranean water: Natl. Speol. Soc. Bull., v. 30, p. 55-74, 17 figs.

Campbell, N., 1973, Scapegoat alpine karst, Montana: Natl. Speol. Soc. Bull., v. 35, p. 49-57, 7 figs.

Cumings, E. R., 1906, On the weathering of the Subcarboniferous Limestones of southern Indiana: Proc. Indiana Acad. Sci for 1905, p. 85-100, 22 pls.

Currens, J. C., 1975, An investigation of Up and Down Cave, Rockcastle County, Kentucky: Natl. Speol. Soc. Bull., v. 37, p. 9-15 6 figs.

Cvijic, J., 1918, Hydrographie souterraine et evolution morphologique du Karst: Recueil des Travaux de l'Inst. de Geogr. Alpine, v. 6, p. 375-426.

Daly, R. A., 1917, Genetic classification of underground volatile agents: Econ. Geol., v. 12, p. 487-504.

Daubree, G. A., 1887, Les eaux souterraines a l'epoque actuelle; v. 1, Paris.

Davies, W. E., 1957, Erosion levels in the Potomac drainage system and their relation to cavern development: Speleo Digest, p. 2-32-36.

Davis, W. M., 1930, Origin of limestone caverns: Geol. Soc. America Bull., v. 41, p. 475-628, 62 figs., 2 pls.

Deike, G. H., III, 1960, Origin and geologic relations of Breathing Cave, Virginia: Natl. Speol. Soc. Bull., v. 22, p. 30-42, 13 figs.

Dicken, S. N., 1935, Kentucky karst landscapes: Jour. Geol., v. 43, p. 708-728, 8 figs.

———————, and Brown, H. B., 1938, Soil erosion in the karst lands of Kentucky: U. S. Dept. Agri., Circular 490, 62 p., 34 figs., 3 tbls.

Dole, R. B., 1906, Use of fluorescein in the study of underground waters: U. S. Geol. Surv., Water-Supply Paper 160, p. 73-85.

Elias, M. K., 1930, Origin of cave-ins in Wallace County, Kansas: American Assoc. Pet. Geol. Bull., v. 14, p. 316-320, 2 figs.

———————, 1931, The geology of Wallace County, Kansas: Kansas Geol. Surv., Bull. 18, 254 p., 7 figs., 42 pls.

Fellows, L. D., 1965, Cutters and pinnacles in Green County, Missouri: Natl. Speol. Soc. Bull., v. 27, p. 143-150, 14 figs.

Finch, J. W., 1904, The circulation of underground aqueous solutions: Proc. Colorado Sci. Soc., v. 7, p. 193-252.

Folk, R. L., Practical petrographic classification of limestones: Bull. American Assoc. Petrol. Geol., v. 43, p. 1-38, 41 figs.

Frey, J. C., 1950, Origin of Kansas Great Plains depressions: Kansas Geol. Surv., Bull. 86, Pt. 1, p. 1-20.

Gardner, J. H., 1935, Origin and development of limestone caverns: Geol. Soc. America Bull., v. 46, p. 1255-1274, 1 fig.

Greene, F. C., 1909, Caves and cave formations of the Mitchell Limestone: Proc. Indiana Acad. Sci., v. 18, p. 175-184, 8 figs.

Grund, A., 1903, Die karsthydrographie: Studien aus Westbosnien; Penck's Geog. Abhandl., v. 7, p. 103-200.

———————, 1910, Beitrage zur morphologie des Dinarischen Gebirges: Penck's Geog. Abhandl., v. 9, p. 144-145.

Haas, J. L., 1959, Evaluation of ground water tracing methods used in speleology: Natl. Speol. Soc. Bull., v. 21, p. 67-76, 4 figs., 2 tbls.

Hack, J. C., and Durloo, L. H., Jr., 1962, Geology of Luray Caverns, Virginia: Virginia Div. Min. Res., Rept. Inves. 3, 43 p., 14 figs., 1 pl.

Hamilton, D. K., 1950, Areas and principles of ground-water occurrence in the Inner Bluegrass Region, Kentucky: Kentucky Geol. Surv., series 9, Bull. 5, 68 p., 13 figs., 15 pls. 4 tbls.

Herak, M., and Stringfield, V. T., 1972, Karst: Important Karst Regions of the Northern Hemisphere: Elsevier, Amsterdam, 551 p., illus.

Hook, J. S., 1914, The brown and blue phosphate deposits of south-central Tennessee: Resources of Tennessee, v. 4, p. 50-86.

Hovey, H. C., 1882, Celebrated American Caverns: R. Clark and Co., Cincinnati, Ohio, 228 p., 50 figs.

Howard, A. D., 1968, Stratigraphic and structural controls on landform development in the central Kentucky karst: Natl. Speol. Soc. Bull., v. 30, p. 95-114, 6 figs., 2·pls.

Hubbert, M. K., 1940, The theory of ground water motion: Jour. Geol., v. 48, p. 785-944.

Imbt, W. C., 1947, Project 7 — Carbonate reservoirs: *in* Fundamental Research in Sedimentology: American Assoc. Petrol. Geol., Research Comm., Houston, Texas, p. 114-132.

Jennings, J. N., 1971, Karst: The M.I.T. Press, Cambridge, Mass., 252 p., 69 figs., 40 pls.

Jones, W. K., 1971, Characteristics of the underground floodplain: Natl. Speol. Soc. Bull., v. 33, p. 105-114, 8 figs.

Keefer, W. J., 1963, Karst topography in the Gros Ventre Mountains, northwestern Wyoming: U. S. Geol. Surv., Prof. Paper 475, p. B 129-130, 2 figs.

Kiersch, G. A., and Hughes, P. W., 1952, Structural localization of ground water in limestones — "Big Bend District," Texas-Mexico: Econ. Geol., v. 47, p. 794-806, 7 figs.

King, F. H., 1899, Principles and conditions of the movement of ground water: U. S. Geol. Surv., 19th Ann. Rept., Pt. 2, p. 59-294.

LaMoreaux, P. E., Raymond, D., and Joiner, T. J., 1970, Annotated Bibliography of carbonate rocks: Geol. Surv. Alabama, Bull. 94, pt. A., 242 p.

Lane, C. F., 1952, Grassy Cove, a uvala in the Cumberland Plateau, Tennessee: Tennessee Acad. Sci., v. 27, p. 291-295, 2 figs.

———————, 1953, The geology of Grassy Cove, Cumberland County, Tennessee: Tennessee Acad. Sci., v. 28, p. 109-117, 3 figs.

———————, 1957, Headward growth of anticlinal valleys in the karst cycle of erosion: Virginia Jour. Sci., v. 8, p. 203-209, 3 figs.

LeGrand, H. E., and Stringfield, V. T., 1973, Karst Hydrology — a review: Jour. Hydrol., v. 20, p. 97-120, 3 figs.

Lobeck, A. K., 1928, The geology and physiography of the Mammoth Cave National Park: Kentucky Geol. Surv., series 6, Pamphlet 21, 69 p., 37 figs.

Logan, W. N., 1918, Certain indicia of dip in rocks: Indiana Acad. Sci. Proc. for 1917, p. 229-234.

Malott, C. A., 1922, The physiography of Indiana: *in* Handbook of Indiana Geology, Indiana Dept. Cons., Pub. 21, p. 59-256, 3 pls., 51 figs., 1 tbl.

———————, 1932, Lost River at Wesley Chapel Gulf, Orange County, Indiana: Proc. Indiana Acad. Sci., v. 41, p. 285-316, 12 figs.

———————, 1952, The swallow-holes of Lost River, Orange County, Indiana: Indiana Acad. Sci., v. 61, p. 187-231, 16 figs.

———————, and Shrock, R. R., 1930, Origin and development of Natural Bridge, Virginia: American Jour. Sci., v. 19, p. 257-273, 5 figs.

Meinzer, O. E., 1923a, Outline of ground-water hydrology, with definitions: U. S. Geol. Surv., Water-Supply Paper 494, 71 p., 35 figs.

———————, 1923b, The occurrence of ground water in the United States: U. S. Geol. Surv., Water-Supply Paper 489, 321 p., 110 figs., 31 pls.

———————, 1927, Geology of large springs: (abstract) Geol. Soc. America Bull., v. 38, p. 213-216.

———————, 1929, Relations of ground water to leakage of reservoirs: American Inst. Min. and Met. Eng., Tech. Pub. 215, p. 19-30, 1 tbl.

———————, 1939, Discussion of question no. 2 of the International Commission on Subterranean Water: Definitions of the different kinds of subterranean water: Trans. American Geophy. Union of 1939, p. 674-677.

Merriam, D. F., and Mann, C. J., 1957, Sinkholes and related geologic features in Kansas: Kansas Acad. Sci., v. 60, p. 207-243, 9 figs.

Miotke, F. D., and Palmer, A. N., 1972, Genetic Relationship Between Caves and Landforms in the Mammoth Cave National Park Area: privately published, 69 p., 58 figs., 1 pl.

———————, and Papenberg, H., 1972, Geomorphology and hydrology of the sinkhole plain and Glasgow Upland, Central Kentucky karst: Caves and Karst, v. 14, p. 25-32, 2 figs., 2 tbls.

Monroe, W. H., 1970, A glossary of karst terminology: U. S. Geol. Surv., Water-Supply Paper 1899-K, 26 p.

Munthe, J., 1974, Scapegoat: Natl. Speol. Soc. News, v. 32, p. 37-42, 10 figs.

Murdock, S. H., and Powell, R. L., 1968, Subterranean drainage routes of Lost River, Orange County, Indiana: Indiana Acad. Sci., v. 77, p. 250-255, 1 fig., 1 tbl.

Palmer, A. N., 1972, Dynamics of a sinking stream system: Onesquethaw Cave: Natl. Speol. Soc. Bull., v. 34, p. 89-107, 11 figs., 1 pl.

———————, 1975, Origin of cave levels in Mammoth Cave National Park: *in* Genetic Relationship Between Caves and Landforms in the Mammoth Cave National Park Area: p. 34-48.

Palmer, M. V., and A. N., 1975, Landform development in the Mitchell Plain of southern Indiana: origin of a partly karsted plain: Zeit. Geomorph., v. 19, p. 2-39, 15 figs., 7 photos., 2 tbls.

Piper, A. M., 1932, Ground water in north-central Tennessee: U. S. Geol. Surv., Water-Supply Paper 640, 238 p., 7 figs., 9 pls.

Pitty, A. F., 1968, The scale and significance of solutional loss from the limestones tract of the southern Pennines: Geol. Assoc. London Proc., v. 79, p. 153-177, 7 figs.

Pluhar, A. and Ford, D. C., 1970, Dolomite karren of the Niagara Escarpment, Ontario, Canada: Zeit. Geomorph., v. 14, p. 392-410, 4 figs., 6 photos., 2 tbls.

Powell, R. L., 1964, Origin of the Mitchell Plain in south-central Indiana: Indiana Acad. Sci., v. 73, p. 177-182, 2 figs.

Posepny, F., 1894, The gensis of ore deposits: Trans. American Inst. Min. Eng., v. 23, p. 197-369, 10 figs.

Purdue, A. H., 1907, On the origin of limestone sinkholes: Science, N. S., v. 26, p. 120-122.

Quinlan, J. F., 1970, Central Kentucky karst: Reunion Inter. Karst. en Languedoc-Provence, 1968, Actes: Mediteranee, Etudes et Travaux: No. 7, p. 235-253, 5 figs.

Rauch, H. W., and White, W. B., 1970, Lithologic controls on the development of solution porosity in carbonate aquifers: Water Resour. Research, v. 6, p. 1175-1192, 14 figs.

Rhodes, R., and Sinacori, M. N., 1941, Pattern of ground-water flow and solution: Jour. Geol., v. 49, p. 785-794, 4 figs.

Russell, W. L., 1929, Local subsidence in western Kansas: American Assoc. Petrol. Geol. Bull., v. 13, p. 605-609.

Sanders, E. M., 1922, The cycle of erosion in a karst region (after Cvijic): Geog. Rev., v. 11, 593-604, 11 figs.

Sauer, C. O., 1927, Geography of the Pennyroyal: Kentucky Geol. Surv., series 6, v. 25, 303 p., 125 figs.

Sellards, E. H., 1907, Origin of sink-holes: Science, N. S., v. 26, p. 417.

———————, 1908, A preliminary report on the underground water supply of central Florida: Florida Geol. Surv., Bull. 1, 103 p.

Smith, J. F., Jr., and Albritton, C. C., Jr., 1941, Solution effects on limestone as a function of slope: Geol. Soc. America Bull., v. 52, p. 61-78, 5 figs., 2 pls.

Stockdale, P. B., 1936, Montlake – an amazing sinkhole: Jour. Geol., v. 34, p. 399-414, 2 figs.

Strahler, A. N., 1944, Valleys and parks of the Kaibab and Coconino Plateaus, Arizona: Jour. Geol., v. 52, p. 361-387, 5 figs.

Stringfield, V. T. and LeGrand, H. E., 1969, Hydrology of carbonate rock terranes – a review: Jour. Hydrol., v. 8, p. 349-417, 5 figs.

Stubbs, S. A., Solution a dominant factor in the geomorphology of peninsular Florida: Florida Acad. Sci., v. 5, p. 148-166, 2 figs.

Sweeting, M. M., 1950, Erosion cycles and limestone caverns in the Ingleborough District: Geog. Jour., v. 115, p. 63-78, 3 figs.

———————, 1973, Karst Landforms: Columbia University Press, New York, 362 p., 127 figs., 57 photos., 16 tbls.

Swinnerton, A. C., 1929, Changes of base level indicated by caves in Kentucky and Bermuda (abstract): Geol. Soc. America Bull., v. 40, p. 194.

———————, 1932, Origin of limestone caverns: Geol. Soc. America Bull., v. 43, p. 663-694, 2 figs.

———————, 1937, Structural control of the form and distribution of sink-holes: Science, N. S., v. 85, p. 218-219.

———————, 1942, Hydrology of limestone terranes: *in* Hydrology, O. E. Meinzer, ed., Nat. Research Council, p. 656-677, 3 figs.

Theis, C. V., 1936, Ground water in south-central Tennessee: U. S. Geol. Surv., Water-Supply Paper 677, 182 p., 2 figs., 7 pls.

Thornbury, W. D., 1969, Principles of Geomorphology: John Wiley and Sons, Inc., New York, 2nd ed., 594 p.

Thrailkill, J. V., 1960, Origin and development of Fulford Cave, Colorado: Natl. Speol. Soc. Bull., v. 22, p. 54-65, 5 figs.

VanHise, C. R., 1901, Some principles controlling the deposition of ores: Trans. American Inst. Min. Engr., p. 27-177, 10 figs.

Varnedoe, W. W., Jr., 1964, The formation of an extensive maze cave in Alabama: Alabama Acad. Sci. Jour., v. 35, p. 143-148, 3 figs.

Waltham, A. C., 1970, Cave development in the limestone of the Ingleborough District: Geog. Jour., v. 136, p. 574-585, 10 figs.

Ward, F., 1930, The role of solution in peneplanation: Jour. Geol., v. 38, p. 262-270, 1 fig.

Warren, W. M. and Moore, J. D., 1975, Annotated bibliography of carbonate rocks: Geol. Sur. Alabama, Bull. 94, pt. E., p. 31-168.

Weller, J. M., 1927, Geology of Edmonson County: Kentucky Geol. Surv., ser. 6, v. 28, 246 p.

Wilson, J. F., 1968, Time-of-travel measurements and other applications of dye tracing: *in* Hydrological aspects of the utilization of water – Gen. Assembly of Bern, 1967: Inter. Assoc. Sci. Hydrol. Pub. 26, p. 252-262, illus.

Woodward, H. P., 1936, Natural Bridge and Natural Tunnel, Virginia: Jour. Geol., v. 44, p. 604-616, 5 figs.

Wray, D. A., 1922, The karstlands of western Yugoslavia: Geol. Mag., v. 59, p. 392-408, 2 figs.

Wright, F. J., 1936, The Natural Bridge of Virginia: Virginia Geol. Sur., Bull. 46-G, p. 53-78, 4 figs., 3 pls.

CHAPTER 13

THE CASE FOR EPISODIC, CONTINENTAL-SCALE EROSION SURFACES: A TENTATIVE GEODYNAMIC MODEL

Wilton N. Melhorn and Dorland E. Edgar

ABSTRACT

Evidence of geomorphic surfaces of deposition, transportation, and erosion are abundant and widespread in the geological record, and should not be ignored in any interpretation of the historical evolution of the continents. Transportational and depositional surfaces are readily visualized because landform families representative of these classes are seen being formed or modified today. Erosion surfaces generally are seen only indirectly as fragmentary remnants of the record of subaerially denuded landscapes of ages past. The precise mode of formation of erosion surfaces is difficult to conceptualize, and thus is difficult to interpret in terms of landscape evolution. Some writers doubt that widespread surfaces exist, though acknowledging that such surfaces of limited extent develop under local controls of climate, base level, *etc.* Terminological ambiguity adds to the confusion.

Lester King in Africa and Soviet geomorphologists in recent years have presented evidence in support of the reality of continental-scale, cyclic erosion surfaces that can be correlated with other aspects of the geologic record. These surfaces are thus an important baseline for geomorphic inference or deduction, and their origin must be explained to achieve any logical reconstruction of former landscapes. A tentative correlation of major erosion surfaces in time and space is outlined in this paper, and is then examined within a framework of current geological concepts in geotectonics, sea-floor spreading, and the cyclical sedimentation record of the cratons.

The stratigraphic record, in terms of paleogeomorphology, shows several erosional and depositional episodes of long duration and wide areal extent. Certain evidence indicates that these events are essentially contemporaneous and correlative. Thus the reality of time-synchronous, world-wide erosion surfaces also seems probable. A geodynamic model is proposed to explain the temporal and spatial attributes of these surfaces. The model depends upon appropriate inputs from geotectonic studies of the scope and periodicity of crustal movement and deformation through time. It is concluded that conditions favorable for net denudation and planation of landscapes existed between approximately +135 m.y. and 110 m.y. B.P. ("Fall Zone Time"); 85 m.y. to 55 m.y. B.P. ("Schooley Time"); 45 m.y. to 20 m.y. B.P. ("Harrisburg Time"), and possibly from about 12 m.y. to 2 m.y. B.P. ("Somerville ? Time"). Considerable uncertainty exists about duration and intensity of the last stage because of conflicting data inputs and eustatic complications arising from late Tertiary events prior to actual continental glaciation. Times listed as available for erosion of uplifted continental masses and concurrent deposition of accepted estimates of volumes of sediment transported to the continental shelves or stored subaerially on continent borderlands agree rather favorably with estimates of average rates of landscape reduction previously described in the literature.

INTRODUCTION

> *Surface*, s.b. 1611. The outermost boundary of any material body, immediately adjacent to the air or any empty space; a magnitude or continuous extent having only two dimensions, without thickness, such as constitutes the boundary of a material body; the upper boundary or top of ground or soil, exposed to the air; the outer boundary of the Earth, 1612.

The foregoing sample definitions are selected from a more extensive listing in the Oxford Universal Dictionary. The definitions given are highly generalized, but they clearly refer to a bidimensional subject without specific geometry, form, or size. The term "surface" also has been used recently in an abstract or symbolic sense by geographers and psychologists, who conceive of social, cultural, and economic surfaces that can be overlaid on and moved across the surface of the solid Earth.

In a purely physical context, however, geomorphology deals with three types or sets of surfaces on the face of the Earth. These surfaces are characterized by the imprint of processes that act on and the landforms that result from the interplay of erosion, transportation, and deposition. However, probably no surface is produced exclusively by any single process acting independently from other processes, nor does any surface of sufficient areal extent display a single landform element. The tendency, rather, is toward development of landform families, dominated perhaps by a single form, but locally including other landform elements as mute evidence of the surface's complex history.

Most geomorphologists readily agree on identification of dominantly transportational and depositional surfaces. Primarily transportation surfaces, such as alluviated pediments, fans, or hamada plains can be observed to undergo change, through flux of transported materials, during the lifetime of a single observer. Likewise, depositional surfaces such as deltas, the tops of volcanic flows, or the floors of playas, are abundantly visible in the modern landscape and changes can be observed or measured. The fact that depositional and transportational forms or surfaces, at various times in their history, may be dissected and thus become part of an erosional episode is easily conceptualized; furthermore, we readily envision these surfaces and solid bodies as parts of some type of wedge or prism of matter resting on an equally solid substrate, and with an upper boundary in contact with the atmosphere. Standard geomorphological texts unfortunately devote little separate treatment to these surfaces, though Thornbury (1969, p. 37) does note that transportation is tacitly assumed to be an integral part of the erosion process (not a separate surface, however). Machatschek (1969) also devotes some commentary to strictly depositional surfaces.

It is the peneplane or erosion surface which has been the subject of scores of papers during the past 85 years and has generated so much dialogue among geomorphologists. Many textbooks, even in introductory geology, marshal arguments for or against the reality of peneplains, pediplains, etchplains, structural plains, straths, *etc.* Regardless of the personal predilection of each writer, an undercurrent of frustration seems to flow beneath the surface of each paper; for despite processes invoked, landforms created, or dynamic mechanisms proposed, in dealing with any more or less planar surface of denudation we study "that which is not, rather than that which is." Instead of a prism, wedge or other solid geometrical body, we are confronted with only fragmentary or nebulous evidence of events that occurred in times long past. Psychologically, perceptive mechanisms and mental sorting devices become confused; it is extremely difficult to receive and conceptualize data output from a system with little tangible input in terms of geometry, composition, or forces acting on a vanished landscape.

Most geomorphologists accept the existence of some relatively smooth geometrical and/or geographical surfaces, and assume that regardless of processes, dynamic controls, or climatic conditions invoked these surfaces are the end product of denudational activity operating upon the Earth's crust during some finite length of time. Such acceptance is particularly universal for local, small-scale beveled surfaces attributed to scour or erosion (rock benches, straths, *etc.*). These are usually ascribed to complex interactions of climate, vegetation, rock structure, and lithology.

Apparent large-scale, broad-area surfaces are quite another matter. The geographical cycle concept of William Morris Davis was geomorphological dogma for nearly 50 years, or roughly from about 1890 to 1940, although serious questions about the validity of the concept were occasionally raised, for example by Trowbridge (1921). The inadequacies of many aspects of the Davisian system are treated in other papers in this volume, so it suffices to note here that the peneplane (or peneplain) was discovered or invented as documentational proof of the end product of the cycle. In recent years, the peneplain has fallen from favor, as the geographical cycle concept itself (at least as proposed by Davis) has succumbed. Furthermore, there are other substantive arguments that challenge the existence of any large-scale erosion surface, for example the assertion that the visual aspect of low-inclination, relatively smooth or planar landscape is largely an illusion. This argument is especially valid in terms of the scalar relation constraints imposed on the optical reference frame of the human observer at ground level; this contention was mostly recently reemphasized by Garner (1974, p. 54).

In summary, it seems possible to safely state that there is good agreement about the existence of local or small-area erosion surfaces, whose development is controlled by a complex interplay of climatic, biologic, and physical forces. However, general agreement is yet lacking about this question: do large-scale, broad-area erosion, planation, or denudation surfaces exist? The objectives of this paper are, therefore, to: 1) review the current status of research on erosion surfaces; 2) demonstrate that a valid case may be made for the possible existence of several more or less time-synchronous, world-wide surfaces in the geologic record; and 3) show that the origin and apparent contemporaneity of these surfaces can be explained reasonably well in terms of some modern ideas about geodynamics, sea-floor spreading rates, and the historical record of cyclical sedimentation of continental platforms.

TERMINOLOGICAL CONFUSION

A lack of clearly defined terms historically has led to complications. The confusion arising from interchangeable use of the words peneplane (a geometric surface) and peneplain (a geographic surface) is so well-known that no review is needed. We will continue use of the latter spelling in the remainder of this paper unless a different meaning is intended.

There appears a current tendency in Europe, particularly in the Soviet Union, to use the term *planation surface* for any surface formed under conditions of prolonged and complete compensation of tectonic movements by exogenetic surface processes, without regard as to whether this surface is an erosional or depositional plain. This usage would, for example, include alluvial fans and terraces, which are not formed by planing. The term *erosion surface* seemingly is reserved for surfaces of uncertain or complex origin. It seems desirable to follow Brown (1968, p. 856) in using planation surface as a general term to describe any geographically plain surface which is the end product of all processes or mechanisms of planing by erosion. It may even be desirable to reinstitute use of the term *denudation surface*, which seems to have fallen into disuse, to unequivocally indicate that we refer to surfaces eroded under subaerial conditions.

GLOBAL CORRELATIONS OF EROSION SURFACES

Lester King's magnificent studies that relate historical geomorphology to structural and tectonic evolution and drifting of continents is well-known to most geomorphologists, and it therefore suffices to briefly summarize his results. A sequence of reports, commencing with the paper on the World's Plainlands (King, 1950) and culminating with *Morphology of the Earth* (King. 1962), led to these general conclusions:

1. Two early Paleozoic supercontinents, Gondwana and Laurasia, after several aborted earlier attempts to sunder, split or "dispersed" in late Mesozoic time.
2. Pre-Mesozoic geography of the globe can be reconstructed by correlation of lithologic, biostratigraphic, and tectonic evidence. As a capstone for this paleogeographic reconstruction, King provides evidence for existence of a world-wide, Jurassic age erosion surface, called the Gondwana surface.
3. Final rifting of the two supercontinents during Cretaceous and later times was followed by development of a series of younger erosion surfaces, but these formed independently on each continent.
4. A final "megacycle" of Geomorphic Terrace Development resulted from escalation of tectonic activity in the Tertiary. The most recent global picture is additionally complicated by periodic large fluctuations of the World Ocean through eustatic change and glaciation.

Russian geomorphologists, apparently somewhat independently, have taken similar approaches to the problem and have reached somewhat similar conclusions. The Soviets have, in recent years, placed great emphasis on the study of planation surfaces as fundamental geomorphic forms, and even greater emphasis on erosion surfaces, because of practical applications in prospecting for mineral deposits in areas where ancient weathering crusts can be identified in the subsurface stratigraphic record. Furthermore, much effort has been devoted to tracing and correlating erosion surfaces over extensive regions. Nearly 20 years has been spent in compiling a Composite Map of Erosion Surfaces in the Soviet Union. As a result of this intensive mapping program, Gorelov *et al* (1970) suggests that all planation surfaces can be classed into two basic groups, each age-dependent and influenced by subsequent tectonic deformations. These groups are:

1. *Ancient denudational surfaces*. These are peneplains or pediplains of Mesozoic or pre-Mesozoic age (and thus by inference are pre-breakup of the supercontinents) that formed under conditions of long-term, relative stable tectonism. This class includes probable late Proterozoic and late Paleozoic planation surfaces identified on the Russian Platform. Each surface developed slowly over a time span that may have embraced several geological periods. These surfaces are now deeply dissected or destroyed where Cenozoic tectonic deformation has been intense, but elsewhere they are preserved essentially intact, for example where buried beneath areas of cratonic downwarping or in downfaulted blocks. The principal criteria for recognition of these peneplains are familiar: maximum beveling of surfaces of gentle

relief, cut across deformed or tilted strata; accordancy of summit levels; and presence of thick kaolinitic, sialitic, or ferruginous weathering crusts (paleosols) wherever preserved beneath a covering blanket of deposits. Similar attributes are ascribed to denudational pediplain surfaces, but presumably these have some characteristic small-scale elements, which resulted in response to dry climatic conditions in contrast to "humid" conditions that produced peneplains.

2. *Geomorphological (relief phase) surfaces*. These developed chiefly in the Cenozoic, under relatively short-lived episodes of tectonic stability or subsidence followed by stages of uplift, so that all surfaces are incompletely planed (the incipient peneplain of Fenneman, 1938, p. 181). These surfaces are areally restricted and lack weathering crusts. Three familiar categories of surfaces are included in this group:

a. *Polygenetic (denudational-depositional) surfaces* formed by exogenetic processes acting during a single geologic stage, chiefly by lateral stream planation, mass wasting of slopes, abrasion, and accumulation of continental, littoral, and marine deposits.

b. *Denudational surfaces*, represented by small-scale, local pediments and local planation in mountainous regions.

c. *Depositional surfaces*, represented by marine, lacustrine and alluvial terraces and plains.

Polygenetic surfaces are presumably characteristic of continental platform environments; depositional surfaces occur in intermontane basins of recent or continuous subsidence; and pedimentation is restricted to the rock-bench forelands of mountain ranges. The timing of relief phase surfaces has been established as late Triassic-early Jurassic; late Jurassic-Cretaceous; Danian-Eocene; Oligocene-Miocene; Miocene-Pliocene (Sarmatian-Pontian); and late Pliocene-early Pleistocene (the present landscape). Mapping of these surfaces reveals synchronous and asynchronous relations of (a), (b), and (c) presumably because of uneven intensity of tectonic activity across a continent in a given geological epoch.

Finally, Gerasimov (1970) proposes three megacycles in the geomorphological development of the Earth during the Mesozoic-Cenozoic. The earliest (Megacycle I) is a Jurassic-Cretaceous basal planation surface surmounted by inselbergs of Paleozoic or older rocks, but which is still present on a global scale. This megacycle took place in a relatively quiet tectonic environment and led to production of the basal or *global peneplain*. Megacycle II starts with a time of tectonic mobility in late Mesozoic or early Paleogene time and climaxed in late Paleogene-early Quaternary time. This megacycle is envisioned as a denudation-deposition cycle, a time of erosion in elevated regions and sediment accumulation in subsidence regions. The erosion-deposition ratio is related to increase or decrease of tectonic activity over short time spans, and the result is widespread, non-completed (partial or incipient) peneplains of only local extent. Megacycle III is triggered by renewed tectonic activity, onset of large-scale glaciation, and rapid fluctuations of sea level during Quaternary time. This megacycle overprints the Cenozoic landscape with thick accumulations of alluvial and fluvioglacial deposits in valleys, and maximized production of regolith deposits on uplands and slopes because of cooler, wetter climates; in areas not directly affected by glaciation, basins tended to develop large accumulation terraces and slope deposits.

Attempts to correlate erosion surfaces are not new, and this was a popular pastime earlier in this century. Correlation charts for regionally named peneplains in North America are abundant, as for example attempts by E. Knopf (1924), Cole (1941), Shaffer (1947), and Thornbury (1965). As the cyclic erosion and peneplain concepts fell into disfavor, the popularity of correlation charts likewise declined, although the general problem was reviewed as late as 1956 by the International Geographical Congress.

If, however, we are to argue affirmatively that time-synchronous regional or continental scale erosion surfaces exist, and to use geodynamic mechanisms to show when and how they were created, it seems useful to reconstruct another world-wide correlation chart of erosion surfaces. This reconstruction (Tables 1 and 2) is based solely on a literature review, and the ultimate rashness of our effort is shown by the inclusion of known or postulated surfaces in the Mesozoic-Cenozoic mobile belts of western North America! The tables thus can represent only such correlations and temporal placement of denudational episodes as is permitted by our interpretations of these older publications. For example, it was common to use the same name for an erosion surface in different regions, while simultaneously assigning the surface to a different part of geologic time. Conversely, surfaces that were presumably of the same age were assigned different names in different regions. Some papers are so purposely vague about chronological placement that it is impossible to determine any chronology at all! It should be remembered, however, that earlier workers lacked the information provided by radiometric dating, and their studies long predated the work of King, the Soviet geomorphologists, or the persuasive concepts of plate tectonics, continental drift, and other elements of the "New Global Geology." These workers relied only on altimetry, the few topographic maps available, cross-cutting relationships of planar surfaces of erosion to rock structure and lithology, and perhaps their long familiarity or "feel" for the historical geology of a particular area or region.

Tables 1 and 2 do suggest that many geomorphologists and geologists, working over a span of many years, were convinced that erosion surfaces are real phenomena, and appear to represent more or less time contemporaneous, regional or continental scale events. Furthermore, the tentative general agreement about the number of major erosion surfaces and their relative spacing in geologic time is too significant to dismiss summarily. Therefore, for the remainder of this paper, we will make the tacit assumption that the surfaces exist, and will present a case for their formation by constructing a

Table 1. Correlation chart of North America erosion surfaces. (Data are compiled from a review of numerous literature sources).

	AGE TO BASE (M.Y.)	APPALACHIANS	INTERIOR LOW PLATEAUS	INTERIOR HIGHLANDS	CENTRAL LOWLANDS	GREAT PLAINS	ROCKY MOUNTAINS	GREAT BASIN	SIERRA/ CASCADE
PLEISTOCENE	2	VALLEY CYCLE	DEEP STAGE	VALLEY CYCLE POST-OSAGE STRATH	DEEP STAGE HAVANNA STRATH	TERRACES	CANYON CYCLE	PEDIMENT CYCLE	KERN RIVER CANYON CYCLE
PLIOCENE	7	SOMERVILLE	PARKER	OSAGE STRATH	CENTRAL ILLINOIS	FLAXVILLE		ANTLER	MOUNTAIN VALLEY (?)
MIOCENE	26	HARRISBURG	LEXINGTON/ HIGHLAND RIM	HOT SPRINGS /OZARK	LANCASTER /CALHOUN		ROCKY MTN./ SUBSUMMIT		CHAGOOPA/ BROAD VALLEY
OLIGOCENE	38								
EOCENE	54					PRAIRIE/ CYPRESS HILLS	FLATTOP/ SUMMIT	"BROKEN HILLS" (?)	SUBSUMMIT/ BOREAL
PALEOCENE	65	SCHOOLEY		OUACHITA/ SPRINGFIELD	DODGEVILLE (?)				SIERRA/SUMMIT (?)
CRETACEOUS	135	FALL ZONE		BOSTON MTN. /SUMMIT (?)		XXX			
JURASSIC	200								

Table 2. Correlation chart of erosion surfaces exclusive of North America. (Data are compiled from a review of numerous literature sources).

	AGE TO BASE (M.Y.)	SOUTH AFRICA	WEST AFRICA	ANGOLA	BRAZIL/ URUGUAY	AUSTRALIA	INDIA	MONGOLIA	CHINA
PLEISTOCENE		CONGO			PARAGUAÇU	WUDINNA	XXX	PANGKIANG	PANCHAIO
						KOONGAWA	JAMDA	GOBI	
———	2								
PLIOCENE		NOSSOB/ COASTAL PLAIN		XXX	VELHAS	MECKERING			
———	7								
MIOCENE			HO-KETA				NAOMUNDI		TANGSHIEN
———	26								
OLIGOCENE						UNO/NONNING	KIRIBIRU	KHANGAI/ MONGOLIAN	PEI-TEI
———	38		ASHANTI	NAMIB					
EOCENE		AFRICAN			SUL-AMERICANA	AUSTRALIAN PEDIPLAIN	INDIAN		
———	54								
PALEOCENE									
———	65	POST-GONDWANA			XXX	SIMMENS/NOTT	POST-GONDWANA	POST-LAURASIAN	POST-LAURASIAN
CRETACEOUS									
———	135		VOLTAIAN	BENGUELLA					
JURASSIC		GONDWANA		PLANALTO	GONDWANA	GONDWANA/ MT. DALE	GONDWANA/ NILGIRI	LAURASIAN	LAURASIAN
———	200		AGU MTN. (?)						
TRIASSIC		SUB-STORMBERG			SUB-BOTUÇATU	LINCOLN			

sedimentational-eustatic-tectonic model that may explain the observed phenomena and justify the subjective deductions of the earlier geomorphologists.

GEODYNAMIC BASIS FOR FORMATION OF EROSION SURFACES

Conditions for Surface Development.

In the foregoing discussion, a considerable body of literature was reviewed which supports the proposition that major erosion surfaces can be identified on a global scale. The review also appears to reflect a general belief in the time-equivalence of certain surfaces. In other words, there seems to have been recurring intervals of optimum conditions for these surfaces to develop during the geologic past. In the sections that follow, additional observations will be presented in support of such a recurrence of events and, in light of recent geological developments, an attempt is made to define this sequence within the framework of geologic and absolute time.

We do not intend to examine specific geomorphic processes which lead to the development of erosion surfaces. Rather, it is simply assumed that in response to a change in geologic conditions, the overall geomorphic regime has been periodically altered to a condition dominated by net erosion and landscape denudation. This is visualized as resulting from three general, but related, sources: tectonic, eustatic, and climatic adjustments. Thus a change in any or all of these factors in the proper direction results in a "lowering" of effective base level and a phase of increased landscape denudation will then occur.

Facts used in deciphering past geomorphic evolution must be obtained from observation of the preserved geologic record. Although the idea of base level is a basically simple and useful concept, many geologic variables and adjustments are involved. Tectonic uplift or eustatic fall of sea level has the effect of lowering base level. However, a climatic change that increases the rate of weathering and erosion would have the same influence on landscape development, although levels of the land surface and sea may remain unchanged. The stratigraphic record reflects many marine transgressions and regressions, but it normally is impossible to state precisely the ultimate cause of such changes. The direction, rate, and location of tectonic movements, rates of sediment production and deposition, volume of free water, isostasy, climatic change, and numerous other factors influence the location and motion of the strand line on local, regional, and world-wide scales. These facts, together with an inability to accurately date an observed phenomenon can introduce significant error into the determination of past base level fluctuations. However, even with these limitations, certain "cyclic phenomena" can be observed to have recurred throughout Phanerozoic time. Thus if the timing and periodicity of climatic, eustatic, and tectonic events can be determined, these factors can serve as input into a system. The output of this system will be determinations about the periodic development of erosion surfaces through time.

Tectonic Cyclicity and Periodicity.

The idea of geologic cycles dates from the very beginnings of geology (see Adams, 1954). Dawson (1866), Hull (1868), Newberry (1873) and many other geologists of the 19th century noted depositional-erosional cycles within the stratigraphic record. A more detailed review of geologic literature shows that the idea of more or less world-wide periodicity in diastrophism and transgression-regression sequences has been a point of discussion and debate for many years. A good review of the literature relating to cyclic sedimentation is given by Weller (1964).

The concept of geologic cycles was the theme of a book written by Umbgrove in 1947 entitled "The Pulse of the Earth". In this text the author presents a significant body of evidence for tectonic, climatic, eustatic, biologic, and extra-terrestrial cycles as based on geologic observations. Umbgrove (1947, p. 86) decided that most geologic phenomena "must be attributed to some common and world-embracing, deep-seated cause, for all the phenomena are related chronologically". In relation to eustatic sea level movements, Umbgrove (1947, p. 93-94) concluded that: 1) world-wide transgressions and regressions are brought about by periodic movements of the continents and/or ocean basins, 2) epochs of folding coincide with periods of world-wide regression, 3) movements of sea level and folding are caused by subcrustal processes which "elapsed periodically, as it were, with a rhythmic cadence", and 4) other phenomena such as magmatic cycles and the formation of basins and domal uplifts are closely related to this rhythm (Fig. 1). Umbgrove also concluded that for climatic change, a periodicity of approximately 250 million years was indicated for alternating periods of glaciation and warmer climates from the Precambrian to Pleistocene time. Umbgrove (1947, p.323) also noted that the same major periodicity of 250 million years is reflected in times of major orogenic activity, and that during intervening periods minor activity occurred with an average cadence of 50 m.y. Interestingly, he observed that our galaxy rotates once in 200 to 300 m.y. Although this relation may be totally coincidental, Steiner (1973) recently presented a galacto-geologic model to correlate transgressive-regressive sequences with cosmic and galactic periodicities.

The timing and coordination of tectonic activity and related eustatic fluctuation has long been, and continues to be, a subject of debate for structural geologists and stratigraphers. The points of debate that are basic to an understanding and interpretation of geomorphic history can be generally categorized as the following: 1) Definite periods of recurring episodes of orogeny separated by extended periods of relative quiescence *vs.* tectonic activity operating somewhere all the time; 2) Times of orogeny are times of marine transgression *vs.* times of orogeny are times of marine regression; 3) Orogeny in mobile belts is directly related, in an orderly fashion, with epeirogenic cratonal movements *vs.* no temporal relation between mobile belts and cratonal tectonism.

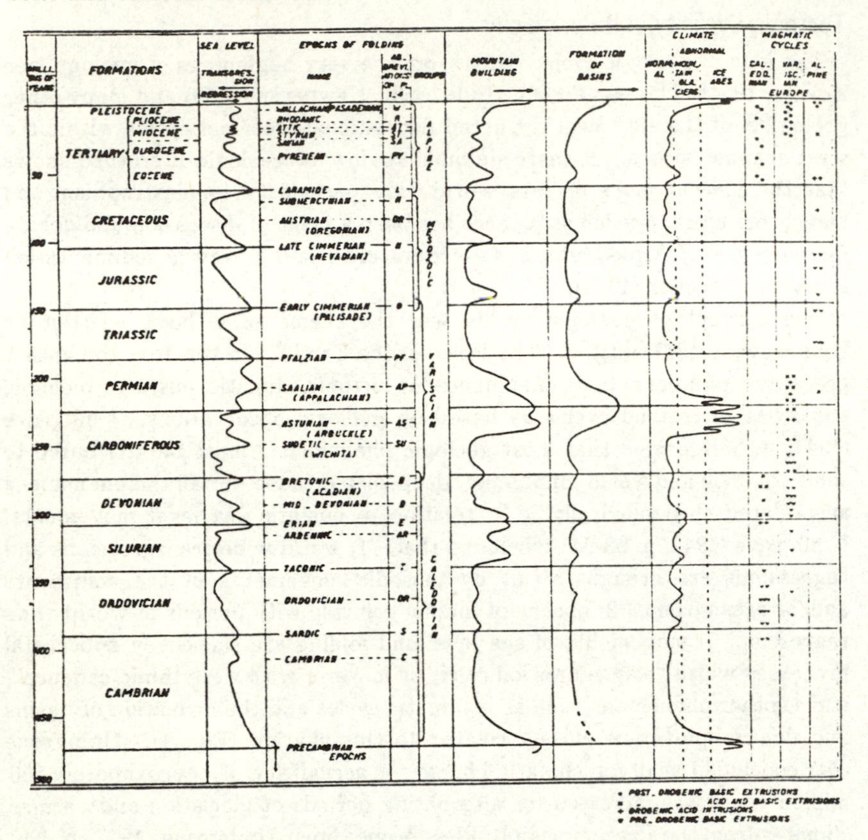

Figure 1. Diagrammatic synopsis of "the pulse of the Earth" (modified from Umbgrove, 1947).

Stille (1924) took the position that "all orogeny is confined to relatively few phases of short duration and of more or less world-wide importance" (Bucher, 1933, p. 393). As noted, Umbgrove also subscribed to this view. Bucher (1933, p. 405) also believed in periodic tectonic activity as evidenced by the statement "At any point on the earth, positive, that is, upward orogenic movements are limited to relatively short epochs separated by much longer epochs of rest or of negative movement. The duration of such an orogenic epoch is measured in terms of hundreds of thousands of years . . .". Bucher also subscribed to the premises of simultaneous orogeny in various parts of the world (his Law 41, Opinion 32); the universal, synchronous nature of sea level change (his Law 44); and that great transgressions occurred during anorogenic intervals (his Law 45). Damon (1972) more recently proposed two tectonic periodicities, an epeirogenic periodicity of about 180 m.y. and an orogenic periodicity of about 36 m.y., and concluded that anorogenic periods are of about the same duration as orogenic periods.

Conversely, Gilluly (1949), Rutten (1949), and others have argued against the concept of short orogenic phases separated by long periods of quiescence. Some geologists agree with Bucher and Umbgrove that times of orogeny are times of regression (Damon, 1971; Grasty, 1967; and Rona, 1973), whereas others (Evernden and Kistler, 1970; Johnson 1971; and Sloss and Speed, 1974) maintain an opposite view. Similarly, expert geologists continue to disagree on other salient aspects of tectonic periodicity.

The recent literature indicates that disagreements on tectonic cyclicity and timing can be resolved, in general terms, by viewing the problem at different scales. Rutten (1949, p. 1768-1769) notes that though alternation of orogenic and anorogenic periods for limited areas of the Earth's crust is an established fact, the Earth as a whole has experienced long periods of tectonic activity characterized by varying and fluctuating phases. In relation to volcanic activity, Gilluly (1973, p. 499) notes that "though magmatism has been episodic at any one locality, taking the Cordillera as a whole, it has been proceeding steadily . . .". In a detailed study of the Antler orogeny, Johnson (1971) notes that although the tectonic activity can be subdivided into separate phases, the orogeny was felt all along the western and Arctic margins of the North American continent during most of Devonian and Mississippian time. Johnson also notes that effects of the orogeny were felt progressively southward through time along the geosyncline. Similar evidence is noted by Gibson (1970, p. 1817) in concluding that the Atlantic Coastal Plain sediments indicate that regional uplift in the Appalachian area proceeded with a north to south progression. This interpretation is substantiated by Mathews (1975) who observes that the amount of sediment (and therefore presumably erosion) derived from the Appalachians increases somewhat from north to south. Similarly, Rona (1973, p. 2867) states that "the migration in space and in time of orogenic activity, rather than its occurrence as a sharp, synchronous, world-wide event, appears consistent with . . . plate tectonics . . .". Consequently, one might visualize discrete periods of orogeny on a smaller scale (local or areal) in response to a migrating focus of more intense activity and/or uplift throughout the orogenic belt or region being uplifted. Tectonism appears therefore to have proceeded in different fashions depending upon the observational scale used.

To summarize, there is considerable evidence, although far from conclusive, to substantiate the premise that tectonic activity may be periodic or episodic. Although geologists disagree on precise timing, distribution, and mechanisms of these events, an operational hypothesis that tectonic activity may be viewed as cyclic or episodic does not seem totally untenable.

Climatic and Stratigraphic Cyclicity and Periodicity.

Cyclicity and periodicity of climate can also be demonstrated. As noted, Umbgrove (1947) proposed a 250 m.y. climatic cycle based on the reoccurrence of glacial conditions. Durham (1959) demonstrated a general cooling trend from late Cretaceous through Tertiary time, with a superimposed cyclic variation of approximately 10° to 15°F. Similar cyclicity in ancient temperatures has been demonstrated by Emiliani (1958, 1966). Alternation of glacial and interglacial climates within the Pleistocene is unquestioned. In attempting to refine the Pleistocene time scale, Evans (1971) found that cyclic climate changes appear to have a somewhat irregular period averaging 40,000 years. Consequently, like tectonic activity, there appears to be an underlying cyclicity of climatic variation through time.

Stratigraphers have long recognized the cyclic nature of sedimentation. Rise and fall of sea level and resulting strand line migration are preserved in the stratigraphic record by transgression-regression sequences. This observation then returns us to the quandary of timing such events, deciding if they are truly world-wide in scope, and what their relationship is to tectonic activity. Approaching this problem with the warning that "one must be specific in discussing particular cyclic patterns, for there are all scales of repetition", Dott (1964) points out that both long and short period cycles are observable in mobile belt sedimentation. Short period cycles usually reflect local geologic pulsation and oscillations, whereas long period cycles, affecting both cratons and mobile belts, reflect universal variations in sea level.

Sloss (1963) demonstrated that the sedimentary record of the North American craton contains six inter-regional unconformities which subdivide the sediments into six stratigraphic sequences (Table 3). Although these unconformities are time-transgressive, as expected, they can be recognized in all parts of the craton interior. Each of the intervening stratigraphic sequences is interpreted as being deposited during an extended transgression-regression cycle. Thus six major circuits of the seas onto and off of the craton are recorded since late Precambrian time. Sloss (1972) subsequently demonstrated the same periodicity of marine transgression and regression on the Russian Platform, and noted that episodes of basin subsidence and relative stability appear to be generally synchronous with North America. Similarly, Dearnley (1966) and Haughton (1963) identified six major rock-stratigraphic units and six intervening major unconformities in Africa. These appear to be broadly correlative with those delineated by Sloss in North America and in Russia (see Fig. 6). Rona (1973) recently showed that sea-level fluctuations and sediment accumulations on the coastal plains and continental shelves of southeastern North America and northwestern Africa exhibit a great degree of correspondence since Cretaceous times. Consequently, it appears that synchronous eustatic sea-level movements are well documented.

In order to justify the interpretation of any present-day landscape as

Table 3. Stratigraphic sequences and bounding unconformities identified across the North American craton (after Sloss, 1972).

TEJAS SEQUENCE
Sub-Tejas unconformity - Paleocene, about 64 m.y.b.p.
ZUNI SEQUENCE
Sub-Zuni unconformity - Early Jurassic, about 172 m.y.b.p.
ABSAROKA SEQUENCE
Sub-Absaroka unconformity - Late Lower Carboniferous, about 325 m.y.b.p.
KASKASKIA SEQUENCE
Sub-Kaskaskia unconformity - Early Devonian, about 385 m.y.b.p.
TIPPECANOE SEQUENCE
Sub-Tippecanoe unconformity - Early Ordovician, about 480 m.y.b.p.
SAUK SEQUENCE
Sub-Sauk unconformity - Late Precambrian, about 600 m.y.b.p.

remnants of former erosion surfaces of great regional or continental extent, it is necessary to demonstrate that conditions have existed in the past to allow formation of such features on a large scale. If several such features are inferred to have formed at different times in the past, it must also be demonstrated that there were recurring intervals of requisite conditions or a recurring sequence of events that produced them. It is presently impossible to prove that these prerequisites have been satisfied. However, the foregoing discussion has encapsuled a significant number of observations that tend to support the premise that tectonic, climatic, and eustatic variations appear cyclic, but with irregular periods if viewed in the proper temporal and geographical perspective. Furthermore, the quantity of sediments found in the Atlantic and Gulf Coastal Plains, the Tertiary basins of the western U. S., and beneath the Great Plains is tangible evidence of the tremendous erosion that occurred during Cretaceous and Tertiary times. Therefore, it seems plausible to assume that during recurring intervals of time, geologic conditions have been such that net erosion occurred concurrently over vast areas of the landscape.

If it is thus assumed that there are periodic recurrences of regional erosion, an underlying causative mechanism must be defined. Sea-floor spreading and plate tectonic theory is appealing in this regard.

Geotectonic Models.

The hypotheses of sea-floor spreading and plate tectonics have become quite popular in recent years because they provide a new approach to many geologic problems. Innumerable reports have been published which attempt to explain diverse geologic phenomena in terms of these new ideas (see Pitman, 1971). However, considerable debate continues about the nature of driving forces, mechanics of motion, plate interactions, and other basic aspects of these concepts. Although several geotectonic models have been proposed, none in vogue seems capable of explaining all observed geologic phenomena. Furthermore, some geologists argue that sea-floor spreading and plate tectonics cannot be applied to pre-Mesozoic Earth history, even though the theory is tenable for later geologic time. Others reason that it is equally valid for earlier history. Bird and Dewey (1970, p. 1032) developed a model to explain late Precambrian to Devonian Appalachian tectonics, stating the opinion that "lithosphere plate theory is the most fundamental paradigm change that has occurred in the earth sciences, and explains the tectonic evolution of the earth during at least the last 1000 m.y." Although this may be an extreme view, it is possible to explain major structural elements such as geosynclines (Wang, 1972), major orogenies (Dewey, *et al*, 1973), magmatism and pluton implacement (Gilluly, 1973), and other tectonic events through the "new global tectonics".

Bird and Dewey (1970, p. 1103) give as the essential tenets of plate tectonic theory: 1) the lithosphere is segmented into rigid plates bounded by the major seismic zones; 2) plates grow by accretion along tensional zones of ocean ridges; 3) plates are consumed in oceanic trenches and return into asthenosphere as cold, seismically active slabs; 4) major displacements of oceanic (magnetic) anomaly patterns occur along transform (strike-slip displacement) faults; 5) global plate accretion and consumption vectors seem to approximate zero; and 6) lithosphere plates move as rigid slabs on the low velocity channel, continents passively ride as superficial parts of plates, and plate boundaries may or may not coincide with continental margins. By analyzing the spacing and distribution of magnetic anomalies adjacent to spreading centers, it is possible to calculate rate and direction of past plate movements (see Pitman and Talwani, 1972). It is then possible to rotate plates about poles to reconstruct continental areas for given times in the past. By incorporating stratigraphic information it is also possible to define those areas beneath epicontinental seas. Thus it is possible to reconstruct paleogeographic and paleogeomorphic conditions and locate contiguous continental areas that were concurrently exposed to subaerial erosion, but which today may be separated by thousands of miles of ocean.

One of the recently proposed geodynamic models that is particularly appealing was developed by Sloss and Speed (1974). This model, based upon the cited tenets of sea-floor spreading and plate tectonics, was formulated to explain the presence of apparently world-wide unconformities and the cratonal tectonic cycles (Sloss, 1964) inferred from the stratigraphic record.

Sloss and Speed deduced that Phanerozoic evolution of continental cratons was marked by repeated global episodes of three types, called emergent, submergent, and oscillatory modes (Fig. 2). The emergent episodes are characterized by slow elevation of cratons with respect to sea level to maxima of more than a kilometer over a period of 10^6 to 10^7 years, followed by equally slow submergence. The uplift is viewed as a smooth uparching without significant structural relief while continental margins remain passive. An emergent episode is recorded as a craton-wide unconformity; therefore, the craton experienced extensive erosion with sediments being deposited in extracratonic sinks. The submergent mode is characterized by progressive, differential subsidence of the craton, where cratonic-interior basins subside more rapidly than separating arches and domes. Most of each episode involves marine transgression, occurring in pulses of a few million to ten million years. The cratonal-continental margins are sites of active orogenic belts formed in response to plate convergence and accompanying subduction and obduction. Finally, the oscillatory episodes (Sloss and Speed, 1974, p. 101-102), are:

> "marked by pulsatory vertical movements leading to general net elevation of cratons with respect to sea level and thus characterized by a sedimentary record of marine regression punctuated by repeated transgressions with a periodicity ranging from less than a million to five million years."

These episodes are further characterized by high-angle faulting, producing mountain blocks and yoked basins, with periods of vertical movements ranging from 10^5 to 10^6 years. Plate convergence occurs along island arc systems, several hundred kilometers from the adjoining craton, with either inactive or active extensional basins lying between the island arc and the continental border.

As a mechanism for these tectonic episodes, Sloss and Speed propose episodic changes in the proportion of melt in the continental asthenosphere. When the concentration of melt increases, the asthenosphere thickens, and continental uplift results. When the fraction of melt in the continental asthenosphere reaches a critical level, melt is extracted and convected to the oceanic asthenosphere where it is discharged at oceanic ridges. This results in increased plate accretion and sea-floor spreading while the cratons undergo subsidence. Oscillatory movements result from "erratic perturbations in the ratio of melt retention to extraction."

To summarize the Sloss and Speed model, the six stratigraphic sequences defined by Sloss (1963) represent deposition during either submergent or oscillatory tectonic modes, whereas the intervening major unconformities represent either the emergent mode or the gradual transition from emergent to oscillatory modes (Fig. 2). Submergent mode deposition occurred during an extended transgression-regression cycle, whereas sediments deposited during oscillatory modes reflect several minor cycles of shorter duration.

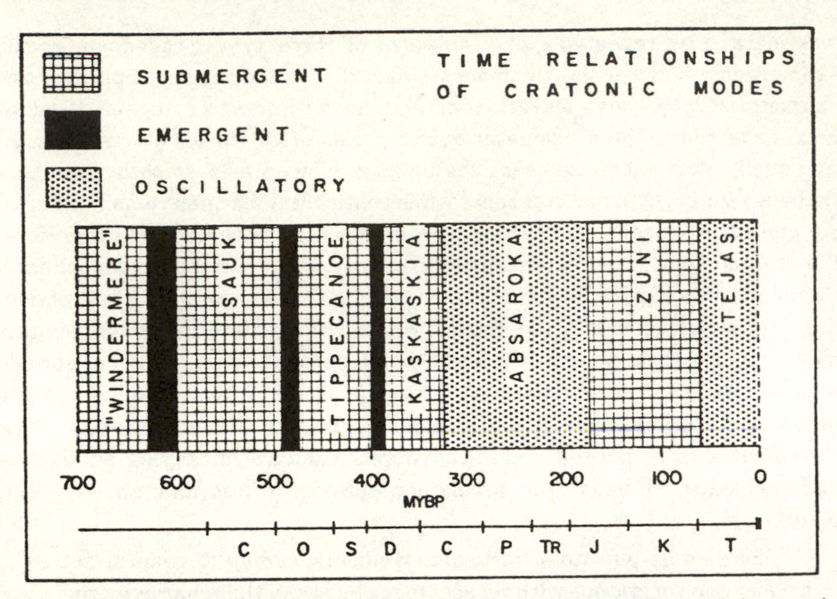

Figure 2. Time distribution of identified cratonic modes, late Precambrian to the present, and relative chronology of episodes of deposition of the North American stratigraphic sequence (modified from Sloss and Speed, 1974).

The cratonal subsidence which results in marine transgression roughly corresponds in time with orogenic activity along cratonal and plate margins and an increased rate of sea-floor spreading and plate accretion along oceanic ridge systems (Fig. 3). During oscillatory modes, cratons experience alternating uplift and subsidence and minor regression and transgression, with an overall uplift-regression tendency or trend. Although sea-floor spreading rates also vary, the correspondence with craton movement may be influenced by the draining of melt by volcanism along faults and rifts.

Support for this model is found in the results of several other studies. A detailed stratigraphic and structural analysis by Johnson (1971, p. 3291) observes that:

> "of the four major orogenies that occurred in North America during the Paleozoic and Mesozoic, three began, reached climactic stages, and went through waning stages during the time epicontinental seas were transgressing to their maximum extent and then regressing to form the great onlap-offlap cycles called sequences."

As Figure 4 indicates, the Taconic, Acadian-Antler, Appalachian-Sonoma, and Nevadian Orogenies were approximately synchronous with deposition of the Tippecanoe, Kaskaskia, Absaroka and Zuni cratonal sequences respectively. This diagram also shows that geosynclinal quiescence is contemporaneous with cratonal erosion during a regressed state. The observation that rapid sea-floor spreading and continental convergence is correlative with

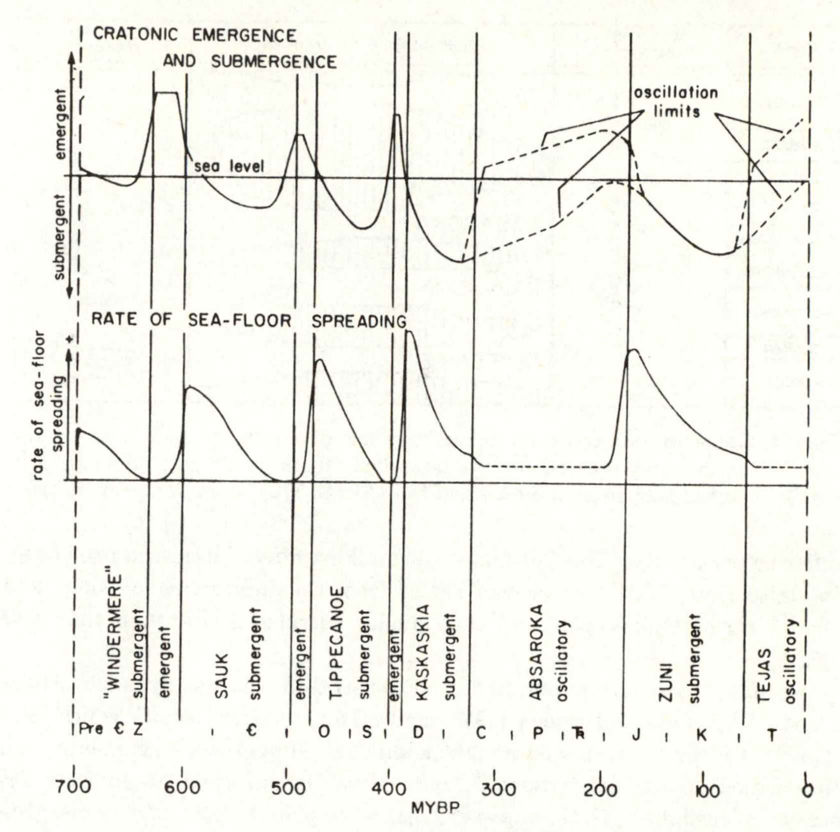

Figure 3. Positions of cratons with respect to sea level (upper graph) and inferred relative rates of sea-floor spreading (lower graph) in terms of geologic time, cratonal tectonic modes, and the periods of deposition of the North American stratigraphic sequences (after Sloss and Speed, 1974).

time of transgression has been made in several studies (Armstrong, 1969; Bird and Dewey, 1970; Johnson, 1971). As an example, Hays and Pitman (1973) showed quantitatively that the world-wide middle to late Cretaceous transgression corresponds with a pulse of rapid sea-floor spreading between 110 m.y. and 85 m.y. B.P. As already noted, Rona (1973) found that areas along and adjacent to the southeastern coast of North America and northwestern coast of Africa have experienced essentially simultaneous alternating periods of maximum and minimum rates of sediment accumulation since late Jurassic times. These fluctuations show a close correspondence to transgression and regression which, in turn, can be correlated with the rate of sea-floor spreading. On these bases, Rona (1973, p. 2864) presented the relations between sediment accumulation, sea-floor spreading, and eustacy shown in Figure 5. Clearly, the relationships depicted in this figure are supportive of the Sloss and Speed model. Note also that column G (Fig. 5) reflects interaction and overprinting of two generalized but discrete cycles of

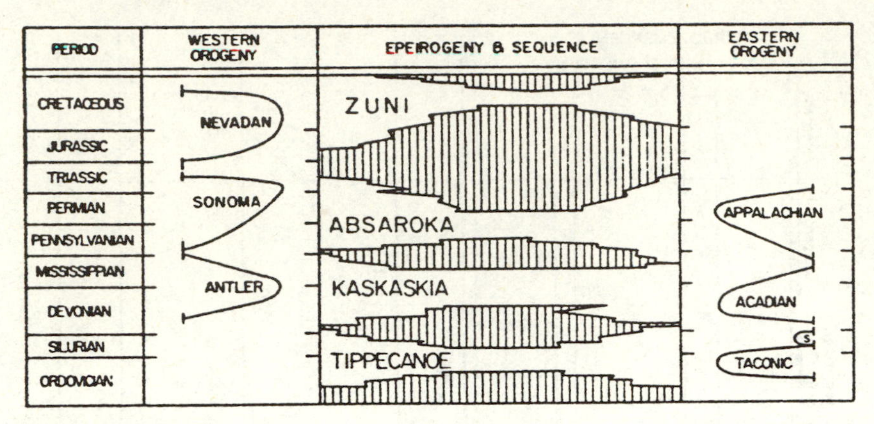

Figure 4. Table of orogenies on opposite sides of North America and of the sedimentary rock sequences deposited on the continental interior. The named sequences are modified from Sloss, 1963 (from Johnson, 1971).

differing periodicity. This illustrates the masking effect, discussed previously (see also Dott, 1964), which can result from the interaction of long- and short-term periodic events and thus hinder correlations through time and space.

We have extensively outlined and illustrated the geodynamic model proposed by Sloss and Speed (1974) and other research results which lend support to this model. Although additional supportive arguments and observations could be presented, our recital is sufficient to indicate its possible plausibility. Note, however, that some plate tectonists may disagree with the hypothesized mechanism driving the crustal elements, and others might argue against the deduced timing and interrelationships of orogenic, eustatic, and epeirogenic events. Deficiencies and apparent discrepancies with local geologic observations can be found in any general geotectonic model yet presented. However, it appears to us that the overall timing, location, and distribution of large-scale geologic phenomena, and their interrelationships as predicted by this model, are in good agreement with those deduced from the geologic record.

Geomorphic Model.

The geodynamic model provides a rationale for deducing and documenting the recurrence of world-wide, penecontemporaneous periods of erosion and net landscape denudation. If it is assumed that activity was approximately synchronous and correlative on all continental plates, as implied in the model and supported by data, and that periods of optimum conditions for net erosion were of sufficient duration to permit the development of a subdued landscape of low relief, the reality of world-wide denudation surfaces seems plausible. The validity of this latter assumption, a crucial point for historical geomorphology, will be discussed in a subsequent section.

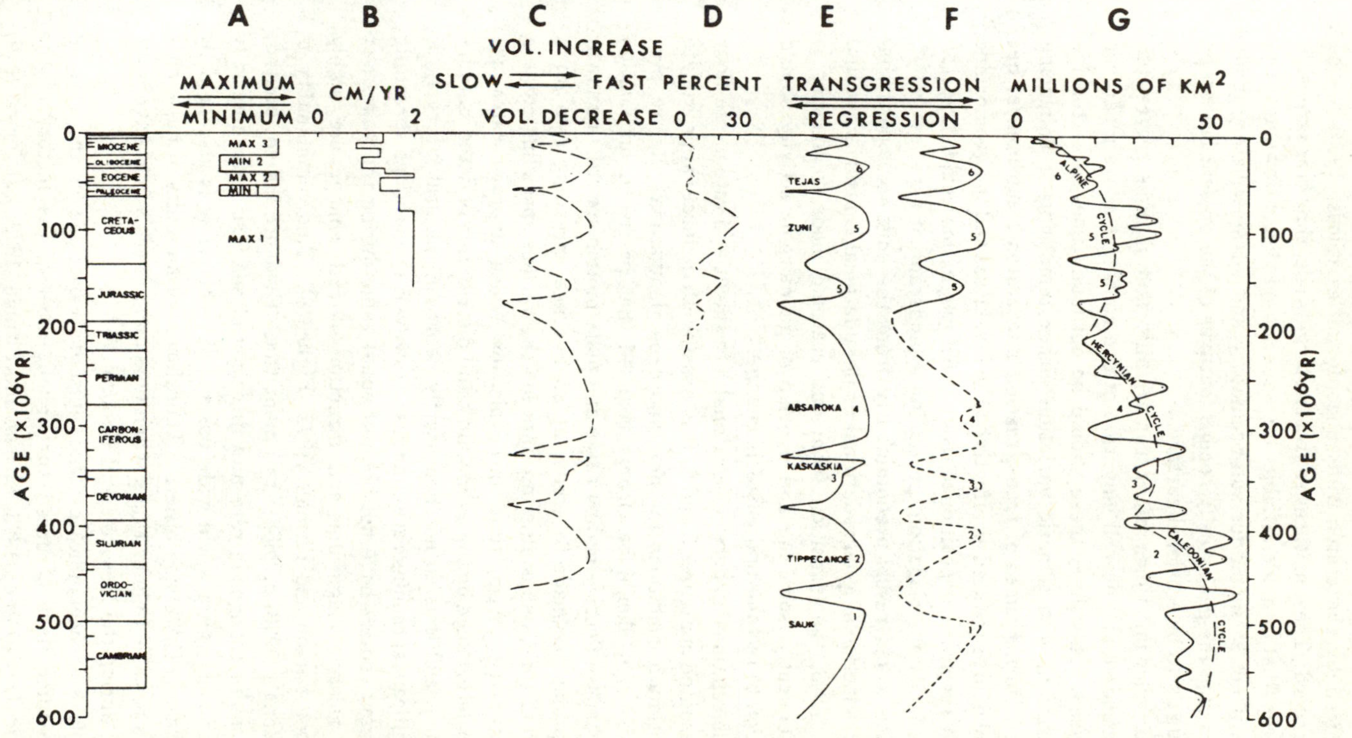

Figure 5. Relations through time between rates of sediment accumulation on continental shelves, sea-floor spreading, and eustacy. A. Average rates of sediment accumulation. B. Half-rates of sea-floor spreading, central North Atlantic. C. Inferred world-wide rate of sea-floor spreading (relatively slow or fast) and inferred state of mid-oceanic ridge system (volume change); these are considered to vary in unison. D. Area of North America covered by sea. E. North American marine transgressions and regressions. F. African marine transgressions and regressions. G. World continental area covered by sea. D through G are from the stratigraphic record. (Data obtained from many sources; figure is from Rona, 1973).

The emergent cratonal mode is a time of widespread and lengthy erosion. This is evidenced by the craton-wide unconformities (Sloss, 1963) in the stratigraphic record. These hiatuses represent large-scale, time-transgressive erosion surfaces that are identifiable on several cratons. However, the interpretation is not so simple for sub-oscillatory and intra-oscillatory mode unconformities, and the geologically recent features of the present landscape that are presumed to be erosional surfaces.

During and following times of cratonal uplift and marine regression, continental freeboard and overall elevation of the continent above mean sea level is increased. Therefore, these should be optimum times for extensive landsurface erosion, with climatic variation either tempering or enhancing this condition. During times of transgression and cratonal subsidence, net erosion rates should be reduced as effective base level rises and the geomorphic regime moves toward a depositional mode. Although this could have the same effect of landscape planation, producing a landscape of very low relief, the landsurface would result predominantly from the shift to depositional conditions. Production of a degradational or denudational landsurface would be retarded. Also, the transition from one cratonal mode to another is gradual; consequently, some lag and overlap of geomorphic activity should occur in relation to tectonic and eustatic change.

Invoking the relationship between cratonal movement and rate of sea-floor spreading presented in Figures 3 and 5, it is possible to delineate approximate periods of optimum conditions for net landscape denudation. Before proceeding, however, it should be clearly realized that precise quantification values for global sea-floor spreading rates and plate motion are still unknown. Many assumptions are made to arrive at an average spreading rate for a given interval of time, and conflicting data for the various spreading centers have been reported. Also, time intervals for which average rates are calculated are determined by the availability and distribution of dated samples of oceanic crust, and thus do not necessarily represent natural intervals or reflect all significant spreading rate changes. However, even though spreading rates at different centers have not been of equal value during the past or at present, the relative magnitude and direction of rate change are highly correlative (see Larson and Pitman, 1972, Figure 8). This communality is reflected in Rona's (1973, p. 2858) statement that "Evidence, as yet limited, suggest that major changes in rate and direction of sea-floor spreading may occur simultaneously on a world-wide basis . . .". This view is also held by Larson and Pitman (1972) and others. Therefore, within limits of basic data precision, it is possible to infer times when erosion surfaces would have most likely formed during the Mesozoic and Cenozoic.

Considerable evidence indicates that episodic variations in the rate of sea-floor spreading have occurred throughout most of the history of the North Atlantic (Dewey *et al*, 1973; Larson and Pitman, 1972; Lattimore *et al*, 1974; and Pitman and Talwani, 1972). Because much more data exist for

Atlantic spreading centers than other centers, geomorphic inference will be made from these observations, with the presumption of analogous activity at other centers.

A schematic presentation of the variation of sea-floor spreading and related phenomena is given in Figure 6. Spreading data were compiled and generalized from Hayes and Pitman (1973), Lattimore *et al*, (1974), Pitman and Talwani (1972), and Rona (1973). The generalized transgression-regression curve was compiled from observations by Hallam (1963), Hayes and Pitman (1973), Rona (1973), and Sloss (1963), whereas the sediment accumulation curve and Atlantic Coastal Plain unconformity data were obtained from Rona (1973) and King (1967) respectively. Corresponding times of landscape erosion, as deduced from the other variations, are our own.

It will be recalled that in the Sloss and Speed (1974) model, times of rapid sea-floor spreading correspond with cratonal subsidence or cessation of uplift and marine transgression. Although the spreading rate data do not extend into the Jurassic, the data trend seems to indicate a possible decreasing rate and thus a more positive cratonal position (see also Fig. 5). This conjecture is in accord with the waxing phase of the proposed submergent mode during Zuni time (Fig. 2). The spreading values appear to generally increase with time up to approximately 85 m.y. B.P., with a distinctively rapid rate from 110 m.y. B.P. to 85 m.y. B.P., and then a rapid decrease through the interval -85 m.y. to -53 m.y. This coincides with a major global transgression, culminating in approximately middle Cretaceous time, followed by more rapid regression and transition into the oscillatory Tejas mode. Spreading rates were low during the Paleocene and the craton was in a regressed, uplifted state with transgression commencing slowly during the latter half of the epoch. It was during this condition of cratonal exposure that the sub-Tejas unconformity formed. This coincides in time with a major unconformity in Coastal Plain sediments below Upper Paleocene strata. The spreading rate data reflect three periods of increase separated by two periods of low values since the beginning of Eocene time (approximately 53 m.y. ago). This is reflected in the transgression-regression curve and is in keeping with the variable vertical motion expected to occur during an oscillatory cratonal mode.

Note that the overall trend of the transgression-regression curve is one of gradual, but continued, regression since the maximum flooding of the Cretaceous. Similarly, the rate of sea-floor spreading appears to have decreased progressively, with periodic rapid phases of comparatively short duration since mid-Cretaceous time. This reflects the overall tendency for cratonal uplift during the waning phase of submergence and through the oscillatory mode. These long-term trends must be incorporated into geomorphic interpretations of base level and continental freeboard changes. Figure 5 indicates a maximum of 30 percent submergence during the Cretaceous, whereas Dunbar (1960, p. 319) states that almost 50 percent of

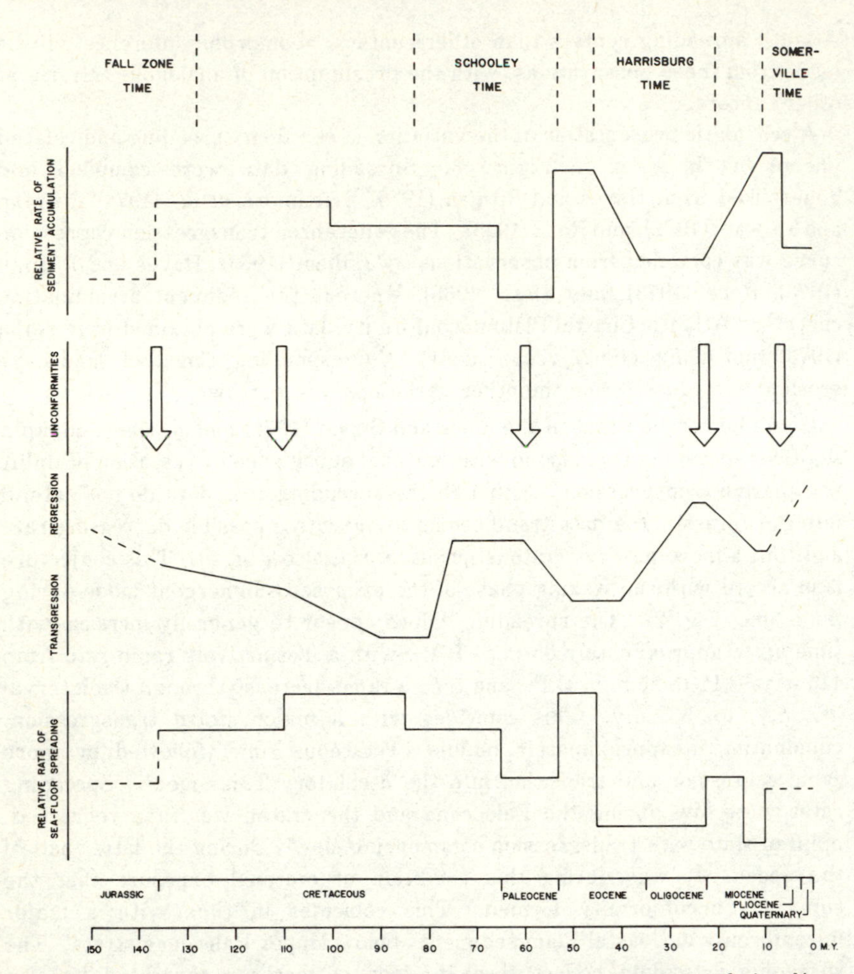

Figure 6. Time available for extensive denudation (peneplanation) of the North American craton. Potential broad-area erosion surfaces are plotted against an absolute time scale and curves of relative rate of sediment accumulation, major unconformities in the stratigraphic record, marine transgression-regression, and relative rates of sea-floor spreading. The figure suggests a cyclic but irregular periodicity of erosive episodes.

the present North American continent was submerged. Regardless of the absolute value, this was a time of marked submergence and rise of base level. The change from the Cretaceous to the Tertiary is marked by a major regression "at least as pronounced as the Upper Cretaceous transgression, after which the sea never returned to its former extent." (Hallam, 1963, p. 406). Maximum Tertiary submergence of the United States never exceeded 10 percent and averaged only about 3 percent (Dunbar, 1960, p. 353). Consequently, although the greatest Tertiary transgression occurred during the early Eocene, the effect upon base level and cratonal erosion rates was considerably less than during the Cretaceous. The same is true for Miocene transgression or base level rise, which was of even less magnitude than during Eocene time. Thus, the lower two curves of Figure 6 indicate a general condition of net accumulation generally occurred during times of transgression, with low rates of deposition being associated with regressional time. This is as expected for the transgressional state favors deposition and preservation of sediments in the near-shore environment.

Interestingly, the relative maxima of sediment accumulation have increased in magnitude since Cretaceous time and the minima have been less extreme since the Paleocene. In other words, there seems to be a long-term trend toward increased sediment production and accumulation since the beginning of the Tertiary. Rona (1973) noted that maximum accumulation rates for the Atlantic Coastal area of the United States during the Eocene and Miocene were respectively 125 percent and 165 percent of the Cretaceous values, whereas the minima during Oligocene and post-Miocene times were respectively 5 to 50 percent and 10 to 40 percent of the Cretaceous values. It will be recalled that the long-term eustatic trend appears to be one of net regression. Therefore, these two observations are not in harmony with the short-term correspondence of maximum accumulation with transgression. This long-term trend of increased sediment production and accumulation probably is an indication of net increase in continental freeboard or effective base level lowering by net cratonal uplift and marine regression since Paleocene time (Tejas oscillatory mode). This implies a tendency for generally increasing erosion rates through Tejas time. Consequently, during a short time span maximum, sediment accumulation appears related to marine transgression but correlates with marine regression over a much longer time period. Such situations are not uncommon in geomorphology, because negative feed back and apparent reversal of independence-dependence or cause-effect relationships can occur when time and space boundaries are crossed (Schumm and Lichty, 1965).

Based on the foregoing discussion, it is possible to infer periods of time during the Mesozoic and Cenozoic when conditions were favorable overall for net landscape erosion. Prevalence of erosive conditions is deduced during portions of the Jurassic, particularly during the regressed cratonal state when the sub-Zuni unconformity (Sloss, 1963) formed and during the early waxing phase of Zuni transgression and submergence. Although several lines of stratigraphic evidence support this reasoning, the data extrapolation presented in Figure 6 is conjectural. Based upon this evidence (see also Fig. 4) and the work of others (see Johnson, 1931), this period of net erosion, herein called Fall Zone time, is tentatively depicted as concluding sometime between -135 m.y. and -110 m.y. with the advent of extensive subsidence and Cretaceous marine transgression. We hypothesize that sediments which accumulated during this Cretaceous transgression were eroded earlier during Fall Zone time and were intermittently stored on and transported across the craton before final deposition in shallow epicontinental seas.

Figure 6 indicates that between 85 m.y. B.P. and 80 m.y. B.P. cratonal uplift began and continued until about 55 m.y. B.P. to 50 m.y. B.P. These data do not give any insight into the rate, magnitude, or timing of uplift; all that may be inferred with reasonable confidence is the general period of positive movement. If our reasoning is sound, one may visualize a more or less continuous general uplift over most of the craton through this time interval with smaller areas, perhaps of regional extent, experiencing a shorter pulse of uplift followed by a period of relative quiescence. Evidence of the magnitude of uplift and regression is given by Hays and Pitman (1973, p. 20) in estimating that between -85 m.y. and -60 m.y. sea level fell more than 250 meters, an amount sufficient to deflood most continental areas. This positive mode of cratonal exposure, marine regression, and lowered base level continued until -55 m.y. to -50 m.y., or approximately the end of Paleocene time. This period of optimum conditions for erosion, tentatively dated between -85 m.y. and -55 m.y., is herein called Schooley time.

From about -53 m.y. to -45 m.y. sea-floor spreading rates were large, reflecting cratonal submergence and marine transgression. Sediment accumulation along the continental margins was great. These sediments are presumed to have been eroded, transported, and temporarily stored during earlier Schooley erosion. In middle to late Eocene time (approximately -45 m.y. to -42 m.y.) uplift and regression recurred and the rate of sediment accumulation began to decrease. These conditions prevailed until approximately -25 m.y. to -20 m.y., or about the Oligocene-Miocene boundary. Thus it seems that ideal conditions for landscape erosion existed for approximately 20 m.y. to 25 m.y., or from late Eocene into earliest Miocene. This interval is named Harrisburg time. The second major Tertiary transgression, marking the end of Harrisburg time, began in the early Miocene, and the contemporaneous increase in sea-floor spreading is presumed to correspond with decreased net uplift or subsidence of the craton. The sediments eroded during Harrisburg time were accumulating in the transgressing seas.

Interpretation of the last 15 m.y. is more difficult because interrelationships as depicted in Figure 6 are not very clear or precise. A marked decrease in sea-floor spreading rate occurred during the approximate interval -15 m.y. to -10 m.y., which should be indicative of a late Miocene time of uplift. The data are somewhat inconclusive for the last 10 m.y., but seem to indicate a relatively uniform spreading rate. Eustatic data are disparate following the early Miocene transgression. Obviously the large glacio-eustatic fluctuations of the late Tertiary and Pleistocene prevent a clear interpretation. Similarly, sediment accumulation data of Rona (1973) are imprecise for this time frame. Accumulation rates for the North American Atlantic Coast, emphasized in Figure 6, correspond quite well with those for Africa until about 6 m.y. B.P., when they diverge (see Rona, 1973, p. 2865). It is probable that various ramifications of glaciation, exemplified through glacio-eustatic, climatic, and isostatic variation, are more evident for this recent period than for the long-term relationships. Obviously, the recent geologic and geomorphic record is more complete and permits more detailed study and interpretation, whereas the more remote record is such that only the large, longer-term variations are easily deciphered. Analogous local and short-term variations logically occurred in the more distant past, but they are not readily discernible because of incompleteness of the geologic record.

One possible interpretation of post-early Miocene time is based upon the apparent period of uplift commencing about -15 m.y. to -12 m.y. Although a transgression is indicated for this approximate time, it was very minor. Thus, although marine encroachment would tend to negate base level lowering through uplift, it would have very little relative effect. This period could therefore reflect a renewed tendency for erosion at a somewhat lower base level, although the length of time and degree of erosion would seemingly be much less than during the preceding erosive intervals. This tentative period of less intense landscape reduction is called Somerville time. It could be reasoned that this condition existed from -12 m.y. until the onset of Pleistocene glaciation approximately 2 m.y. ago, or that it continued mid-way into the Pleistocene. One could also argue that we are still in Somerville time and that climatic, glacio-eustatic, and isostatic adjustments have alternately enhanced and retarded the overall erosive condition. Furthermore, a case could be presented for extending Harrisburg time into the Pleistocene or even to the present, omitting the Somerville designation entirely, because of the minor extent and presumed short duration of Miocene transgression. The data are not yet sufficiently detailed and refined to solve this dilemma.

Most of the presumed erosional elements of our landscape which are inferred to have formed during latest Tertiary and Pleistocene, such as Parker Strath, Scottsburg Lowland, Deep Stage Valleys, and their analogs (see Table 1), and the many local pediments and terraces are viewed as being preliminary and temporary stages in the reduction of the present cratonal surface during post-Harrisburg or post-Somerville time. Their formation, when viewed in the longer term, is interpreted as resulting from continued but intermittent net cratonal elevation during oscillatory Tejas time. If a shorter time span reference is used, these features represent geomorphic response to glacio-eustatic, isostatic and climatic fluctuations related to glaciation, and to local lithologic, tectonic, and base level variations. Although these landscape components yield valuable information for deciphering the more recent geomorphic history of an area, they represent local, short-term variations and are insignificant, transient features compared with geomorphic activity over a longer time span as depicted in Figure 6.

Discussion.

Relationships depicted in Figure 6 seem, in general, logical and reasonable. However, several salient aspects of the proposed interpretations are somewhat tenuous. One major problem is the inability to accurately and precisely date geologic events. Furthermore, geologic processes require time to adjust to new controls. This condition results in a lag time and some overlap between stimulus and response, or cause and effect. Interpretation is hindered by seemingly anomalous relationships and complex interactions as time and space boundaries are crossed. The data used in constructing Figure 6 are heavily weighted toward the Atlantic Coast-Appalachian region. If our reasoning and interpretations are valid, and if such an approach to historical geomorphology is to be useful, similar analysis and interpretation should be attempted for other areas. The geodynamic model which serves as the interpretative base is presumably valid for all cratons (Sloss, 1975, personal communication); thus this approach may be extended to other plate trailing-edge situations and subsequently, with additional refinement, to other plate boundaries and cratonal interiors.

Finally, the assumption is made that the periods of erosion designated in Figure 6 were of sufficient duration to reduce large areas of the landscape to such a condition of subdued relief as to merit the connotation of "erosion surface" or "peneplain". Although several efforts (for example, Gilluly, *et al*, 1970; Judson and Ritter, 1964; Menard, 1961; and Schumm, 1963) have been made to determine a regional average rate of landscape denudation, results have been disparate. Several factors, each being difficult to quantify, must be incorporated into any such analysis if results are to be meaningful. Judson and Ritter (1964) found that landscape lowering varied significantly between the major drainage basins of the United States. Gilluly *et al*, (1970, p. 367) suggest that this results principally from the difference in geologic materials comprising the watershed surfaces; thus, lithologic variations are reflected in different rates of lowering. A second major source of error in such estimates is failure to incorporate the effect of climatic changes. Although interrelationships between temperature, precipitation, vegetation, weathering, and erosion are complex, Langbein and Schumm (1958) and Schumm (1965) have demonstrated that runoff and sediment yield are at a maximum under certain climatic conditions, but are considerably less under other temperature-precipitation combinations. Schumm (1968, Fig. 2) also has pointed out that the relationship between mean annual precipitation and mean annual runoff has changed through geologic time as terrestrial vegetation evolved. Furthermore, several recent studies (for example, Douglas, 1967; Edgar and Melhorn, 1974; and Wolman, 1967) of present drainage basin stability and erosion indicate that present erosion rates may be inflated as a result of human influences. Consequently, simply extending present erosion values into the past, when different climatic and vegetative conditions prevailed, can cause significant errors.

Because of this problem of defining a regional landscape denudation rate that is valid for different increments of the geologic past, it is difficult to decipher the actual effect of the erosional periods noted in Figure 6.

Somerville, Harrisburg, and Schooley times are tentatively judged as spanning approximately 10 m.y., 25 m.y., and 50 m.y. respectively. Judson and Ritter (1964, p. 3400) concluded that if the average elevation of the United States is 2300 feet, "it would take 11 to 12 million years to move to the oceans a volume equivalent to that of the United States lying above sea level." Using a rate of 0.25 feet per 1,000 years (Schumm, 1963, p. 12) it would take slightly more than 9 million years to perform the same task. If these estimates are reasonably accurate, ample time has existed in each of the above erosive intervals to produce extensive denudational surfaces. Similarly, Schumm (1963, p. 12) concluded that "relatively rapid rates of denudation make peneplanation a very likely event under conditions which were probably common in the geologic past."

In addition to other possible criticism, the preceding reasoning requires the assumption of uplift followed by extended tectonic stability as erosion progresses. This assumption is questionable. In fact, Schumm (1963, p. 12) noted that present rates of orogeny are significantly greater than denudation rates. This observation is in agreement with the Sloss and Speed (1974) oscillatory mode wherein the cratonal areas have been experiencing net uplift since the beginning of Tertiary times. Schumm (1963) also noted that isostatic adjustment will interrupt erosion cycles and can partly explain the existence of multiple or warped erosion surfaces. Thus an imbalance between uplift and denudation can account for the preservation of remnants of ancient erosion surfaces that would likely be destroyed under a different geologic regimen.

SUMMARY AND CONCLUSIONS

This paper has deliberately presented a positive argument for the possibility of recurring episodes of net landscape reduction by erosion throughout geologic time. Stratigraphic and tectonic concepts are discussed for the purpose of demonstrating that it is possible to interpret certain geologic activities and processes as occurring on a cyclic or periodic basis. Also, plate tectonic and sea-floor spreading concepts and the Sloss-Speed sedimentational model have been invoked to serve as a base for deducing Mesozoic and Cenozoic structural, eustatic, and geomorphic history. The principle conclusions reached in our paper are as follows:

1. Abundant geologic literature exists to suggest that there is a periodicity of tectonic, eustatic, and climatic activity and variation throughout geologic time. The long-term pattern, however, is overprinted by shorter-term and local variations that hamper interpretation. The problem is further complicated by dating techniques and time scales that are not sufficiently precise or refined to allow complete knowledge about true synchroneity and/or overlapping of events.

2. Long-term stratigraphic sequences, as outlined by Sloss for the North

American, African, and Eurasian cratonic areas, developed primarily in response to tectonic and eustatic controls, and appear to be equivalent in time and mode of formation. The same stratigraphic pattern may occur elsewhere but we find no results reported in the literature. The current Sloss-Speed stratigraphic model identifies six major unconformities in the post-Precambrian record, and thus there are six hiatuses of long enough duration for erosion surfaces to develop on a cratonal or possibly continental scale.

3. The Sloss-Speed model, sea-floor spreading rates, cratonal tectonics, and marine transgressive-regressive relations can be used to identify and temporally refine rapid oscillatory movements occurring during Tejas (post-Paleocene) time. A reasonable correlation then is possible between tectonic and erosional events.

4. Optimum conditions existed during Mesozoic and Cenozoic times for development of major, continental-scale erosion surfaces. The suggested erosion intervals are:

a. Late Jurassic-early Cretaceous: development of the Fall Zone (Gondwana-Laurasian) surface between 135 m.y. B.P. and 110 m.y. B.P.
b. Very late Cretaceous (Danian ?) through Paleocene: formation of the Schooley (post-Gondwana) erosion surface during the interval from 85 m.y. B.P. to about 50 m.y. B.P.
c. Latest Eocene through Oligocene: construction of the Harrisburg (African) surface, dating from about 45 m.y. B.P.
d. Upper Miocene-early Pleistocene: possible formation of the Somerville-Parker strath surfaces during the interval from 12 m.y. B.P. to 2 m.y. B.P. This interpretation is extremely tenuous because of overprinting by rapid, short-term climatic and glacio-eustatic changes, and inadequacy of the sea-floor spreading data.

The information available to us suggests that the lack of exact synchroneity of surfaces is not really a problem. Smaller-scale areal or regional overprints as a result of climatic variations, differences in lithology, and local base level controls have always tended to influence the rate and manner of denudation and formation of erosion surfaces on the continents throughout geologic time. However, these short-term variations are obscured in the older record by the major global events, evidence of which is still imprinted on the landscape.

All interpretations presented in this paper clearly should be considered tentative; they represent only a preliminary attempt to use a new and potentially useful data base from other branches of geology as a tool for geomorphic inference. We merely have tried to determine when optimal conditions have existed for formation of large-scale erosion surfaces, and have shown that millions of years were available for them to develop.

Finally, this paper argues neither for nor against any of the so-called process response models; no bias is demonstrated for peneplanation, pediplanation, *etc.*, any or all of which, acting over the time available, could produce the observed surfaces.

REFERENCES CITED

Adams, F. D., 1954, The Birth and Development of the Geological Sciences: Dover Press, New York, 506 p.

Armstrong, R. L., 1969, Control of sea level relative to the continents: Nature, v. 221, p. 1042-1043.

Bird, J. M., and Dewey, J. F., 1970, Lithosphere plate-continental margin tectonics and the evolution of the Appalachian orogen: Geol. Soc. America Bull., v. 81, no. 4, p. 1031-1960.

Brown, E. H., 1968, Planation surface; *in* Fairbridge, R. W., Ed., The Encyclopedia of Geomorphology: Reinhold Book Corp., New York, p. 856.

Bucher, W. H., 1933, The Deformation of the Earth's Crust: Princeton Univ. Press, Princeton, 518 p.

Cole, W. S., 1941, Nomenclature and correlation of Appalachian erosion surfaces: Jour. Geol., v. 49, p. 129-148.

Damon, P. E., 1971, The relationship between late Cenozoic volcanism and tectonism and orogenic-epeirogenic periodicity, *in* Turekian, K. K., ed., The late Cenozoic glacial ages (Flint volume): Yale Univ. Press, New Haven and London, p. 15-35.

———————, 1972, Orogenic-epeirogenic periodicity (abs.): 24th Internat. Geol. Cong. Rept., sec. 3, p. 462.

Dawson, J. W., 1866, On the conditions of the deposition of coal, more especially as illustrated by the Coal Formation of Nova Scotia and New Brunswick: Geol. Soc. London Quart. Jour., v. 22, p. 95-169.

Dearnley, R., 1966, Orogenic fold-belts and a hypothesis of earth evolution, *in* Ahrens, L. H., Press, F., Runcorn, S. K., and Urey, H. C., eds., Physics and Chemistry of the Earth: Pergamon Press, v. 7, p. 1-114.

Dewey, J. F., Pitman, W. C., III, Ryan, W. B., and Bonnin, J., 1973, Plate tectonics and the evolution of the Alpine system: Geol. Soc. America Bull., v. 84, no. 10, p. 3137-3180.

Dott, R. H., Jr., 1964, Superimposed rhythmic stratigraphic patterns in mobile belts, *in* Merriam, D. F., ed., Symposium on cyclic sedimentation: Kansas Geol. Survey Bull., 169, v. I, p. 69-85.

Douglas, I., 1967, Man, vegetation and the sediment yields of rivers: Nature, v. 215, p. 925-928.

Dunbar, C. O., 1960, Historical Geology, 2nd ed.: New York, John Wiley and Sons, 500 p.

Durham, J. W., 1959, Paleoclimates, *in* Ahrens, L. H., Press, F., Runcorn, S. K., and Urey, H. C., eds., Physics and Chemistry of the Earth: Pergamon Press, London, v. 3, p. 1-16.

Edgar, D. E., and Melhorn, W. N., 1974, Drainage basin response — documented historical change and theoretical considerations: Purdue Univ. Water Resour. Research Center, Tech. Rept. 52, 196 p.

Emiliani, C., 1958, Ancient temperatures: Scientific American, v. 198, no. 2, p. 54-63.

Emiliani, C., 1966, Paleotemperature analysis of Caribbean cores P6304-8 and P6304-9 and a generalized temperature curve for the past 425,000 years: Jour. Geol., v. 74, no. 2, p. 109-126.

Evans, P., 1971, Towards a Pleistocene time-scale, *in* The Phanerozoic time-scale — a supplement: Geol. Soc. of London Spec. Pub. 5, p. 123-356.

Evernden, J. F., and Kistler, R. W., 1970, Chronology of emplacement of Mesozoic batholithic complexes in California and western Nevada: U. S. Geol. Survey Prof. Paper 623, 42 p.

Fenneman, N. M., 1938, Physiography of the Eastern United States: McGraw-Hill, New York, 714 p.

Garner, H. F., 1974, The Origin of Landscapes: Oxford Univ. Press, 734 p.

Gerasimov, I. P., 1970, Three major cycles in the history of a geomorphological stage of the development of the Earth: Geomorphology, v. 1, no. 1, pp. 12-17.

Gibson, T. G., 1970, Late Mesozoic-Cenozoic tectonic aspects of the Atlantic coastal margin: Geol. Soc. America Bull., v. 81, no. 6, p. 1813-1822.

Gilluly, J., 1949, Distribution of mountain building in geologic time: Geol. Soc. America Bull., v. 60, no. 4, p. 561-590.

———————, 1973, Steady plate motion and episodic orogeny and magmatism: Geol. Soc. America Bull., v. 84, no. 2, p. 499-514.

———————, Reed, J. C., Jr., and Cady, W. M., 1970, Sedimentary volumes and their significance: Geol. Soc. America Bull., v. 81, no. 2, p. 353-375.

Gorelov, S. K., Drenev, N. V., Meschcheryakov, Yu. A., Tikanov, N. A., and Fridland, V. M., 1970, Planation surfaces of the USSR: Geomorphology, v. 1, no. 1, pp. 18-29.

Grasty, R. L., 1967, Orogeny, a cause of world-wide regression of the seas: Nature, v. 216, p. 779-780.

Hallam, A., 1963, Major epeirogenic and eustatic changes since the Cretaceous and their possible relationship to crustal structure: Am. Jour. Sci., v. 261, no. 5, p. 397-423.

Haughton, S. H., 1963, The Stratigraphic History of Africa South of the Sahara: Oliver and Boyd, Edinburg and London, 365 p.

Hays, J. D., and Pitman, W. C., III, 1973, Lithospheric plate motion, sea level changes and climatic and ecological consequences: Nature, v. 246, p. 18-22.

Hull, E., 1868, On the physical causes which seem to have regulated the distribution of calcareous and sedimentary strata of Great Britain, with special reference to the Carboniferous Formation (summary): Geol. Mag., v. 5, p. 143-146.

International Geographical Congress, Eighteenth, 1956, Eighth report of the commission for the study and correlation of erosion surfaces around the Atlantic: UNESCO, New York, 64 p.

Johnson, D. W., 1931, Stream Sculpture on the Atlantic Slope, a Study in the Evolution of Appalachian Rivers: Columbia Univ. Press, New York, 142 p.

Johnson, J. G., 1971, Timing and coordination of orogenic, epeirogenic, and eustatic events: Geol. Soc. America Bull., v. 82, no. 12, p. 3263-3298.

Judson, S., and Ritter, D. F., 1964, Rates of regional denudation in the United States: Jour. Geophys. Research, v. 69, no. 16, p. 3395-3401.

King, L. C., 1950, The study of the World's plainlands: a new approach in geomorphology: Geol. Soc. London Quart. Jour., v. 106, pt. 1, pp. 101-131.

———————, 1967, The Morphology of the Earth, 2nd ed.: Oliver and Boyd, Edinburg and London, 726 p.

Knopf, E. B., 1924, Correlation of residual erosion surfaces in the eastern Appalachian Highlands: Geol. Soc. Am. Bull., v. 55, p. 633-668.

Langbein, W. B., and Schumm, S. A., 1958, Yield of sediment in relation to mean annual precipitation: Amer. Geophys. Union Trans., v. 30, no. 6, p. 1076-1084.

Larson, R. L., and Pitman, W. C., III, 1972, World-wide correlation of Mesozoic magnetic anomalies, and its implications: Geol. Soc. America Bull., v. 83, no. 12, p. 3645-3662.

Lattimore, R. K., Rona, P. A., and DeWald, O. E., 1974, Magnetic anomaly sequence in the central North Atlantic: Jour. Geophys. Research, v. 79, no. 8, p. 1207-1209.

Machatschek, F., 1969, Geomorphology (translation of Geomorphologie, 9th Edition): Oliver and Boyd, Edinburg, 212 p.

Mathews, W. H., 1975, Cenozoic erosion and erosion surfaces of eastern North America: Am. Jour. Sci., v. 275, no. 7, p. 818-824.

Menard, H. W., 1961, Some rates of regional erosion: Jour. Geology: v. 69, no. 2, p. 154-161.

Newberry, J. S., 1873, The geological relations of Ohio: Ohio Geol. Survey Rept., v. 1, p. 50-88.

Pitman, W. C., III, 1971, Sea-floor spreading and plate tectonics: EOS (Am. Geophys. Union Trans.), v. 52, no. 5, p. 130-135.

———————, and Talwani, M., 1972, Sea-floor spreading in the North Atlantic: Geol. Soc. America Bull., v. 83, no. 3, p. 619-646.

Rona, P. A., 1973, Relations between rates of sediment accumulation on continental shelves, sea-floor spreading, and eustacy inferred from the central North Atlantic: Geol. Soc. America Bull., v. 84, no. 9, p. 2851-2872.

Rutten, L. M. R., 1949, Frequency and periodicity of orgenetic movements: Geol. Soc. America Bull., v. 60, no. 11, p. 1755-1770.

Schumm, S. A., 1963, The disparity between present rates of denudation and orogeny: U. S. Geol. Survey Prof. Paper 454-H, 13 p.

———————, 1965, Quaternary paleohydrology *in* Wright, H. E. Jr., and Frey, D. G., eds., The Quaternary of the United States: Princeton Univ. Press, Princeton, p. 783-794.

———————, 1968, Speculations concerning paleohydrologic controls of terrestrial sedimentation: Geol. Soc. America Bull., v. 79, no. 11, p. 1573-1588.

———————, and Lichty, R. W., 1965, Time, space, and causality in geomorphology: Am. Jour. Sci., v. 263, no. 2, p. 110-119.

Shaffer, P. R., 1947, Correlation of erosion surfaces of the southern Appalachians: Jour. Geol., v. 55, p. 343-352.

Sloss, L. L., 1963, Sequences in the cratonic interior of North America: Geol. Soc. America Bull., v. 74, no. 2, p. 93-114.

———————, 1964, Tectonic cycles of the North American craton, *in* Merriam, D. F., ed., Symposium on cyclic sedimentation: Kansas Geol. Survey Bull. 169, v. II, p. 449-460.

Sloss, L. L., 1972, Synchrony of Phanerozoic sedimentary — tectonic events of the North American craton and the Russian platform: 24th Internat. Geol. Congress Rept., sec. 6, p. 24-32.

———————, and Speed, R. C., 1974, Relationships of cratonic and continental-margin tectonic episodes, *in* Dickinson, W. R., ed., Tectonics and sedimentation: Soc. Econ. Paleontologists and Mineralogists Spec. Pub. 22, p. 98-119.

Steiner, J., 1973, Possible galactic causes for synchronous sedimentation sequences of the North American and eastern European cratons: Geology, v. 1, no. 2, p. 89-92.

Thornbury, W. D., 1965, Regional Geomorphology of the United States: John Wiley and Sons, New York, 609 p.

———————, 1969, Principles of Geomorphology, 2nd Edition: John Wiley and Sons, New York, 594 p.

Trowbridge, A. C., 1921, The erosional history of the Driftless Area: Univ. Iowa Studies in Natural History, v. 9, no. 3, 127 p.

Umbgrove, J. H. F., 1947, The Pulse of the Earth: Martinus Nijhoff, The Hague, 358 p.

Wang, C. S., 1972, Geosynclines in the new global tectonics: Geol. Soc. America Bull., v. 83, no. 7, p. 2105-2110.

Weller, J. M., 1964, Development of the concept and interpretation of cyclic sedimentation, *in* Merriam, D. F., ed., Symposium on cyclic sedimentation: Kansas Geol. Survey Bull. 169, v. II, p. 607-621.

Wolman, M. G., 1967, A cycle of sedimentation and erosion in urban river channels: Geog. Annaler, v. 49A, p. 385-395.

CHAPTER 14

THE MIND OF GROVE KARL GILBERT

Stephen Pyne

INTRODUCTION

Grove Karl Gilbert was born in 1843 — ten years after Charles Lyell published the final volume to his *Principles of Geology*; he died in 1918 — ten years before the published proceedings of the first American symposium on continental drift. The seventy-five years of his life consequently spanned the heroic age of American geology. He knew personally most of its grand figures — Hall, Dana, Powell, Dutton, Newberry, Thomas Chamberlin and William Morris Davis. He contributed impressively to the geology of the heroic age in its several cultural functions. Geology was a mechanism for resolving certain intellectual problems, particularly the consciousness of a landscape whose spatial and temporal scales were rapidly expanding; Gilbert addressed those central questions in a series of brilliant, now classic monographs. Geology served as a frontier institution, an economic and intellectual subsidy to the westward migration; in the course of his career Gilbert made the West a field of special interest, pursuing problems of significance to frontier settlement and conservation. Finally geology simply expressed one of man's most elementary impulses — his curiosity, manifest in his determination to explore; and as a member of the Ohio, Wheeler, and Powell Surveys, as well as the U. S. Geological Survey from the time of its founding, Gilbert earned a reputation as one of America's premier exploring geologists. Yet the most extraordinary aspect of Gilbert's distinguished career may well be the fact that he ever became a geologist at all.

GILBERT'S YOUTH

One can triangulate this improbable career from two circumstances of Gilbert's youth: one is his family environment and the other his education. The progenitor of the Gilberts arrived in Massachusetts in 1630, but if we accept a joking remark by G. K. we might thrust the family line back further. Thanking a paleontologist friend for naming another fossil after him, Gilbert threatened to construct a "paleogenealogy" and "gloat over the DARs with their brief historical pedigrees." But the significant fact is not the longevity of the family, but the unorthodox behavior of one of its members, Grove

Sheldon Gilbert, Karl's father. Breaking with family tradition, he left New England for Rochester, New York; he abandoned the trades of coopers and machinists to become a self-taught portrait painter; and he gave up formal religion for speculations more to his own liking. The family relied heavily on its own resources — frugality was a necessity, and entertainment consisted largely of the "intellectual game," as Karl called it. It was a small environment — the family called its home "The Nutshell" — a world whose axis was the dynamic, opinionated, unorthodox father; and it became the archetype for all of Karl's later associations. Throughout his life he tried to recreate it whenever possible, usually found it in scientific clubs, his own homelife, and frequent visits to his sister and brother. And he gravitated around men who resembled his father — men like John Newberry, John Wesley Powell, and C. Hart Merriam. Entertainment consisted of the intellectual game — round-robin poetry composition, riddles, and reading aloud to small, family-sized groups (1).

His education was equally informative. At substantial cost to his family, he matriculated at the nearby University of Rochester — though, as a means of protecting the family investment, with the stipulation that he shore up his frequently frail health with a program of outdoor exercise. Exploring geology easily extended this pattern of education throughout his adult life. What he received at the university was a traditional, classical education. Biased toward language, it nevertheless introduced Gilbert to science — though of a classical variety; a blend of physics, astronomy, and especially mathematics collectively known as natural philosophy. At this time physics was defining the laws of thermodynamics, but as John Merz has observed the chief discoverers tried to bring the new "energetics" into "harmony and continuity with the older Newtonian laws." The *Principia* of Newton was again studied, and re-edited in the unabridged form, and an interpretation and amplification of the third law of Motion — so as to embrace the principle of energy — was made the key to the science of mechanics. It was in this context, and probably through the medium of the *Manual of Applied Mechanics* by William Rankine, the Scotch physicist and civil engineer, that Gilbert learned the fundamental sciences.

When he graduated his largest block of credits was in mathematics, but he carried away an award for his Greek and with it an indelible bias of classicism which stamped his intellectual career. That impression appears in his prose style, his scientific concepts, his philosophy of method, and his metaphysical assumptions about nature; it merged easily with the scholarly temperament of a man who was already reserved, imperturbable, serious, famous for his self-control and his often self-deprecating humor (2).

Fortunately for geology Gilbert had difficulty deciding on a career or on a way to leave the Nutshell. A decisive moment came when, despite the fact that he was qualified for little else, he failed at school-teaching. He finally landed a job at Henry Ward's Cosmos Hall, a factory for processing natural history specimens located on the university campus. He worked at Cosmos Hall for five years. Unlike many of his contemporaries, however, he was not overcome by a Humboldtean fever at the sight of mountains of specimens, but there were diversions from the routine that made the job palatable, such as the occasions when Cosmos Hall sent representatives to reconstruct the fossil remains of Pleistocene mammals. It was on one such expedition to Cohoes Falls near Albany that Gilbert became a geologist.

Amid a channel of gravel-filled potholes, James Hall was supervising the excavation of a mastodon. The discovery of a mastodon could be a major cultural event, like finding the ruins of a lost civilization. The exhumation of exotic specimens could almost redeem a state geological survey, for example, from its otherwise fatal inability to locate gold or coal. At any rate Hall fell into the pit, injured himself, and Gilbert took over. He did a creditable job, and was rewarded by later assignments with Irish elks. But when he completed the reconstruction of the Cohoes mastodon, he returned to the site to examine the nature of something that intrigued him far more than the lost bones — he studied the potholes. In fact he surveyed all of them in the vicinity, measured them, counted them, collected materials, and then, dangling by ropes to where he could cut trees and count growth rings, he estimated the rate of recession for the falls. That is, rather than reconstruct the evolution of the mastodon, he preferred to reconstruct its physical environment. He tried to convert a problem of geography and history into one of geometry and physics. And it was in the course of this investigation that he became seriously attracted to geology. Recalling his graduation from the university, he later remarked that, having been exposed to both, he had found engineering more interesting than geology. Amid the Cohoes potholes, however, he realized that the two could be combined. He never deviated from that lesson (3).

The study of the Cohoes potholes prefaces all of his mature work. It addressed problems posed by the environment of upstate New York; it showed a reserved classicist responding to discoveries that made enthusiastic romantics out of most of his associates; and it turned fossils into physics, transforming an artifact from the past into a mechanism of the present. It also, not incidentally, sparked a lifelong fascination with potholes, which he pursued with equal enthusiasm in the gorge of the Grand Canyon and on the slopes of the High Sierras; he even imagined Niagara Falls — the inheritor of the Cohoes study — as a pothole on its side. When Gilbert later termed Niagara a "physiographic engine," when he conceived of the Henry Mountains as a gigantic piston, when he gave an address on "The Use of the Colorado Canyons for Weighing the Earth," he was elaborating on the lesson of the Cohoes mastodon. So also when, during the San Francisco earthquake of 1906, he calmly measured the burning time for wooden buildings and when, during the hard-pressed Wheeler Survey expedition up the Colorado River, he imperturbably took astronomical sightings on Venus from the gorge. The long search for mathematical forms like the potholes perhaps climaxed in 1908 when he led a visiting geologist into the sequoia groves of the Sierras. Oblivious to the trees, symbols though they were of the wonder and transcendence of the organic world, Gilbert labored for hours looking for a spider which made its web in the shape of a paraboloid. The greatest wonder of all was that Nature was a geometer.

GILBERT — THE WESTERN EXPLORER

In 1869 Gilbert acquired a position as voluntary assistant on the Ohio State Geological Survey under John Strong Newberry. He was 26 years old, rather late in life, he felt, to begin a career; and in a sense he had begun in a hole. The position was an apprenticeship, and for two years under the tutelage of Newberry, Gilbert learned the basics of geologic field work and the elementary themes of the science. Newberry was not only an exceptional teacher, but one of the great pre-Civil War explorers — his experience with the Ives Expedition into the Grand Canyon had given him an international reputation. So when the Army planned to repeat that expedition, they asked Newberry to recommend a replacement. He named his promising protege, G. K. Gilbert.

Thus began Gilbert's tour of duty as a Western explorer. With the Wheeler Survey Gilbert thoroughly learned the trade. For three years he toured Nevada, Utah, Arizona, and New Mexico; distinguishing himself on the 1871 expedition, a marathon which marched over 6000 miles including a crippling July traverse of Death Valley and a boat trip up the Colorado River into the Grand Canyon. A fourth year of frustration he spent writing his report. But the Wheeler Survey, like its 1871 itinerary for the Grand Canyon, travelled conceptually, no less than geographically, against the mainstream. Its rigid schedules and production standards annoyed Gilbert, he likened its travels to reading a book while someone else turned the pages. In 1875 he deserted Lt. Wheeler in favor of Major Powell and transferred to the U. S. Geographical and Geological Survey of the Rocky Mountain Region.

Davis has aptly described the relationship to Powell as the "greatest determining factor of his mature life." It would be hard to find two men so close, or so different. Gilbert was a classicist and a scholar, a man who abhorred controversy; Powell was a romantic, a reformer, and a scientific soldier of fortune who had lost one arm at Shiloh and risked the other in the trench warfare which the Gilded Age witnessed annually in Congress. Like exploration it was a moral equivalent to war that the Major relished. Yet in Powell's intellectual education with respect to the Colorado Plateau, it is clearly Gilbert who is the nexus between the Ives Expedition and that by Powell. It was a connection preserved in later years with the Geological Survey, as Bailey Willis reminisced when he called Gilbert:

> "Powell's better half. Perhaps no one else ever thought of them in that way, but in constant relations with the two I learned to know how much Gilbert, the true scientist, contributed to the Geological thinking of Powell, the man of action. I do not think that they themselves were conscious of the degree to which the latter absorbed and gave out as his own the ideas that the former had silently passed through. But as Gilbert's assistant, I was sometimes jealous for that generous soul and devoted friend."

Gilbert's first task with Powell was to arrange his fossil collection, and that bit of scientific housecleaning typified his official function for twenty years as adjutant to the Major (4).

For four seasons he worked in Utah. Wheeler had made him a professional, but Powell cashed that training in for the reward. In 1877 Gilbert published the *Report on the Geology of the Henry Mountains*, and his scientific reputation was assured. The following year he supplied the scientific muscle for Powell's *Arid Lands* report. But most of his time was spent resurveying areas that Powell's original mapping crews had botched.

In 1879 the U. S. Geological Survey consolidated western exploration. Gilbert assumed control of the Division of the Great Basin stationed at Salt Lake City. At one point, in exasperation, he pointed out to Clarence King, the Survey director, that while he had already purchased considerable office furniture, mules, and equipment, he had no authorization to rent an office. King soon rectified that error and sent him off to study Lake Bonneville; most of Gilbert's text was ready by 1881. Powell, of course, took over the directorship that year, and for the next decade Gilbert's original research virtually died. It took nine anticlimatic years to publish *Lake Bonneville*. The Major needed someone to arrange his administrative fossils, and the chore fell to his loyal friend. Gilbert could do little more than issue a series of short papers. For a while he tried to initiate research for an eastern companion to the Bonneville study, focusing on the Great Lakes. That fell through, and desperate to capitalize on his knowledge of lake forms, he spent his evenings at the Naval Observatory ("congressmen and clouds," he remarked, "were equally obstructive") scrutinizing the lunar marias. Those evenings culminated in "The Moon's Face: A Study of the Origin of its Features" in which he proposed a mechanism for fusing "moonlets" that Chamberlin used a decade later in his planetismal hypothesis. For many of these short papers Gilbert took leave without pay. But his reputation was secure. As a friend from the Ohio Survey (5) wrote:

> "Your Henry Mountains and Lake Bonneville monographs have more genuine thoughts in them than one finds in a cord (wood measure) of the [USGS] bulletins." (5)

It was not until Powell was driven from office that Gilbert returned to the field, before he again satisfied his longing to "be astride the occidental mule." For three years he studied artesian wells on the High Plains, literally an inversion of his classic studies and his hopes seemed to be as buried as his rivers and lakes. Yet it brought him to the field again, and thrilling to the grandeur of the Rocky Mountains on the horizon, he could freely speculate on the question: "What is happiness?"

> "'The soul's calm sunshine.' True enough, but too abstract and metaphoric; give us something specific. Well, specifically, happiness is sitting under a tent with walls uplifted, just after a brief shower, when most of the flies have quit lightning on the lobster-red wrists burnt during the morning ride, and gone off to see what the cook is going to do next, and when the thirsty air is rapidly exchanging its heat for moisture left by the shower. It is rising at 4:30, while Jupiter is still palely visible, but there is no longer any temptation to hunt for the comet, taking a sponge bath in the open, breakfasting from off a box lid gaudily decked by a painted table cloth, and then sallying forth on the white horse Frank to study the limits of the alluvial veneering on the base-level mesas, measure the dips of rows of rusty nodules, sketch problematic buttes, and gather the houses of Ammonite, Scaphite, and Hamite. It is going to bed by early candle light in the midst of a grove of *Rhus tox*, hunting the double stars near Lyrae and Cygni among the branches of overhanging cotton woods, moralizing on the development of character through the trying associations of camp life, congratulating yourself that you are not a pessimist, and finally dropping off to sleep" (6).

In 1899 he joined the Harriman Expedition to Alaska. If the Plains recalled his years on the Ohio Survey, the Harriman Expedition harked back to his years with Wheeler, offering little more than a rapid reconnaissance of tidal glaciers. In 1901 he returned to the Basin Range, but his map notes were destroyed in a fire. That same year a proposal to the Carnegie Institution to drill into the earth's crust was rejected. The informal cloth of Gilbert's social life began to unravel as fully as the intellectual fabric. His wife, brother, and Powell were dead. His last ties to the great period of exploration and federal science were broken. But Gilbert was no more inclined to long lament the passing of his youth than he did the passing of the Earth's. It was well, for at age 60 he seemed to have eased into a career as an elder statesman of geology. For several decades, out of loyalty to Powell, he had replaced original research with administrative chores; he had assumed, at substantial cost in time, the offices of a great many scientific societies; instead of producing a Great Lakes companion to *Bonneville*, he appeared to have dissipated his energies and entered a long, graceful denouement — writing elegant articles for encyclopedias; interpreting the field for younger geologists; looking philosophically at his early undertakings; touring as a sort of scientific celebrity with International Geological Congresses and the

Harriman Expedition; perfecting a methodology that had little substantive meat.

So it seemed. But like the great masses beneath Lake Bonneville, so long depressed with its overburden of water, there remained a plasticity. Like his model for the Earth's crust, Gilbert's personality could be locally rigid and broadly plastic. When the climate changed, when the burdens of the past evaporated, Gilbert slowly rebounded. His final years, even though racked by illness, were a ringing testimony to Gilbert's continued powers, his blend of patience and intelligence. While he became more dignified as the years progressed, he never became a simple dignitary. His reddish-brown hair ripened to white, but his blue eyes burned as clearly as ever. If *Lake Bonneville* had been a culmination of American geomorphology of the nineteenth century, his hydraulic studies in California were an inauguration of the twentieth.

By 1903 Gilbert had reconstructed his life — he took up residence in the house of C. Hart Merriam and discovered a new region of study in California. In 1905 those investigations acquired a theme when the USGS assigned him to evaluate the effects of hydraulic mining. The study was twice interrupted by disasters — one natural, the San Francisco earthquake; the other, personal, a near-fatal illness. From his general reconnaissances of the Sierras came brief but brilliant essays on exfoliation, the convexity of hilltops, and the mechanisms of glacial erosion. From the earthquake came reports for both the state and national commissions. From hydraulic mining came two classic studies: a laboratory report on the mechanisms of fluvial transport and a field report on the distribution and natural processing of hydraulic mining debris, completing a subject he began with the *Henry Mountains* in 1877. But by 1917 and 1918 at age 74 Gilbert was already back in the field, this time to complete a cycle begun under Wheeler — the conundrum of Basin Range structure. He died before it was finished; the incomplete text was published posthumously in 1928.

GILBERT — THE BOOKKEEPER

In his personality no less than in his scientific thought the man who measured the Cohoes potholes stands as a somewhat isolated, even unique figure. He came from an urban rather than a rural environment, so that he failed to share that fascination with the organic world that typified most of his associates. He dealt with physics rather than fossils; he sympathized with an inorganic nature whose hieroglyphics were mathematical forms rather than an organic nature alive with transcendental symbols. He accepted a mechanical metaphor rather than an organic one; he saw equilibrium rather than evolution; his intellectual mentor was William Rankine rather than Charles Darwin. He came to geological science from a formal education in the classics, rather than as a self-taught naturalist. His father was not a minister, nor was his family famous for its moral fervor. Grove Sheldon

Gilbert, on the contrary, had apostatized, and any lingering religious sentiment in his son vaporized when G. K.'s daughter, Bessie, died of diptheria at age 6. He was more of a scholar than an adventurer, a conservative rather than a reformer; he passed through the Civil War without seeing value to the concept "struggle for existence," and without becoming a scientific captain of industry. He was an "investigator," a mental state he contrasted with that of the teacher. Consequently he never accepted an academic position or government podium or wrote a definitive textbook, so that despite his years of intellectual leadership he never established a "school" of followers or articulated a distinctly Gilbertian creed. He was an unassuming maverick — his reserve rather than his enthusiasm set him apart. The contrast to Newberry, LeConte, Powell, Chamberlin, and Davis is almost perfect.

His friend, W. C. Mendenhall, perhaps best characterized Gilbert when he wrote (7):

> "In sheer balanced mental power Gilbert was probably unsurpassed by any geologist of his time. Fundamental among the qualities of his mind were self-knowledge and self-control. These qualities he possessed in a degree equalled by few. That mind which he knew and controlled so well was a quiet, efficient, powerful instrument, which functioned perfectly. Thus he was the very antithesis of the brilliant, temperamental, erratic genius. He recognized both his powers and his limitations, and did not undertake that which he was not equipped to do." (7)

He kept to private entertainment, shunning the theatre in particular, because, during sentimental melodramas, he could lose control and begin to weep — much to the ladies' amusement and his mortification. He presents a personality almost gyroscopic in its internal workings. But if one wanted to reduce that temperament to a single image, there is the portrait of Gilbert the bookkeeper.

The sense of accounting, or accountability, runs like a motif throughout Gilbert's life. In part it was instinctive; he counted and kept double-entry books of various sorts by impulse, as a beaver might build dams or a wasp its nest. In part it harmonized with his conception of science, which unlike the arts, dealt with quantities not qualities. This belief was aided by his thermodynamic model for geologic processes. "Whenever a tentative theory involves the application of force or the expenditure of energy," he observed, "the investigator (or his critic) habitually asks whether the assumed cause affords a sufficient amount of force or of energy." He added that "quantitative tests of this particular type are among the most familiar resources of investigation." And after ultimately identifying "mankind, collectively, through the agency of its men of science and inventors . . . as an investigator, slowly unravelling the complex of Nature and weaving from the disentangled threads the fabric of civilization," he phrased his conclusions in the imagery of accounting: "Knowledge of Nature is an account at bank, where each dividend is added to the principal and the interest is ever compounded; and hence it is that human progress, founded on natural knowledge, advances with ever increasing speed." (8).

Nor was it simply a quirk of his financial pressures; it informed his whole outlook on nature. During his solitary trip to Europe, while guided about Paris, the whole complex surfaced. When the Arch of Triumph and the Cathedral of the Sacred Heart were proudly pointed out to him, the bookkeeper, classicist, and engineer in him rebelled. He could only exclaim: "What a worthless use of money!" His attitude toward Nature was almost identical. Like Gilbert, Nature was expected to balance her books, and as though a mirror to Gilbert's temperament, she would never act in excess. She would always rely on her prodigious powers of self-control.

Where most of his colleagues could glamorize the marvels of ancient geologic worlds, Gilbert conceived of Nature in the present tense. Like him Nature was not given to reminiscence but lived in a continual present. Hence his analytical tools were the timeless forms of mathematics, rather than the time-designating forms of fossils. Newberry, on the other hand, would speak for a generation of geologists when he portrayed the "poetry" of coal (9):

> "It has been formed under the stimulus of the sunshine of long past ages, and the light and power it holds are nothing else than such sunshine stored in this black casket, to wait the coming and serve the purposes of man. In this process of formation it composed the tissues of those strange trees that lifted their scaled trunks and waved their feathery foliage over the marshy shores of the carboniferous continent, where not only no man was, but gigantic salamanders and mail-clad fishes were the monarchs of the animated world." (9)

It was a romantic vision, and one multiplied by Powell, Dutton, LeConte, Chamberlin, and Davis as they celebrated fossil-equivalents like mountains, rivers, the Silurian, peneplains, and even the "dynamic vestiges" of the solar system. The ruined castles that piqued the romantic imagination found geologic analogues in the skeletons of extinct animals and the devasted remains of ancient geologic empires.

At the same time most of the data which the culture asked geology to process concerned evidence from the past. Organizing it into evolutionary sequences gave it order, and allowed for the confident predictions of a grander, progressively higher future which men like Powell and Chamberlin loved to prophesy. Yet for the man who at Cohoes found more meaning in the geometry of potholes than the poetry of mastodons, the present was a more compelling theme. At Lake Bonneville, for example, what excited the most wonder in him was the freshness of the record. Eyeing the bars and deltas along the Wasatch, he exulted in the glistening pebbles of the beach, so recent one could imagine the slap and wash of waves still upon them.

Gilbert's theoretical understanding of Nature has two concepts at its foundation: one is equilibrium and the other rhythmic time. Both derived from his readings in natural philosophy. They merged well; as with his prose style, Gilbert was fundamentally interested in the preservation of form. Consider the Henry Mountains. In analyzing the process of laccolith formation, Gilbert conceived that the sum of the forces acting on the final form equalled zero. The driving force of the rising magma continued until it was countered by a resisting force of equal magnitude; in reaching this equilibrium the magma obeyed the principle of least force. Describing the action of rivers on the landscape, Gilbert used exactly the same formula. The downward force of the river, acting under gravity, was resisted by the channel; the energy of the system, expressed as velocity, adjusted itself to this resistance. The resulting river profile was a profile of equilibrium — an "equilibrium of action," he termed it; and the adjusted river he referred to as "graded." When he depicted one of the streams draining the slopes of the Henrys as shifting its channel laterally, the river, like the laccolith, was only solving the problem of least force. What Gilbert had done was to apply thermodynamics to the physical geology of the Henrys exactly as Willard Gibbs was doing for physical chemistry; it is interesting to note that the "graded river" concept and the "phase rule" were published almost simultaneously.

The graded river has equivalents in practically all of Gilbert's writings. At Bonneville, one finds the graded beach; in the Sierras, the graded hillslope; and outside the Golden Gate, a graded tidal bar. The perception of equilibrium was a philosophical conviction. Where he could not name specific forces, he nevertheless held that the landscape was a product of two competing tendencies: one created diversity and the other uniformity. In the case of the Great Lakes, for example, he began his essay with the statement that there were lake-creating and lake-destroying forces and proceeded from there. It was a vision that had more in common with Newton's depiction of planets, whose orbits inscribed a compromise pattern between two forces, than with Darwin's struggle for existence, in which one force overcame another.

> (In fact Gilbert even conceived organic evolution in identical terms. "Thus the secular evolution of species," he wrote, "combined with the secular and kaleidoscopic revolution of land areas, leads to two antagonistic tendencies, one toward diversity of life on different parts of the globe, the other towards its uniformity. The tendency toward uniformity affords the basis for the correlation of terranes by comparisons of fossils; the tendency toward diversity limits the possibility of correlation." The apparent progressiveness evident in the fossil record was an illusion.)

The philosophical difference between Newton and Darwin is that between Gilbert and Davis (10).

So is the difference in their conception of time. Operating again out of a Newtonian paradigm, Gilbert argued that geologic time was rhythmic. Natural processes beat to particular rhythms, and any event represented, as he termed it, a "plexus" of particular rhythms. The basic rhythms were astrophysical — the motion of the Earth, for this could affect climate and thereby geologic processes. Gilbert contrasted this perception with that derived from a progressive process like evolution or entropy; in doing so he criticized physicists and geologists equally for their addition to continual decay or growth. On the contrary, Gilbert felt that every swing of the pendulum was countered by a swing in the opposite direction, that every credit had an equal deficit. He recommended that geologists try to correlate some geologic event to an astrophysical cycle (he suggested the precessional wobble) and then construct a time scale by applying ratios from that fixed point. The precession, he claimed, was the ideal natural timepiece, a frictionless pendulum" (11).

Insofar as they influenced climate, these cycles could fashion a vaguely rhythmic or rippled texture to the Earth. But Gilbert offered another relationship more directly influential. The action of force and resistance generated friction, and this friction created a rhythm, a pulsing action across the Earth's surface. One could find such frictional rhythms in the motion of earthquakes, rivers, and glaciers; and by analogy to all the dynamic processes that sculptured the Earth's surface. Gilbert's conception of time was uniformitarian, but in the sense that eighteenth century neo-classicists defined it: as unchanging time, a continual present.

In sum, the landform, at least those which attracted Gilbert, existed in a state of dynamic equilibrium; its history was rhythmic; and its shape was rippled. The problem of geomorphology was to identify these forces (that is, processes), quantify them, and show their dynamic competition. Almost all of Gilbert's writings conform to this mold, but to illustrate the assumptions underwriting it, consider his almost compulsive concern with Niagara Falls, the immediate heir to the study of the Cohoes potholes. Gilbert never wrote a comprehensive monography on Niagara; but during his many visits to the Nutshell he frequently stopped by it, and pondered its meaning. The more he studied it, the more complex it became. Before long he abandoned any effort to measure its recession, which was its major geologic theme, as a regular progression. Instead he began documenting qualifying circumstances until he decided that the calculation of recession demanded a regional synthesis of the Pleistocene. Still the definiteness of the problem attracted him. The age of the cataract had a definite beginning, when its work of gorge cutting initiated, and a definite ending, when it had passed from Erie to Ontario. In this respect it resembles the theme of his work on hydraulic mining debris, in which one studied the passage of a rhythm, the debris wave, through the Sacramento River system. Gilbert's analysis of Niagara hardly constituted one of his great studies, but it shares similar assumptions with them. In particular Gilbert emphatically rejected all attempts to make the recession of the falls uniform.

He expressed this interpretation best in a critical review of a book by Joseph Spencer titled "The Falls of Niagara: Their Evolution and Varying Relations to the Great Lakes." Spencer assumed that the energy of the falls had been constant, and therefore its rate of erosion uniform. Gilbert objected by remarking:

> "Rate of recession is proportional to the height of the falls and the discharge of the river. As the energy of the cataract (per unit time) is measured by the product of the height, or head, into discharge, it is implied that the rate of recession is proportional to the energy of the cataract."

He denied that head and discharge had been constant.

Putting the erosional formula into this form was not accidental. It was integral to the intellectual and aesthetic awe with which Gilbert beheld the spectacle. "I put the law into this form," he explained, "for the sake of comparing it with the experience of mechanical engineers. The cataract is a natural engine, and the erosion and recession correspond to what Rankine calls 'useful work' in the discussion of artificial engines." Arguing by analogy to the efficiency of heat machines, Gilbert attacked the hypothesis of Spencer that the efficiency (*i.e.*, recression) of the falls had been constant. On the contrary, "not merely does Spencer's supposed law fail to find support in engineering experience; it is contradicted by it." In short Gilbert employed the same thermodynamic model to explain the action of Niagara Falls that he had used with such effect at the Henry Mountains and Lake Bonneville. It was a model he hoped to extend generally to all landforms. "It would perhaps be more pertinent to compare the Niagara engine with other physiographic engines," he lectured, "but in general the efficiencies of such engines have not been investigated." That was a defect in the understanding of Niagara he boldly intended to correct through his flume experiments in California. "The solitary exception" to the absence of this form of analysis "is that of running water regarded as a carrier of detritus, and it happens that the unpublished results of a study of this engine are in my possession." It is not surprising that when he applied those figures to the problem of hydraulic mining debris in the Sacramento River system, the organization of the text had more in common with a chapter from Rankine's *Manual* than with the Davisian cycle of erosion (12).

The scientific wonder of Niagara was that it was both complex and informative, as though the panelling to a steam engine had been removed and the interior machinery plainly exposed. Not surprisingly the same elements piqued Gilbert's sense of aesthetic wonder. "The great cataract," he marvelled, "is the embodiment of power. In every second, unceasingly, seven thousand tons of water leap from a cliff one hundred and sixty feet high, and the continuous blow they strike makes the earth tremble. It is a spectacle of great beauty. The clear, green, pouring stream, forced with growing speed against the air, parts into rhythmic jets which burst and spread till all the green is lost in a white cloud of spray, one which the rainbow floats." (13).

This rare burst of rhapsody is revealing vis-a-vis his contemporaries. Where his colleague Clarence Dutton saw the sublimity of the Grand Canyon in terms of an artistic ensemble of immense time and space, Gilbert sensed the sublimity of Niagara in the spectacle of mechanical power, the apotheosis of the thermodynamic meaning of energy. Where Dutton and Powell saw ensembles organically related and progressively evolving, Gilbert saw a mechanical system naturally guided toward thermodynamic efficiency. What Dutton perceived as wondrous in the Canyon was akin to the romantic sensations discovered in exhumed mastodons, exotic scenery, and the decaying ruins of lost civilizations. What Gilbert found awesome in the great cataract of Niagara resembled the amazement that hushed crowds standing before the Corliss engine at the Philadelphia Centennial. His enthusiasm for Nature was no less than Dutton's or Powell's or Chamberlin's or Davis', it only sprang from a different sense of its unity and meaning. A subtle difference, perhaps, but one which drove him to search for paraboloid webs amid the giant sequoias and to measure potholes among the burial grounds of extinct mastodons.

As with Niagara Falls so it was with virtually all landforms. Each was distinct, and beyond broad philosophical considerations, there existed no single formula to explain particular shapes. Gilbert's philosophy of scientific method was a response to this problem. It was also the program of a classicist and the creed of a man who worked along the interface of geology and physics. His most famous essays on the subject are "The Inculcation of Scientific Method by Example" (1886) and "The Origin of Hypotheses" (1896). Yet his most succinct statement is contained in an 1885 book review. He wrote (14):

> "The principles which distinguish modern scientific research are not easily communicated by precept, and it is by no means certain that they have yet been correctly formulated. However it may be in the future it is certain that in the past they have been imparted, and for the present they must be imparted, from master to pupil by example; and whoever in publishing the result of a scientific inquiry sets forth at the same time the process by which it was attained, contributes doubly to the cause of science." (14)

There was no precept for geology, only the proven example of other sciences. Scientific thinking was a blend of close observation and analogic reasoning. Since this thinking was hardly peculiar to science, scientific thought was distinguished only by its "tendencies", in particular the effort to erase the ego of the observer from the scene. One learns this pattern of thought by imitating the great scientists, in a way reminiscent of that neo-classical doctrine by which artists were urged to imitate the great masters in painting, prose, and sculpture. The same impulse that led Horatio Greenough to sculpture the new hero George Washington in a Zeus-like pose draped in a toga led Gilbert to explain newly discovered scenery like the Henry Mountains with hydrostatics, the language of Archimedes. Most of the great scientists that Gilbert admired were physicists, and the process of analogic thought was a means to translate their style of thought, their concepts, and their example into geology.

Which is not to say that geology was merely a derivative science. Rather the two were appositives or analogues. One could no more derive geology *a priori* from physics than he could Greek from Sanskrit. The crux of the problem was that there was no simple formula to guide the translation, only the example of successful translators; hence, one needed a very broad background in the sciences as well as an inventiveness in generating analogies. The test on such hypothetical analogues was almost always quantitative. Gilbert criticized physics and evolutionary geology equally when, on the basis of their respective systems, they tried to tell him what he saw in Nature.

Gilbert was a liaison between the two sciences, a hybrid position from which he criticized each equally and yet tried to harmonize them. His lifelong fascination with questions of geologic time and isostasy illustrate this perfectly. The physicists claimed that the Earth was highly rigid, the geologists that it was highly plastic; Gilbert argued that it was broadly plastic and locally rigid, and tried to define that boundary quantitatively. In the case of geologic time he criticized both camps for assuming that natural time was progressive. The crucial discrimination between his interpretation and theirs was not simply a difference in the mode of inquiry, but in the object of inquiry. The reason geologists and physicists quarreled during the late nineteenth century was for the same reason theologians and geologists quarreled in the early part of the century; they addressed the same questions, in particular the age of the Earth, but with different techniques. Gilbert blended the two sciences because he asked different questions about the earth.

But perhaps the best illustration of Gilbert's method is to see it at work. And rather than the example of his monographs, consider the story told about an incident that occurred on the High Plains. It was past sunset, and one member of the party, a novitiate geologist that Gilbert was training, had not returned to camp. There was no wood for a signal fire, so a search had to be organized. Orienting himself by the stars, Gilbert walked out a fixed number of paces, turned 120° and paced off the same number of steps, then turned 120° again. He returned to camp having inscribed an equilateral triangle. When many explorers went West, they found themselves awash amid a cascade of new information, exotic landforms, and natural marvels; there seemed to be no familiar landmarks to direct them. Yet what Gilbert did scientifically was exactly what he did that night on the Plains. He inscribed a nearly mathematical order on a bizarre landscape, guided by the "frictionless pendulum" of the stars (15).

DAVIS AND GILBERT

It was not a method William Morris Davis found comfortable. And the contrast between the two men is instructive. Davis envisioned himself as the synthesizer and heir to the American school of geology. In terms of his personality and conceptual apparatus he was right. He was a confirmed evolutionist, this was what made his geographical cycle creditable. Like his contemporaries Henry Adams scanning the history of a civilization or Sigmund Freud imagining the psychological growth of an individual, Davis saw the natural landscape as an evolutionary system passing through progressive stages. Where Gilbert accepted a mechanical metaphor, Davis, climaxing a century of geologic speculation, accepted an organic one. For Davis explanation followed a genetic taxonomy based on evolution; for Gilbert a rational theory based on mathematical-mechanical principles. Philosophically the difference between them is that between Hegel and Kant. The geographical cycle was rapidly promulgated. If Darwin had demonstrated how paleontology could be organized by the principle of natural selection, Davis showed how evolution could unify the earth-fossils of geomorphology by the operation of universal denudation. His concept fashioned a "denudation chronology" to match the "depositional chronology" of stratigraphy. It is revealing to see how both men also applied the techniques of landform analysis to features far removed from the American West; but where Gilbert, on the example of Newton, puzzled over the Moon, Davis, on the example of Darwin, studied coral reefs.

The geographical cycle was not only good pedagoguism, it was also a culturally meaningful concept. Consider a Currier & Ives print published in 1869, the year which witnessed Powell's descent down the Colorado River and the completion of the transcontinental railroad. One can see the stages of civilization marching West. The evolutionary interpretation of Western settlement made sense, and so did the evolutionary interpretation of the Western landscape. Yet by 1905 there were cracks in the smooth masonry of the cycle; by 1920 major revisions were underway. The reason is simple. The normal cycle was breaking down when applied to more exotic regions; or to put it another way, the information base of geology continued to expand exponentially as it had from the time of Werner and Hutton. On the average its doubling time, as measured by its publications, was about 15 years. Geology responded in the hands of Thomas Chamberlin by shoring up the cosmological foundations of evolutionism and in the hands of Davis by multiplying the normal cycle of geomorphic evolution to account for environments different from that of the Appalachians. With nothing to take its place, abandonment of evolutionism meant intellectual anarchy (17).

In all this it is easy to detect an unmistakable parallel to simultaneous events in physics. The effort to preserve the natural laws of mechanics through the transformations of relativity theory was exactly correlative to the Davisian effort to "relativize" the geographic cycle. Just as systems of physical laws were seen to vary according to the relative velocity of the system and its observer, so systems of evolutionary laws varied according to the climates and erosive rates of the system and the role of the observer, man, in modifying it. And as with Einstein, so the heroic stature of Davis and Chamberlin warded off challengers. Each would have found the "chance universe" of the quantum as unpalatable as had Einstein. When challenged by Walther Penck in Germany or with instances of exceptional landforms, Davis sought to incorporate them as special cases of his general laws in a way analogous to Einstein's futile search for a unified field theory.

The relationship of Gilbert to Davis, then, is of some concern. Not only was Davis the acknowledged master of the field in which Gilbert produced some of his most notable essays, but it was Davis who, at the request of Charles Walcott, wrote the National Academy of Sciences biographical memoir on Gilbert. It revealed what is confirmed by other sources: the relationship between the two men was cordial, frank, and of genuine intellectual power. Yet it is doubtful that Davis understood Gilbert any more than Gilbert did Davis. Davis claimed that Gilbert instructed him "in method," and Gilbert publicly defended Davis's organic terminology as "apt." Yet neither used the borrowed ideas in anything like the way their author did. The two men had more mutual respect than mutual understanding (18).

The differences between them are striking. Both wanted to make their work practical, but for Davis this meant packaging geologic concepts in ways useful to teachers; for Gilbert, to engineers. At one point, defending his cycle as elastic rather than rigid, Davis pointed out that the system had far more categories than, at the present, there were facts to fill it. That would never have happened in a Gilbert monograph. When he said that, Davis resembled a scholastic debating the great chain of being, plotting categories of creatures not known to exist except by force of logic. Gilbert's monographs distilled, rather than elaborated. In stark contrast to Davis's defense of the geographic cycle concept as flexible, Gilbert's apology for his failure to mathematically define stream transport took it as "a matter of scientific faith" that "if our data were so precise as to substitute definite quantitative relations for the fascicle of trends and indefinite parallelisms they have actually furnished, some way would be found leading from complexity to simplicity." He stubbornly clung to his hope that his actual conclusions did "not necessarily follow. Just as a highly complex mathematical expression may be the exact equivalent of a fairly simple expression of a different type, so a physical law may defy formulation when approached in a certain way yet yield readily when the best method of approach has been discovered." Though the end product was highly complex, it was the impulse to simplify that motivated the investigation. If Davis's scheme had the design of an ornamented Gothic cathedral, Gilbert's monographs have the clear firmness of the Parthenon.

The differences are even apparent in their use of language. To the onslaught of new information, the romantic Davis responded by coining new terms, new concepts, even a new science, as though science, like organic evolution, spawned new genera and species in the course of adapting to novel conditions. Gilbert invented terms reluctantly, preferring to take an old word an give it a new meaning. He did not find radical new landforms in the West, but analogues of scenes he knew in New York, just as he did not see new principles acting on that landscape but analogues of old ones. Davis wrote a flexible, elastic, limpid prose; sentences flow from one to another, arbitrarily terminated by punctuation marks. It is an adaptable, opportunistic, noncommittal style, well adjusted to an evolutionary world of perpetual change. Gilbert, in contrast, saw a world whose forms were less fluid, a world of almost classical structure which he imitated in his prose. To compare their language is to compare the poetry of Walt Whitman to that of Alexander Pope. Yet to accuse Davis of mere metaphor is to ignore the concepts his language carried; his life cycle, after all, was no more metaphorical than Gilbert's physiographic engine.

Inevitably the Davisian system collapsed, both from internal and external pressures. His students filled the blanks in the master's maps and then began a new overlay until the map became a virtual palimpsest. To cope with new situations they carved and compounded the Davisian schema, until their historical landscape resembled the maps of Ptolemaic astronomy. Instead of the clean lines of the geographical cycle, there was a melange of epicycles and equants overlying basic orbits in a virtual maze. At times scholastic debate replaced scientific discussion until the logic of the Davisian methodology seemed perilously close to asking how many peneplains could sit on the head of a mountain.

At the same time the cultural functions of geology changed, and brought with them new questions, new criteria for geologic explanation, even, if you will, a new metaphysic. Geology is no longer a frontier institution, interceding between settlement and a natural wilderness, but a mechanism to cope with a synthetic environment. Geology interfaces increasingly with engineering because it is asked increasingly to deal with an engineered landscape. As Gilbert discovered along the Sacramento, the question about how a pure natural river might behave is less relevant that the problem of a river modified by dams, levees, and the tailings of industry and agriculture.

Insofar as geology has been history, it witnesses the same changes as in other branches of historical writings. Since the 1950s the progressive histories of Turner, Adams, and Parrington have succumbed to a movement which does not interpret the American past as providential, progressive, or evolutionary, but as contradictor, paradoxical, and understandable only in terms of particular social-cultural environments. This has afflicted even the history of science. Regardless of whether one accepts Thomas Kuhn's *Structure of Scientific Revolutions* as valid or not, it seems appropriate that the early 1960s should re-evaluate science and decide that it is not really progressive, but only replaces one paradigm by another without getting closer to truth. The depiction of the Earth's past as a series of more or less discontinuous systems is exactly analogous to the history of science as a succession of paradigms.

It seems to me that a fundamental phenomenon underwrites this entire complex of thought. Since the discovery of America, Western Civilization has confronted a problem of growth — in all dimensions of its existence, including intellectual. William Goetzmann has aptly termed this event "The Big Bang." Geology developed in response to one phase of the Big Bang — the information explosion, in particular the expansion in the knowledge of geography and history. But the Big Bang may have ended with the atomic bomb. Since then Western Civilization has faced a different problem — a contraction. While the mid-nineteenth century was impressed with the growth of organisms, the mid-twentieth in the form of systems theory and cynernetic models, is attracted to the machinery of control and self-regulation in organisms (*i.e.*, feedback). It conceives of Nature on the pattern of the human and electronic brains, as an information-processing system. Where the mid-nineteenth century was fascinated with evolution and entropy, the progressive irreversible decay and growth of systems, the mid-twentieth is concerned with information — defined mathematically as negative entropy, a measure of the amount of structure in a system. In like manner Europe has reversed four centuries of historical experience. It is losing rather than acquiring colonies, just as biology and geology face the problem of a contraction in our appreciation of space and time, and a loss of natural species and landscapes rather than the discovery of new ones.

Scientific thought has responded to this condition. Plate tectonics, for example, like ecology, emphasizes a recycling of materials rather than their growth; for its time scale it relies on rhythmic oscillations of magnetic reversals rather than an expanding tree of evolution. One finds this shift even in art, as it obliterates the sense of perspective that had characterized it since the Renaissance. If Davis' science may be likened to relativity theory, his illustrations occasionally resemble its artistic counterpart, cubism. Consider his coral reef drawings. Contrast it with an aerial photograph, perhaps the dominant form of geologic illustration today. All perspective is flattened. Compare in like way the flattening of cubism into minimal art and pop art. The replacement of spatially and historically vast canvases by local, even trivial subjects can be found in a Warhol painting of a Campbell's soup can, a LANDSAT infrared photograph, or perhaps in a geomorphic analysis from photogrammetry.

Consider once again Niagara Falls. In the nineteenth century Frederick

Church could paint an enormous dramatic canvas of the brink, but the view from space lab has trivialized, almost vaporized the scene. Or take the Grand Canyon, perhaps the epitome of nineteenth century evolutionary geology. William Holmes captured the enormity of the scene, its expanding texture with this panorama from Point Sublime, but the LANDSAT photo has reduced that complexity to something that resembles a knife carving on a wooden picnic table. In the same way, with evolutionary geology replaced by plate tectonics, the Grand Canyon holds less interest for geologic science than something like the Marianas Trench.

It is not a situation William Morris Davis could have predicted, but it is one he could have easily understood. His sense of time was tremendous. Nothing in the natural landscape could endure it unscathed. In his memoir of Gilbert he declaimed rhetorically:

> "Would that the pencilled outlines in the little pocket diaries have been written out elsewhere more at length; and yet how short would have been their endurance as the centuries roll by, even had they been engraved on tablets of stone with an iron quill."

As in the natural landscape, so in the human. No one would have appreciated better than William Morris Davis the erosion time had wrought on his own works (20).

Yet Gilbert's writings, like his landscape, have preserved their forms well after half a century of critical weathering. The reason it seems to me is a happy accident, perhaps a paradox. In many ways for his own time Gilbert was less an anticipation of the twentieth century than a throwback to the eighteenth. His interpretive designs have a strangely classical shape; coming upon them in the age of evolutionism — it was, after all, Darwin's Century — is like discovering some of those early railroad depots built as a Greek temple. Yet his reliance on elementary physics, his urge to quantify, and his emphasis on equilibrium has given him continuing significance to moderns.

Such considerations again type Gilbert as the antithesis of the romantic artist, the Melville or the Alfred Wegener, the wasted, revolutionary genius, forgotten until the future should rediscover him; rather Gilbert was more the Mark Twain, celebrated in his own time, but subject to continual reinterpretations by each subsequent generation. If it is less his geophysics than his metaphysics which strikes a modern chord today, it is nevertheless his example, which is how he felt science was communicated, which yet has meaning. That fact would have pleased even the scrupulously modest Gilbert.

And while it may be paradoxical that our accidental geologist should be honored for qualities that appeal to the present more than as a monument of the past, that fact is not inappropriate for one who preferred the glistening pebbles of Bonneville to the exotic bones of a mastodon. His concepts continue to speak in the present tense. That, too, Gilbert would have liked.

SUMMARY

He maintained his particular vision to the end. When his hydraulic studies were finally published in 1917, he was 74 years old. Repeatedly those last monographs had been threatened by crippling illness. Yet with hardly more than a pause for the galley sheets, Gilbert prepared to begin a new life. He arranged to remarry, and with his new wife to raise his grandson. He would return to the desert mountains of the Basin Range province to finish the work he began on the Wheeler Survey. But en route West, laying over at the last vestige of the Nutshell, his sister's house in Jackson, Michigan, his health collapsed. He spent several weeks slowly recovering at the local hospital. On April 11 he wrote his son Arch to come. He wrote his other son, Roy, soon afterwards. Yet from his sick bed Gilbert labored over his accounts. It demanded considerable effort, but he persevered until ultimately the books balanced. In truth it had taken some seven decades and the span of a continent before than final tally could be made. But what began in the Nutshell on the shores of Lake Ontario and ended at the tidal bar outside the Golden Gate squared (21).

The question of what Nature was and how it should behave had merged inextricably with the problem of what man was and how he should conduct his life; the problems posed by a boy's birth in 1843 proved inseparable from the riddles of the Rochester landscape. The two were solved simultaneously. Nearby Ontario became the Great Lakes of the Holocene, the igneous "lake rocks" of the Henrys, the Pleistocene Bonneville, the buried reservoirs beneath the High Plains, the lunar marias, and San Francisco Bay. The Genessee River became the Escalante, the Colorado, the Niagara, and the Sacramento. The warped shorelines around Rochester led him to those at Erie, at Bonneville, and to the crustal undulations of the planet. The glacial topography of Rochester brought him to the tumbling moraines of Ohio and the arid shore bars etched into the Wasatch; the enigma of glacial climate, to the raw fjords of Alaska and the jeweled lakes of the High Sierras. The logic of his mathematics merged with his inexhaustible reserves of kindness, patience, and good humor; he generated himself the sympathy he saw abundantly in nature, and regulated himself with the same self-control, equally serene and demanding. Now the accounts squared. Gilbert drew a small box to indicate that fact at the bottom of the page.

A few days later he packed to leave. There were more studies beckoning in the Utah desert; with Alice Eastwood, there was a new wife; with his grandson, a new family. There was a new life promised at San Francisco to replace that he had known at Rochester and Washington. But the old life had run its course. His heart failed. He died on May 1, five days before his seventy-fifth birthday.

Arch telegraphed the news to Merriam. The stack of letters congratulating him on his birthday, an honor roll of international geology, stayed unread.

On May 27 he was cremated. His ashes were brought to the family plot. There they were placed over his wife's grave.

TEXT NOTES

(1) Gilbert's early life is best summarized in William Morris Davis: "Biographical Memoir of Grove Karl Gilbert, 1843-1918," *National Academy of Sciences Biographical Memoirs*, Volume XXI, No. 5 (Washington, 1922). Some evidence is also present in materials (mostly letters) now in the possession of Gilbert's grandson, Mr. Karl Palmer. Quotation is taken from a letter by Gilbert to William Dall dated 8-1-02; the letter is in the William Dall Papers, Record Unit 7073, Smithsonian Archives.

(2) John Merz: *A History of European Scientific Thought in the Nineteenth Century,* Volume 1 (Dover Reprint: New York, 1965), p. 144.

(3) The Cohoes episode is told in Davis, *op. cit.*, p. 6-8. Gilbert published two articles out of it. A popular article appeared in Moore's "Rural New Yorker" in March, 1867 (a copy of the article is preserved in a bound collection of Gilbert's works now stored at the USGS National Center library). A formal article appeared as "Notes of Investigations at Cohoes with Reference to the Circumstances of the Deposition of the Skeleton of the Mastodon," 21st Annual Report, State Cabinet of Natural History (Albany, 1871), p. 137-139.

(4) Davis, *op. cit.*, p. 71. Bailey Willis: *A Yanqui in Patagonia* (Stanford, 1947), p. 33.

(5) G. K. Gilbert: "The Moon's Face: A Study of the Origin of its Features," *Philosophical Society of Washington Bulletin*, v. 12 (1893), p. 241. Letter from J. J. Stevenson to Gilbert dated 10-12-12, on file with the Field Records of the USGS at Denver.

(6) Davis, *op. cit.*, p. 206.

(7) W. C. Mendenhall: "Memorial of Grove Karl Gilbert," *Geological Society of America Bulletin*, v. 31 (March, 1920), p. 42.

(8) G. K. Gilbert: "The Origin of Hypotheses," *Science*, n.s., v. 3, (1896), p. 1-13.

(9) J. S. Newberry: "Report of Progress in 1870," *Geologic Survey of Ohio* (Columbus, 1871), p. 32.

(10) G. K. Gilbert: "The Work of the International Congress of Geologists," *American Journal of Science,* 3d series, v. 34 (1887), p. 437.

(11) G. K. Gilbert: "Rhythms and Geologic Time," *Science*, n.s., v. 11, (1900), p. 1001-1012.

(12) G. K. Gilbert: "Review of Spencer, *The Falls of Niagara*," *Science*, n.s., v. 28 (1908), p. 148-151.

(13) G. K. Gilbert: "Niagara Falls and Their History," *National Geographic Monograph*, v. 1 (1895), p. 333.

(14) G. K. Gilbert: "Review of Geikie," *Nature*, v. 27 (1885), p. 237-239.

(15) Davis, *op. cit.*, p. 206.

(16) For a biography of Davis, see Richard Chorley, et al: *The History of the Study of Landforms; or, The Development of Geomorphology*. Volume II: The Life of William Morris Davis (London, 1973).

(17) For data on the exponential growth of scientific literature, see Derek Price: *Little Science, Big Science* (New York, 1965) and Henry Menard: *Science: Growth and Change* (Cambridge, 1971).

(18) The remark by Davis belongs on an inscribed note to a reprint of his "Structure of the Triassic Formation of the Connecticut Valley" now in the possession of

Mr. Karl Palmer; the Gilbert commentary belongs in "Style in Scientific Composition," *Science*, n.s., v. 21 (1905), p. 29.

(19) The Davis quotation belongs in his essay "Complications of the Geographic Cycle," included in *Geographical Essays* (Dover reprint: New York, 1954), p. 286. Gilbert's remark comes from "The Transportation of Debris by Running Water," USGS *Professional Paper 26* (Washington, 1914), p. 109, 190, 129.

(20) Davis, *op. cit.*, p. 10.

(21) Gilbert's death is best recorded in the journals of C. Hart Merriam located in the archives of the Library of Congress. Other materials can be found in letters now in the possession of Mr. Karl Palmer.

CHAPTER 15

"HOW MANY PENEPLAINS CAN SIT ON THE TOP OF A MOUNTAIN?"

Sherwood D. Tuttle

I have chosen as the title of my "summing up" this question tossed out by Steve Pyne during his address on "The Mind of Grove Karl Gilbert" (one of the most interesting presentations that I can recall in many a postprandial session) because Pyne's apt paraphrase of the ultimate in academic eggheadism seems to epitomize the futility and the charm of a symposium on theories of landscape development. And yet, after all has been said, my mood is one of hope for the vigor and future of our science.

In the past, personal and individual philosophies have influenced aspects of this complex and somewhat ambiguous scientific discipline. Some of us who are geologists have tended to be uniformitarian (perhaps too much so) and sequential in our thinking, mainly because we were taught that way. Some geomorphologists visualize approaches to natural processes and resulting changes only in quantitative terms, feeling that a historical or time dimension is not a cogent factor. Some view landscape development as dynamic and subject to equilibrium, with processes testable by equations and models of various sorts. As Charles Higgins said in the first paper, our scientific forebears who did the pioneer work in geomorphology regarded their investigations as shedding light on the workings of an orderly world. More recently we have been influenced by theories suggesting that patterns in the universe may be random, not subject to a higher law or organization.

The papers in this volume reflect diverse points of view and may be looked at in terms of a reader's conscious or unconscious bias about how landscape-shaping processes operate. However, most of us have come to accept the principle that geomorphic hypotheses and their applications vary with the geographical settings of areas under study and with the length of geologic time involved. Higgins points out that there may be no definitive theory or geomorphic system that can fit all landscapes. Therefore, we need multiple theories or different theories for different purposes. He suggests that as scientists we may all be seeking a "correct" or complete rational answer to landform origins, but if the natural world is irrational, no internally complete and substantive theory or system would work.

Geomorphology in the past has been an integral part of geology and its principal usefulness was to explain the evolutionary sequences that areas have undergone, as was done in investigations of the development of topography and drainage in the Appalachians and other mountain ranges

(Bryan, 1941). These explanatory and sequential regional histories became a major part of historical geology and are still found in most historical geology textbooks. However, because much of this material was based on the uplifted and dissected peneplain story, it has been subject to criticism. Modern studies of late Tertiary and Quaternary time have drawn upon new findings in paleontology, sedimentology, and geochronology for insights into the historical geology of the last ten million years — rather than depending on accordance of summits, evidence of rejuvenation, and other traditional aspects of landscape study.

Today's applications of geomorphology involve problems of periods of time shorter than those used 50 or 75 years ago. Older theories are not directly useful in looking at changes in Holocene processes, especially in small areas. Procedures involving small, open or closed systems, quantitative descriptions, and statistical methods are more helpful, particularly when one is working with investigators in related disciplines. One can't wait for Davisian maturity to arrive or pediplanation to occur. The need is to understand current process, not past geologic history. Thus it is that changes in the uses and practice of geology and geomorphology have brought about changes in the theoretical framework in which landscape development is studied. What is evident is that the demands of modern environmental geology require more in terms of theory than Davis's geographical cycle can provide. More profuse and more discriminating detail is needed for small areas being investigated or evaluated for human purposes.

Geologists who are classroom teachers have to deal with the problem of what geomorphic system to use in order to best convey scientific concepts in a meaningful and relevant way. Some instructors start elementary students with Davis's geographical cycle, in part or in total, while others are so doubtful about the validity of the cycle, they refuse to mention it. An instructor can try to comment briefly on weaknesses and criticism of the theory, generated during the more than 75 years of its use, but to do so may take up more time than he feels he can spare. He will need to bring in ideas of downwastage versus backwastage, shifting equilibrium in place of sequential change, and then review strengths and weaknesses of other theories. If he spends very much time with criticisms and alternative theories, his students may accuse him of teaching something as the "truth" and then unteaching it later. Perhaps a teacher of elementary students should emphasize only process, omit discussion of landform theory and use the time for expounding on plate tectonics, metamorphism, or environmental problems.

As many of us have found out, students' search for the truth and for simple correct answers to put on final examinations can create great frustration in the testee and uncertainty in the mind of the professor. Is the peneplain concept an invalid postulation? If so, can one fairly grade a multiple choice question about peneplains? Can a teacher teach something he may not believe in, and can or should he finish an explanation of how landscapes are formed by admitting confusion and uncertainty?

I have found, over a long span of teaching elementary students, that a pessimistic point of view is self-defeating and disheartening to students. It seems far better to look at the various theories of landscape development in terms of the strengths and inconsistencies that reflect the philosophy and intellectual milieu of the author. If the ultimate truth is not yet known (and it may never be known!), why not enjoy playing with all suggested possibilities? Assume different points of view and scientifically demonstrate an eclectic approach. Build composite and multiple theories and enjoy with students the academic fun of trying out a variety of ideas.

As an undergraduate, I became a Davisian by direct order — in a sense. When George W. White, my instructor (that was in the mid-thirties), explained the relationship of peneplain and monadnock in central New England in terms of the geographical cycle, I couldn't believe this visualization. My boyhood home was a few miles from Mt. Monadnock, and the ranges of southern New Hampshire comprised the skyline I knew (or thought I knew) so well. When George White described Mt. Monadnock as a knob on a flat surface, I was indignant! His response was to assign me to read Davis in the original. So I learned about the geographical cycle, believed it, accepted it, and taught it for a while before realizing that there were objections to it, as well as some alternative theories. Later, as a graduate student, I participated in a month-long series of discussions about the differences between a peneplain and a pediment (or pediplain), where neither the instructor nor the students could agree on how to answer the question.

Few textbooks of physical geology or geomorphology give a balanced, complete discussion of theories of landscape development. Some writers ignore geomorphic theory altogether. Are students expected to find out about these theories by accident? Some who teach geomorphology do not use a text that contains a discussion of the different approaches to landform study because of disagreement with (*i.e.*, disapproval of) a particular theory or because in their opinion not enough coverage is given to a favored theory. I myself was upbraided by a critic of a small volume I had written (for elementary students and nonscience majors) because I had included more paragraphs about one theory than another — ergo, I was biased and unfair!

In talking to students in elementary geology classes about the development of landscape theories, I have at times suggested that probably about 50 per cent of what I was teaching wasn't true, the trouble being that I didn't know which 50 per cent was which! Have we not an obligation to our students to help them understand that absolute truth is not the grail we are pursuing when we teach, speculate about, or discuss these hypotheses? A geomorphic theory is, in a sense, like a garment. We use one that best fits our data or best serves a particular need or purpose in the given circumstances. We can be familiar with the proposed hypotheses and their methods of application;

but when I am asked, what do I believe, I have to say I don't know.

In the "real world" — as understood by geologists involved in oilfield and mining operations, engineering studies, environmental evaluations, and the like — theories of landscape development do not seem to be of major significance. In a report surveying geologic literature in Iowa pertaining to geomorphology between 1942 and 1968 (65 papers by 109 authors), it was noted that there was a total absence of discussion of theories of landscape development, classification, and geomorphic philosophy. Emphasis in these papers was on what could be called "process geomorphology" (Tuttle and Prior, 1968). Theories of landscape development were not of burning importance, it would seem, to Iowa geologists.

In academia, however, fascination with and controversy about these theories and their application continue to be a matter of considerable scientific interest and some ego involvement. Having to deal with unsolved theoretical problems attracts some people to geology and, in particular, geomorphology. After years of observing students, I have concluded that some simply do not feel comfortable in scientific areas where lines of investigation are well defined and readily predictable or where orthodoxy rules. Some minds appear to draw away from any form of dogmatism. Ambiguity presents a challenge. Such a student can admit almost with enjoyment how little he knows, and he willingly spends hours reading, investigating, and scrambling over terrain and then more hours arguing about what he has perceived and how he hopes to deal with what he has found.

I have mentioned earlier that students at different levels can gain an appreciation of geomorphic systems, ranging from simple to complex. I would suggest the following schema:

1. Beginning students and nonscience majors. Included in this group are individuals who are turned off by mathematical concepts and some who "don't like science." And yet as inhabitants of the natural world, they can respond to a challenge to cultivate and deepen their appreciation of the physical environment. A first step is to improve visual acuity and naturalistic observation through use of photographs, slides, maps, sketches, landscape paintings, etc. Soon they become "car window geologists," on the lookout for symmetry, accordance, similarity, and diversity in the landscape. (They start picking up on authentic or unauthentic landscape backgrounds in television and movies!) They can begin to recognize effects of geomorphic processes and some relationships of structure to the changes that are taking place. Acquiring some understanding of at least one geomorphic system can help them in observing, visualizing, and evaluating landscapes and the physical environment wherever they may live, work, or travel.

2. Undergraduate students in a geomorphology course. They can accept the idea of long-time change, polygenetic topography, structural control of topography, and perhaps multicyclical topography. They have had some exposure to various geomorphic systems but can recognize that there is no definitive geomorphic system or single correct way of explaining and accounting for landscapes. They understand the influence of baselevel and equilibrium in the evolution of landscapes and are beginning to use quantitative methods to describe and analyze topography. They are able to do simple visual field work, relating form to lithology and structure, etc.

3. Geology or geography graduate students. In addition to having the understandings outlined above, students at this level are aware of divergent approaches and points at which different geomorphic systems impinge or conflict. They are able to make evaluations of these systems in terms of their strengths and weaknesses. (They can discuss the peneplain argument, explain terrace formation, classify coastlines, etc.) They can understand and deal with quantitative approaches to landforms and landscapes and are becoming competent in the field, capable of mapping geomorphic features as well as relating surficial materials to geomorphic processes and history.

4. Geology or geography graduate students in geomorphology. These students can be expected to have a more intensive understanding of the levels listed above. They should be able to discuss the significant aspects of all geomorphic systems, historic and current, and know about the applications of these, as well as arguments brought forward in objection to the applications. However, if they are not ready, from personal observation and conviction, to make an intellectual commitment to a particular approach, theory, or system, they should not be expected to do so. (Sometimes we need to be reminded of Emerson's caution: "Don't make the student another you; one's enough.") Students at this level of training can map all types of surficial features and materials, use soil profiles, collect pollen, fossils, and samples for radiocarbon dating, etc. They need to be geomorphically and mathematically competent to use, understand, and evaluate modern procedures of morphometric analysis.

5. Professional geomorphologists, geologists, geographers, and engineers involved in landscape or surficial studies and environmental projects. A professional may use any geomorphic system or procedure that he wishes. However, his peers expect him to have collected field evidence to support his hypothesis and his conclusions, and he has to be able to rebut possible criticisms of

his applications. When he disagrees with a predecessor, he can carefully marshal evidence to support the disagreement and his own proposals. I would like to add a plea at this point that professionals try, as much as they are able, to develop lines of communication with nonprofessionals and with professionals working in related areas, particularly in fields having to do with environmental evaluations.

Several points have seemed to me significant after having listened to the papers presented at this Sixth Annual Geomorphology Symposium and having reflected on them. All of us should be careful in distinguishing between what W. M. Davis said and wrote himself and how others have used and expanded on Davisian ideas. Several of our speakers brought out specific points in this vein.

Moreover, after a number of years and many a quadrangle of field work, we are still faced with the question of alleged accordance of summits in the Pennsylvanian Appalachians. Are the ridge-tops smooth and flat, and are they reasonably accordant in elevation with adjacent ridges? Does a postulated accordance reflect structural and lithologic control, or do these features represent uplift and dissected surfaces that are accordant in elevation and discordant across structures? My eye can see accordance in the field and on maps, but as a geomorphologist, I experience problems with any theory involving multiple peneplains. Better field studies dealing with problems of summit accordance should help us find the answers we look for.

For additional comments, I would say, by way of appreciation, that Higgins most capably delineated the significant aspects of scientific problems relating to the origin and development of landscapes. Meyerhoff and Crickmay brought to the group the benefit of their years of dedicated study and field work. Meyerhoff has investigated a series of surfaces located *en echelon* at higher and higher elevations inland in Massachusetts and Vermont. He postulated earlier that these surfaces were produced by fluvial processes and slope retreat rather than peneplanation (see Meyerhoff and Hubbell, 1929). At that time he was not aware, he said, of the Penckian models and hypotheses, but came to the conclusion that the New England peneplain was "like Santa Claus" (*i.e.*, not existing in the real world). Crickmay described how his ideas on lateral planation, as a significant process in producing erosion surfaces, had developed and expanded through the years (see also Crickmay, 1933).

Hack and Peltier reiterated points of view that have for some years stimulated fellow professionals and students in geomorphology. Hack remarked that when he first published his Appalachian studies, his purpose was "not to start a revolution" but rather to open up some possibilities that would allow for some more divergent or diverse ideas about landscape evolution. Judson, applying data from plate tectonics, seismology, gravity, and precision leveling, suggested that W. M. Davis' explanation of the origins of drainage and topography in the Appalachians is incompatible, in terms of dating, and inconsistent in terms of episodes of uplift and still-stand. Davis himself, if he were alive and active today would not be satisfied with his 19th century model of deformation and mountain-building but would doubtless have come up with several revised and updated hypotheses utilizing new data.

Knox and Schumm have been investigating very recent geomorphic changes, some within less than a hundred years. Their studies suggest that variations in erosion have been due to climatic fluctuations and man's activities.

Morisawa and Palmquist gave examples of how structure and lithology have controlled the patterns of landscapes they have studied. Melhorn showed intriguing examples of how erosion surfaces may be correlated on a worldwide basis. Morisawa's and Palmquist's material suggests the unimportance of any particular geomorphic system; and Melhorn's data, interestingly enough, point to a similarity of landscape development at more or less the same time all over the world.

Any application of principles of allometry and visualization of a steady state in open or closed systems has to be done with great care, Bull warns us. The model of allometric change is useful in understanding adjustments between materials, processes, and landforms that may be operating in a disequilibrium state. Powell in his paper commented that landform theory regarding areas of karst topography is not very far along and confusion exists because of diverse terminology in that field.

All in all, the papers varied considerably, with some authors using definite theories or systems and others offering explanations to fit specific situations. Parameters of time and space ranged widely — from under a hundred years to most of post-Precambrian time and from small basins to areas of worldwide extent. The differing ages and experiences of the speakers are reflected in the patterns of the papers, with youthful writers tending to be more quantitative and experimental. Wide-ranging presentations such as these are particularly worthwhile, I feel, for students coming along in the field, and as a teacher, I found it gratifying that so many were there and participating in the discussions.

Geomorphology, it is evident, still ranges from the extremes of the school of arm-wavers to the practitioners of geomorphysics. But, more importantly, this Sixth Annual Symposium has shown, as did previous sessions, that geomorphology is alive and well and carrying on in Binghamton, New York! Thank you.

REFERENCES CITED

Bryan, Kirk, 1941, Physiography. Geol. Soc. Am. 50th Anniversary Volume, p. 1-5.

Crickmay, C. H., 1933, The later stages of the cycle of erosion. Geol, Mog. 830, v. 70, p. 337-347.

Meyerhoff, H. A., and Hubbell, M., 1929, The erosional landforms of eastern and central Vermont. Vermont State Geologist, 16th Rept., p. 315-381.

Tuttle, S. D., and Prior, S. J., 1968, Geomorphology in Iowa, 1943-1968, An annotated bibliography of the literature. Proc. Iowa Acad. Sci., v. 75, p. 253-267.